Ionic Liquids:
Current State and Future Directions

ACS SYMPOSIUM SERIES **1250**

Ionic Liquids: Current State and Future Directions

Mark B. Shiflett, Editor
The University of Kansas
Lawrence, Kansas

Aaron M. Scurto, Editor
The University of Kansas
Lawrence, Kansas

Sponsored by the
ACS Division of Industrial and Engineering Chemistry

American Chemical Society, Washington, DC

Distributed in print by Oxford University Press

Library of Congress Cataloging-in-Publication Data

Names: Shiflett, Mark B., editor. | Scurto, Aaron M., editor. | American Chemical Society. Division of Industrial and Engineering Chemistry.
Title: Ionic liquids : current state and future directions / Mark B. Shiflett, editor, The University of Kansas, Lawrence, Kansas, Aaron M. Scurto, editor, The University of Kansas, Lawrence, Kansas ; sponsored by the ACS Division of Industrial and Engineering Chemistry.
Description: Washington, DC : American Chemical Society, [2017] | Series: ACS symposium series ; 1250 | Includes bibliographical references and index.
Identifiers: LCCN 2017035029 (print) | LCCN 2017039058 (ebook) | ISBN 9780841232129 (ebook) | ISBN 9780841232136
Subjects: LCSH: Ionic solutions.
Classification: LCC QD561 (ebook) | LCC QD561 .I5687 2017 (print) | DDC 541/.372--dc23
LC record available at https://lccn.loc.gov/2017035029

The paper used in this publication meets the minimum requirements of American National Standard for Information Sciences—Permanence of Paper for Printed Library Materials, ANSI Z39.48n1984.

Distributed in print by Oxford University Press

Foreword

The ACS Symposium Series was first published in 1974 to provide a mechanism for publishing symposia quickly in book form. The purpose of the series is to publish timely, comprehensive books developed from the ACS sponsored symposia based on current scientific research. Occasionally, books are developed from symposia sponsored by other organizations when the topic is of keen interest to the chemistry audience.

Before agreeing to publish a book, the proposed table of contents is reviewed for appropriate and comprehensive coverage and for interest to the audience. Some papers may be excluded to better focus the book; others may be added to provide comprehensiveness. When appropriate, overview or introductory chapters are added. Drafts of chapters are peer-reviewed prior to final acceptance or rejection, and manuscripts are prepared in camera-ready format.

As a rule, only original research papers and original review papers are included in the volumes. Verbatim reproductions of previous published papers are not accepted.

ACS Books Department

Contents

Fundamentals

Indexes

Preface

The purpose of this book is to provide an update on some of the latest research and applications in the broad field of ionic liquids. This volume spans research and development activities ranging from fundamental and experimental investigations to commercial applications. A brief history of the field is included, as well as both new developments and reviews organized in the general topical areas of applications, materials, biomass processing, and fundamental studies.

This text was developed from a selection of papers presented in a two-session symposium entitled *Ionic Liquids: Current and Future Trends* at the 251st American Chemical Society (ACS) National Meeting which was held in San Diego, California on March 14, 2016. The symposium was organized by Aaron Scurto, a professor from the University of Kansas, Department of Chemical and Petroleum Engineering. The symposium was a virtual "who's who" in the field of ionic liquids and thus the book chapters are written by some of the leading experts in the field.

The symposium was held in honor of Dr. Mark B. Shiflett who was named a Division Fellow by the Industrial and Engineering Chemistry (I&EC) Division of the American Chemical Society. He received the award in recognition of his research at the DuPont Company which has impacted both applied chemistry and chemical engineering.

Dr. Shiflett was honored specifically for:

- Working with his team to invent and patent several energy-efficient refrigerant mixtures based on hydrofluorocarbons that replaced ozone-depleting chlorofluorocarbons (CFCs). These products saved the refrigeration industry hundreds of millions of dollars in retrofit costs and accelerated the transition away from CFCs which has led to the healing of the Earth's ozone layer.
- Developing novel materials for hydrogen storage, a next-generation technology for the green manufacturing of titanium dioxide and pioneering work to study the interaction of fluorinated ionic liquids and fluorochemicals.
- Publishing over 70 papers in peer-reviewed journals and being an inventor on 44 U.S. patents.
- Teaching and advising the next generation of chemical engineering students at the University of Delaware in the Department of Chemical and Biomolecular Engineering.

Dr. Shiflett was a Technical Fellow in DuPont Central Research and Development (CR&D) located at the Experimental Station in Wilmington, Delaware. Dr. Shiflett retired from the DuPont Company in September of 2016 and joined the faculty in the Department of Chemical and Petroleum Engineering at the University of Kansas to continue his research and teaching in the field of ionic liquids.

Acknowledgments

The editors would like to thank first the authors for writing the excellent chapters contained in this book. We would also like to thank the following reviewers for their time to carefully read and provide constructive feedback on each chapter.

Prof. Paschalis Alexandridis, University at Buffalo, Buffalo, New York, United States
Prof. Ewa Andrzejewska, Poznan University, Poxnan, Poland
Prof. Gary Baker, University of Missouri, Columbia, Missouri, United States
Prof. Jason Bara, University of Alabama, Tuscaloosa, Alabama, United States
Prof. Jason Clyburne, Saint Mary's University, Halifax, Nova Scotia, Canada
Prof. Zhifeng Ding, Western University, London, Ontario, Canada
Prof. Rasmus Fehrmann, Technical University of Denmark, Lyngby, Denmark
Dr. Clotilde Gaillard, Université de Lyon, Lyon, France
Prof. Rosa Espinosa-Marzal, University of Illinois, Urbana-Champaign, Illinois, United States
Prof. João Coutinho, University of Aveiro, Aveiro, Portugal
Dr. Joe Magee, National Institute of Standards and Technology, Boulder, Colorado, United States
Prof. Edward Maginn, University of Notre Dame, Notre Dame, Indiana, United States
Prof. David Mecerreyes, University of the Basque Country, Donostia-San Sebastián, Spain
Prof. Luís Santos, University of Porto, Porto, Portugal
Prof. Kenji Takahashi, Kanazawa University, Kanazawa, Japan

The editors would also like to thank Dr. Kristina Davis and Prof. Ed Maginn at the University of Notre Dame for providing the cover art. The image is from a simulation of an ionic liquid (1-ethyl-3-methylimidazolium bis(trifluoromethylsulfonyl)imide, $[C_2C_1Im][Tf_2N]$) vaporizing.

Finally, we would like to thank the team at ACS books, in particular Elizabeth Hernandez and Aimee Greene, for their support. It is always a pleasure to work with such a professional group of people.

Mark B. Shiflett
Foundation Distinguished Professor
Department of Chemical and Petroleum Engineering
Center for Environmentally Beneficial Catalysis
The University of Kansas
Life Sciences Research Laboratory
Building A, Suite 110
1501 Wakarusa Drive
Lawrence, KS 66047-1803
(785) 864-6719 (telephone)
mark.b.shiflett@ku.edu (e-mail)

Aaron M. Scurto
Associate Professor
Department of Chemical and Petroleum Engineering
Center for Environmentally Beneficial Catalysis
The University of Kansas
4132 Learned Hall
1530 West 15th Street
Lawrence, KS 66045-7609
(785) 864-4947 (telephone)
ascurto@ku.edu (e-mail)

Chapter 1

Ionic Liquids: Current State and Future Directions

Mark B. Shiflett* and Aaron M. Scurto

Chemical and Petroleum Engineering, Center for Environmentally Beneficial Catalysis, University of Kansas, Lawrence, Kansas 66047, United States
***Tel.: 785-864-6719; E-mail: mark.b.shiflett@ku.edu**

Twenty years ago, research involving ionic liquids was a minor field of interest. Only a few chemists and even fewer engineers were interested in salts with melting points near room temperature. In April 2000, the first NATO advanced research workshop on ionic liquids was held in Heraklion, Crete. The conference was the first international meeting devoted to ionic liquids and attracted most of the active researchers at that time. Following that meeting, activity in the field began to flourish and the first books and international conferences devoted to ionic liquids began to appear. Today, over 75,000 publications and 12,000 patents have been published involving ionic liquids!

This symposium series book based on the ACS conference, Ionic Liquids: Current State and Future Directions held in San Diego, California in 2016 attempts to propel the field forward by bringing together contributions from some of the foremost researchers on ionic liquids. Recent products and new large-scale processes using ionic liquids, both in operation and being announced, indicate that an exciting new chapter in this field is about to begin.

This introductory chapter summarizes some of the history, applications, conferences, books, databases, issues related to data quality and toxicity for researchers working in the field of ionic liquids and includes an overview for each proceeding chapter with an introduction about the authors.

Introduction

The first book dedicated to room temperature ionic liquids (RTILs) was edited by Kenneth Seddon and Robin Rogers in 2002 entitled "Ionic Liquids, Industrial Applications to Green Chemistry" (*1*). The book contained the key papers from the American Chemical Society (ACS) National Meeting symposium, "Green (or Greener) Industrial Applications of Ionic Liquids" held in San Diego, California, April 1-5, 2001. The symposium was the first "open" international meeting on the fundamentals and applications of ionic liquids (an invited NATO advanced research workshop (ARW) meeting was held in Crete in April 2000). Another seminal text that appeared in 2002 edited by Peter Wasserscheid and Tom Welton was entitled "Ionic Liquids in Synthesis" (*2*). The book was designed to take the reader from little or no knowledge of ionic liquids (ILs) to an understanding reflecting the best knowledge at that time. These authors are now eminent scholars in the field and their texts are must-read books for anyone beginning to study the exciting and rapidly growing field of ionic liquids. Today, a number of excellent books have been written about ionic liquids, some which now specialize in particular fields of use such as organic synthesis, electrochemistry, bio-processing, pharmaceuticals, catalysis, separations, and industrial applications (*3–31*).

The first observations of materials we now recognize as ionic liquids is believed to date back as far as the mid-19th century. However, many consider Paul Walden the grandfather of ionic liquids who published in 1914 the synthesis of the first room-temperature ionic liquid, ethylammonium nitrate, with a melting point of 12.5°C (*32*). However, it was not until John Wilkes and Michael Zaworotko at the U.S. Air Force Academy in 1992 reported the synthesis of the first "air and water stable" imidazolium ionic liquids, such as 1-ethyl-3-methylimidazolium hexafluorophosphate and 1-ethyl-3-methylimidazolium tetrafluoroborate, did the field of ionic liquids start to rapidly expand. Many consider John Wilkes the modern day father of ionic liquids. John wrote one of the first review papers on ionic liquids entitled "A short history of ionic liquids – from molten salts to neoteric solvents" (*33*). A must read for anyone working in the field.

Today, thousands of papers and patents are being published every year with the latest discoveries using ionic liquids. A recent literature search found over 300 "review" articles have been published since 2010 on the use of ionic liquids! Despite the title for this chapter, "Ionic Liquids: Current State and Future Directions", it is literally impossible to summarize everything in a single chapter, let alone a single book. Therefore, this book presents a selection of papers from the ACS symposium "Ionic Liquids, Current State and Future Directions" held in San Diego, California on March 13-17, 2016, almost 15 years after the first ACS symposium was held on ionic liquids (*1*). The symposium brought together some of the most preeminent authors in the field of ionic liquids.

Organization

The symposium as well as this book are organized into four sections: Applications, Materials, Biomass Processing, and Fundamentals. We were honored to have both Professor Kenneth Seddon and Professor Robin Rogers as our two keynote speakers for the morning and afternoon sessions at the ACS Symposium.

Professor Kenneth Seddon and his group conduct ionic liquids research at the Queens University of Belfast (*34*). Under Prof. Seddon's leadership the Queens University Ionic Liquids Laboratory (QUILL) was founded in 1999 to explore, develop and understand the role of ionic liquids and focuses on their synthesis, characterization and applications (*35*). Professor Seddon is Chair of Inorganic Chemistry at Queen's University of Belfast and director of QUILL Research Centre, the world-leading industrial-academic consortium that was awarded the 2006 Queen's Anniversary Prize for Higher and Further Education.

Professor Robin Rogers and his groups at McGill University and the University of Alabama have provided Chapter 2 in the first section under Applications entitled, "Translational Research from Academia to Industry: Following the Pathway of George Washington Carver". Their chapter focuses on the challenges of translating technology from academia to industry and provides several excellent case studies as examples. They correctly point out that academia needs to take basic research one step further from beyond the laboratory to a commercial readiness that allows industry to properly assess for commercialization. Professor Rogers holds the Canada Excellence Research Chair in Green Chemistry and Green Chemicals (*36*).

Professor Rogers and Professor Seddon have both had a profound impact and played an influential role in the expansion of interest and research in the field of ionic liquids. Together they have edited six books and organized numerous conferences on the topic of ionic liquids (*1*, *3*, *5*, *6*, *13*, *21*).

In addition, to some of the leading academic researchers in the field, we also had a presentation by Dr. Boyan Iliev from the IoLiTec Company, a leading supplier of ionic liquids for a variety of applications including synthesis, catalysis, analytics, electroplating, heat and refrigeration engineering, biotechnology and sensor technology (*37*). Dr. Thomas Schubert, CEO and Founder of IoLiTec and his colleagues have written Chapter 3, "Current and Future Ionic Liquid Markets" with some keen insights into where the field is headed.

Section two covers the area of "Materials" and begins with Chapter 4, "Photopolymerization of Alkyl- and Ether- Functionalized Coordinated Ionic Liquid Monomers" by Professor Jason Bara and his group at the University of Alabama. In this chapter, they describe the synthesis of coordinated ionic liquid monomers by photopolymerization of small organic monomers with lithium bistriflimide ($LiTf_2N$). They describe the ability to select from many types of organic species in the formulation of coordinated ionic liquid monomers with $LiTf_2N$ which enables the synthesis of a vast array of polymer-inorganic composites via photopolymerization. Professor Bara's group is focused on the development of processes for clean energy generation, new polymer and composite materials with highly tunable nanostructures for separations and applications utilizing solvents such as ionic liquids (*38*).

Chapter 5, "Self-assembly of Block Copolymers in Ionic Liquids" was prepared by Professor Norm Wagner's group at the University of Delaware. This chapter provides an up-to-date review on the field of amphiphilic block copolymer (ABC) self-assembly in ionic liquids. The group provides perspectives on the current understanding, characterization techniques, challenges, opportunities and new applications to assist in better formulation of ABCs in ionic liquids. Professor Wagner holds the Robert L. Pigford Chair in Chemical Engineering and his group focuses on developing a fundamental understanding of the molecular and nanoscale structure and dynamics of complex materials, especially during flow and processing, which falls under the broader disciplines of rheology, nonequilibrium thermodynamics, complex fluids and soft matter (*39*).

Professor Gary Baker and his group at the University of Missouri along with his collaborators at the Oak Ridge National Laboratory have written Chapter 6, "Multi-Purpose Cellulosic Ionogels". Their work describes the immobilization and characterization of ionic liquids in the form of ionogels prepared from bacterial cellulose alcogels for chemosensory applications. Professor Baker's research is motivated by problem-solving using sustainable nanoscience and task-specific solvents such as ionic liquids for engineering approaches (*40*).

Chapter 7, "Liquid-liquid Extraction of f-block Elements using Ionic Liquids" was written by Dr. Sheng Dai and colleagues at Oak Ridge National Laboratory and the University of Tennessee. This chapter focuses on the liquid-liquid extraction of f-block elements such as lanthanides for recovery of rare earth elements (REEs) and removal of actinides from spent nuclear fuel and waste using task-specific ionic liquids. Dr. Dai and his group are internationally recognized for their research in designing and synthesizing functional porous materials, nanomaterials and ionic liquids for solutions to energy-relevant problems and he holds a joint-faculty appointment at Oak Ridge National Laboratory and in the Department of Chemistry at the University of Tennessee (*41*).

Section three covers the topic of "Biomass Processing" and begins with Chapter 8, "Viscosity and Rheology of Mixtures of Cellulose, Ionic Liquid and Cosolvents for Advanced Processing" by Professor Aaron Scurto and his group at the University of Kansas. Professor Scurto and his group have demonstrated that by adding an aprotic cosolvent such as dimethyl sulfoxide (DMSO), etc. to mixtures of ionic liquids, such as 1-ethyl-3-methylimdiazolium diethyl phosphate [EMIm][DEP], that the thermodynamic cellulose solubility increases, the viscosity of the mixture significantly decreases and the overall economics of the process improves. Professor Scurto's group focuses on sustainable chemistry and engineering, biomass processing, and alternative solvents such as ionic liquids and compressed CO_2 for catalysis and separation applications (*42*).

Chapter 9, "Ultra-low Cost Ionic Liquids for the Delignification of Biomass" was written by Dr. Jason Hallett and his group at Imperial College in London. Dr. Hallett has pioneered the use of ultra-low cost ionic liquids using aqueous acids such as sulfuric acid and amine bases which dissolve the lignin and hemicellulose leaving a cellulose-rich pulp ready for saccharification. Dr. Hallett's research interests include the solvation behavior of ionic liquids and the use of ionic liquids in the production of lignocellulosic biofuels and sustainable chemical feedstocks (*43*).

Section four covers the topic of "Fundamentals" and begins with Chapter 10, "Water at Ionic Liquid Interfaces" written by Professor John Newberg and his group at the University of Delaware. They have summarized experiments using microscopy, spectroscopy (high-pressure XPS) and scattering techniques with molecular dynamic (MD) simulations involving water at ionic liquid (IL-vacuum, IL-gas and IL-solid) interfaces. Professor Newberg's research interests include metal oxide surface chemistry relevant to atmospheric and catalytic processes, ionic liquid surface chemistry relevant to sequestration and catalytic processes, and the chemistry of ice relevant to atmospheric and polar region environments (*44*).

Dr. Jim Wishart and his group at Brookhaven National Laboratory have prepared Chapter 11, "Radiation and Radical Chemistry of Ionic Liquids for Energy Applications". His work presents current progress in understanding radiation chemistry of ionic liquids especially for photoelectrochemical solar cells, batteries, and processing nuclear fuels. The properties of ILs, especially slower dynamics, allow a unique window into their radiation chemistry on short timescales. Radiation chemistry and pulse radiolysis are excellent tools for studying general chemical reactions in ILs. In addition, ILs may be advantageous materials for separations in the nuclear fuel cycle. Dr. Wishart is an expert in radiation chemistry and the effects of radiation on ionic liquids (*45*).

Chapter 12, "Experimental Study of the Interactions of Fullerene with Ionic Liquids" was written by Professor Margarida Costa Gomes and her group at the Institut de Chimie de Clermont-Ferrand. In this work, they present an experimental study on the energy required to replace a good solvent for fullerene (C_{60}) such as 1,2-dichlorobenzene (DCB) with a room temperature ionic liquid (RTIL) such as 1-decyl-3-methylimidazolium bis(trifluoromethylsulfonyl)imide [$C_{10}C_1$Im][NTf_2]. Professor Costa Gomes and her group study the thermodynamic, thermophysical, and phase equilibrium properties of solutions such as the solvation and transport in ionic liquids and the interactions between ionic liquids and solid materials using experimental and theoretical techniques (*46*).

Professor Rico Del Sesto and his group at Dixie State University in collaboration with researchers at Los Alamos National Laboratory, Sandia National Laboratory and Northern Arizona University have prepared Chapter 13, "Biphasic Extraction and Identification of Organic and Inorganic Compounds with Ionic Liquids". Their work demonstrates that room temperature ionic liquids can be used for biphasic extraction and recovery processes from aqueous solutions for a broad range of compounds including organics, biomolecules, and metal salts with high efficiency. Professor Del Sesto's expertise includes research on ionic liquids for energy and materials applications at the U.S. Air Force Academy and using water immiscible ionic liquids to extract organic, inorganic and highly water-soluble compounds from aqueous solutions and solid materials through biphasic separation (*47*).

A few additional authors who presented at the conference but that were unable to provide chapters should also be mentioned here. Dr. Joe Magee at the National Institute of Technology (NIST) in Boulder, Colorado presented, "Vapor + Liquid Phase Equilibrium for [C_6mim][Tf_2N]". This presentation highlighted the vapor liquid equilibrium for 1-hexyl-3-methylimidazolium bis(trifluoromethyl)sulfonylimide [C_6mim][Tf_2N] which serves as a reference ionic liquid for researchers to compare and verify their results and methods. Dr. Magee's research interests are measurements, models, databases and dynamic data evaluation for thermodynamic and transport properties, with a focus on ionic liquids, natural gas, hydrogen, aqueous mixtures, biological systems and biofuels. He also has a key interest in engineering education with a focus on research internships for undergraduate, graduate and postdoctoral engineering students at NIST (*48*).

Dr. Olga Kuzmina presented for Professor Tom Welton's group at Imperial College London on the "Physicochemical Properties of Cellulose-dissolving Ionic Liquids and their cellulose solutions". Professor Welton and his group work in the field of sustainable chemistry and study the properties of ionic liquids, their interactions with solutes, and the resulting effects on chemical reactions (*49*).

Professor James Davis presented his work on "Ionic Liquids as 'Ionic Solids' and their Design for Separations, Catalysis and more". Professor Davis is a Professor of Chemistry at the University of South Alabama and his group has made many contributions to the field including the introduction of functionalized or "task-specific" ionic liquids (*50*).

Professor Richard Noble at the University of Colorado in Boulder presented "Poly(ionic liquid)/ionic liquid Composite Membranes for High Temperature Ion Conductance". Professor Noble and his group have pioneered the development of polymerizing ionic liquids to create unique membrane materials (*51*).

Professor Dr. Hans-Peter Steinrück presented "Advances in Surface and Interface Science of Ionic Liquids". Professor Dr. Steinrück works at the Friedrich-Alexander Universitat (FAU) in Erlangen, Germany and uses high-pressure XPS to study surfaces of ionic liquids (*52*).

Commercialization

Several commercial products have been developed using ionic liquids in the past decade. We highlight only a few here which have made a substantial impact. Professor Daniel Armstrong at the University of Texas in Arlington has developed a new class of capillary gas chromatography (GC) columns with stationary phases based on ionic liquids. His group has synthesized dicationic and polycationic ionic liquids which are stable to water and oxygen even at high temperatures (*53*). A variety of capillary GC columns are now available based on IL technology. The technique can also be used for detection of water using a thermal conductivity detector (TCD) at extremely low limits of detection (LOD). Compared with the standard Karl Fischer titration (KFT) method used today with a lower limit of detection (LOD) of 10 μg (1 ppm), the ionic liquid method has a lower LOD of ~2 ng (0.0002 ppm) (*53*). The columns are now commercially available from Supelco/Sigma Aldrich (*54*).

In the past couple of years (2015-2016) we have seen some of the largest scale ionic liquid processes announced. For example, the ionic liquids process developed by Queens University Ionic Liquids Laboratory (QUILL) in collaboration with the Malaysian oil and gas company, Petronas for the efficient scrubbing of mercury vapor from natural gas (*35*, *55*). The process is now operating on an industrial-scale using chlorocuprate (II) ionic liquids impregnated on high surface area porous solid supports. The supported ionic liquid phase (SILP) approach to heterogenize the ionic liquid allowed the material to be used in standard industrial-scale mercury removal equipment and the rapid commercialization of the process. The SILP containing ionic liquid outperformed the incumbent activated carbon and better manages process upsets such as spikes in mercury concentration.

Chevron recently announced in October, 2016 the development of a new chloroaluminate ionic liquid alkylation catalyst (*56*). The chloroaluminate ionic liquids provide high activity, selectivity and catalyst stability for C_4 alkylation and provide Chevron with an alternative to using corrosive and toxic hydrofluoric acid (HF) as a catalyst. Chevron stated that they began developing the technology in 1999 and have operated a demonstration unit for the past five years. They plan to start construction in 2017 on a full-scale alkylation plant at their Salt Lake City refinery. After the plant is complete in 2020, Chevron plans to remove all HF specific equipment and its inventory of HF from the site. Honeywell UOP will license and market the new IL technology called Isoalky to the refining industry (*57*, *58*).

There continues to be a flurry of activity in the number of patents published involving ionic liquids. As of March 2017, there were over 3300 U.S. patent applications pending with the term "ionic liquid" just in the claims. This is approximately double the number of currently granted U.S. patents!

Conferences

A number of excellent conferences have been held since the original NATO-ARW was held in Crete in April, 2000. A few conferences which we would like to specifically mention that have occurred now for a number of years and focused entirely on the field of ionic liquids include: Congress On Ionic Liquids (COIL 2005, 2007, 2009, 2011, 2013, 2015) (*59*), International Conference on Ionic Liquids in Separation and Purification Technology (ILSEPT 2011, 2014, 2017) (*60*), and the Gordon Research Conference on Ionic Liquids (GRC 2014, 2016, 2018) (*61*). In addition, hundreds of sessions too numerous to count have been held world-wide at a variety of scientific meetings.

Books

A set of three books were recently published (2013, 2014, 2015) by some of the leading authors in the field which provide overviews on a variety of current topics and future trends. The books, "Ionic Liquids UnCOILed", "Ionic Liquids Further UnCOILed" and "Ionic Liquids Completely UnCOILed" were edited by Natalia Plechkova and Professor Ken Seddon (*23*, *27*, *29*). We point the reader to this set of books in particular because we (MBS and AMS) have found many errors in the papers being published (over 100 per week), and some errors are being propagated from one paper to the next. These critical reviews provide honest feedback and insights from some of the leading authors in the field and provide important corrections in some areas.

Database

Ionic Liquids Database (ILThermo v2.0) (*62*, *63*) is a free web research tool that allows users worldwide to access an up-to-date data collection from the publications on experimental studies of thermodynamic and transport properties of ionic liquids. The database also includes binary and ternary mixtures containing ionic liquids with other solutes. ILThermo contains information on hundreds of ions and ionic liquids. The collected data cover the relevant literature from 1982 to 2016. The experimental data stored in the database include phase transitions, transport, volumetric, and thermal properties as well as electrical conductivity, surface tension, refractive index, speed of sound, vapor pressures, and activity coefficients. ILThermo also includes information on chemical identification, sample purity, details of experimental methods and numerical data uncertainty.

Data Quality and IUPAC Reference Ionic Liquid

The quality of some of the ionic liquid data available in the open literature is of concern. This issue led to the establishment of a task force group to systematically study and publish thermophysical and thermodynamic results for the ionic liquid, 1-hexyl-3-methylimidazolium bis(trifluoromethylsulfonyl)imide ($[C_6C_1Im][NTf_2]$) that research groups could use for reference to validate their experimental methods. The International Union of Pure and Applied Chemistry (IUPAC) sponsored the project to make physical property and gas solubility measurements available as part of Project 2002-005-1-100 (Thermodynamics of Ionic Liquids, Ionic Liquid Mixtures, and the Development of Standardized Systems) (*64*, *65*). The editors (MBS and AMS) strongly encourage any research group making measurements with ionic liquids to check their methods by first testing $[C_6C_1Im][NTf_2]$ to make sure their results agree within the experimental uncertainty. A thorough error analysis needs to be conducted for any measurement of a pure IL or IL mixture property (or any substance for that matter). The inherent uncertainty (accuracy of measurements, propagation of error for calculated properties, etc.) and the run-to-run and statistical error (precision, average, standard deviation) over multiple measurements needs to be pursued, explained in the experimental methodology, and documented in the Tables and Figures (error bars). A good reference is the NIST Technical Note (*66*).

Part of the discrepancies in the literature can be attributed to IL sample quality. Basic measurements should be performed and reported for the IL, such as water content (usually Coulometric Karl-Fischer analysis); residual halide content for ILs synthesized through anion exchange from a halide precursor (anion specific electrodes, etc.); IL purity measurements from such techniques as quantitative NMR, HPLC, and elemental analysis. Ionic liquid stability is sometimes another reason for discrepancies. It has been well-documented that some of the fluorinated anions such as tetrafluoroborate (BF_4^-) and hexafluorophosphate (PF_6^-) have low hydrolytic chemical stability and in the presence of water can form HF (*67*). We recommend that future publications avoid the use of these anions for applications where water is likely going to be present.

Toxicity

There is a general lack of data on the toxicology of many ionic liquids and their impact on human health and the environment. The data that has been collected to date such as properties on the toxicity and biodegradability varies immensely from one ionic liquid to another. Minor structural differences, such as the length of the alkyl chain on an imidazolium cation can have a pronounced positive or negative impact on toxicity. The large number of existing ionic liquids provides opportunity for toxicologists to study these materials. Furthermore, the property of toxicity such as antimicrobial activity can be put to use in applications such as antiseptics, disinfectants and anti-fouling agents (*15*). In addition, the toxicity (or biological activity) of ionic liquids can be designed for use as pharmaceutical ingredients or agricultural intermediates. The approach is to combine the activity of two active pharmaceutical ingredients (APIs) as a cation and anion in an ionic liquid (*36*). Ionic liquids as APIs has been referred to as the "third evolution of ionic liquids" and is an exciting new area where ILs will lead to new products with improved efficacy (*68*).

Conclusions

The tremendous research effort over the past two decades focused on the field of ionic liquids has yielded a wealth of fundamental knowledge. Many applications in chemistry, engineering, and material science are being studied, thousands of papers and patents are being published, and new products and processes utilizing ionic liquids are being commercialized. This ACS symposium series book has brought together contributions by some of the foremost researchers to help both current and new researchers in the area obtain knowledge and understanding of the field's broad interest and potential. This introduction chapter has summarized some of the history, applications, conferences, books, databases, issues related to data quality and toxicity of ionic liquids.

In this age where digital information is available in the palm of one's hand, the editors still experience the joy and serendipity of reading a book. We hope the readers will delve into all the chapters!

References

1. Rogers, R. D.; Seddon, K. R. *Ionic Liquids: Industrial Applications to Green Chemistry*; American Chemical Society: 2003.
2. Wasserscheid, P.; Welton, T. *Ionic liquids in Synthesis*; Wiley-VCH: 2003.
3. Rogers, R. D.; Seddon, K. R. *Ionic Liquids as Green Solvents Progress and Prospects*; American Chemical Society: 2003; Vol. 856.
4. Brazel, C. S.; Rogers, R. D. *Ionic liquids in Polymer Systems: Solvents, Additives, and Novel Applications*;American Chemical Society: 2005.
5. Rogers, R. D.; Seddon, K. R. *Ionic Liquids III A: Fundamentals, Progress, Challenges, and Opportunities*; American Chemical Society: 2005; Vol. 901.

6. Rogers, R. D.; Seddon, K. R. *Ionic Liquids IIIB: Transformations and Processes*; OUP USA: 2005.
7. Ohno, H. *Electrochemical Aspects of Ionic Liquids*; John Wiley & Sons: 2005.
8. Letcher, T. M. *Development and Applications in Solubility*; Royal Society of Chemistry: 2007.
9. Wasserscheid, P.; Welton, T. *Ionic liquids in Synthesis*; Wiley Online Library: 2008; Vol. 1.
10. Wasserscheid, P.; Welton, T. *Ionic liquids in Synthesis*; Wiley Online Library: 2008; Vol. 2.
11. Koel, M. *Ionic liquids in Chemical Analysis*; CRC Press: 2008
12. Endres, F.; MacFarlane, D.; Abbott, A. *Electrodeposition from Ionic Liquids*; John Wiley & Sons: 2008.
13. Plechkova, N. V.; Rogers, R. D.; Seddon, K. R. *Ionic Liquids: from Knowledge to Application*; American Chemical Society: 2009; Vol. 1030.
14. Kirchner, B. *Ionic liquids*; Springer: 2009; Vol. 290.
15. Freemantle, M. *An Introduction to Ionic Liquids*; Royal Society of Chemistry: 2010.
16. Gaune-Escard, M., Seddon, K. R., Eds. *Molten Salts and Ionic Liquids: Never The Twain?* John Wiley & Sons, Inc.: 2010.
17. Malhotra, S. V., Ed. *Ionic Liquid Applications: Pharmaceuticals, Therapeutics, And Biotechnology*; American Chemical Society: 2010.
18. Trulove, P., Mantz, R., De Long, H. C., Eds. *Physical and Analytical Electrochemistry in Ionic Liquids*; Electrochemical Society: 2010.
19. Ohno, H. *Electrochemical Aspects of Ionic Liquids*; John Wiley & Sons: 2011.
20. Kokorin, A., Ed. *Ionic Liquids: Applications and Perspectives*; InTech: 2011.
21. Rogers, R. D.; Seddon, K. R.; Volkov, S. *Green Industrial Applications of Ionic Liquids*; Springer Science & Business Media: 2012; Vol. 92.
22. Dominguez de Maria, P., Ed. *Ionic Liquids in Biotransformations and Organocatalysis: Solvents and Beyond*; John Wiley & Sons, Inc.: 2012.
23. Plechkova, N. V; Seddon, K. R. *Ionic Liquids UnCOILed: Critical Expert Overviews*; John Wiley & Sons: 2013.
24. Fehrmann, R.; Riisager, A.; Haumann, M. *Supported Ionic Liquids: Fundamentals and Applications*; John Wiley & Sons: 2013.
25. Fang, Z.; Smith, R. L., Jr.; Qi, X. *Production of Biofuels and Chemicals with Ionic Liquids*; Springer Science & Business Media: 2013; Vol. 1.
26. Kadokawa, J.-i., Ed. *Ionic Liquids-New Aspects for the Future*; InTech: 2013.
27. Plechkova, N. V.; Seddon, K. R. *Ionic Liquids further UnCOILed: Critical Expert Overviews*; John Wiley & Sons: 2014.
28. De Los Rios, A. P.; Fernandez, F. J. H. *Ionic Liquids in Separation Technology*; Elsevier: 2014.
29. Plechkova, N. V; Seddon, K. R. *Ionic Liquids completely UnCOILed: Critical Expert Overviews*; John Wiley & Sons: 2015.

30. Dupont, J., Itoh, T., Lozano, P., Malhotra, S. V., Eds. *Environmentally Friendly Syntheses Using Ionic Liquids*; CRC Press: 2015.
31. Bogel-Lukasik, R., Ed. *Ionic Liquids in the Biorefinery Concept: Challenges and Perspectives*; Royal Chemical Society: 2016; Vol. 36.
32. Walden, P. *Izv. Imp. Akad. Nauk (Bull. Acad. Imp. Sci. St. Petersburg)* **1914**, *8*, 405.
33. Wilkes, J. S. A Short History of Ionic Liquids – From Molten Salts to Neoteric Solvents. *Green Chem.* **2002**, *4*, 73–80.
34. Professor Kenneth R. Seddon, Queen's University Belfast. http://pure.qub.ac.uk/portal/en/persons/kenneth-seddon(ab6b1fc0-694e-4118-bd5f-d333cbbe2af1).html.
35. QUILL, Queen's University Ionic Liquids Laboratory. http://www.qub.ac.uk/schools/SchoolofChemistryandChemicalEngineering/Research/ResearchCentres/.
36. Professor Robin D. Rogers, McGill University. https://www.mcgill.ca/chemistry/faculty/robin-d-rogers.
37. Iolitec, Ionic Liquid Technologies. http://www.iolitec.de/.
38. Professor Jason E. Bara, University of Alabama. http://jbara.eng.ua.edu/.
39. Professor Norman J. Wagner, University of Delaware. http://sites.udel.edu/wagnergroup/.
40. Professor Gary A. Baker. University of Missouri. https://chemistry.missouri.edu/people/baker.
41. Dr. Sheng Dai, Oak Ridge National Laboratory. https://www.ornl.gov/our-people/sheng-dai.
42. Professor Aaron M. Scurto, University of Kansas. https://cpe.ku.edu/aaron-scurto.
43. Dr. Jason P. Hallett, Imperial College London. http://www.imperial.ac.uk/people/j.hallett/research.html.
44. Professor John T. Newberg, University of Delaware. https://sites.google.com/site/newberglab/.
45. Dr. James F. Wishart, Brookhaven National Laboratory. http://www.chemistry.bnl.gov/SciandTech/PRC/wishart/wishart.html.
46. Professor Margarida Costa Gomes, Institut de Chimie de Clermont-Ferrand. http://tim.univ-bpclermont.fr/guida/.
47. Professor Rico Del Sesto, Dixie State University. https://science.dixie.edu/faculty/rico-del-sesto/.
48. Dr. Joe W. Magee, National Institute of Standards, Boulder. https://www.nist.gov/people/joe-w-magee.
49. Professor Tom Welton, Imperial College London. https://www.imperial.ac.uk/people/t.welton.
50. Professor James H. Davis. Jr., University of South Alabama. http://www.southalabama.edu/colleges/artsandsci/chemistry/jameshdavisjr/index.html.
51. Professor Richard D. Noble, University of Colorado. http://www.colorado.edu/chbe/richard-d-noble.
52. Prof. Dr. Hans-Peter Steinrück, Friedrich-Alexander Universitat (FAU). https://www.chemie.nat.fau.de/person/hans-peter-steinrueck/.

53. Zheng, Y.; Wang, C.; Armstrong, D. W.; Woods, R. M.; Jayawardhana, D. A. Rapid, Efficient Quantification of Water in Solvents and Solvents in Water Using an Ionic Liquid-Based GC Column. *LCGC* **2012**, *30* (2), 142–158.
54. Sigma-Aldrich Ionic Liquid Gas Chromatography Columns. http://www.sigmaaldrich.com/analytical-chromatography/analytical-products.html.
55. Mahpuzah, A.; Atkins, M. P.; Hassan, A.; Holbrey, J. D.; Kuah, Y.; Nockemann, P.; Oliferenko, A. A.; Plechkova, N. V.; Rafeen, S.; Rahman, A. A.; Ramli, R.; Shariff, S. M.; Seddon, K. R.; Srinivasan, G.; Zou, Y. An Ionic Liquid Process for Mercury Removal from Natural Gas. *Dalton Trans.* **2015**, *44*, 8617–8624.
56. McCoy, M. Chevron Embraces Ionic Liquids. *Chem. Eng. News* **2016**, *94* (39), 16.
57. Timken, H. K. C.; Elomari, S.; Trumbull, S.; Cleverdon, R. Integrated Alkylation Process Using Ionic Liquid Catalysts. U.S. Patent 7,432,408, Oct. 7, 2008.
58. Elomari, S.; Trumbull, S.; Timken, H. K. C.; Cleverdon, R. Alkylation Process Using Chloroaluminate Ionic Liquid Catalysts. U.S. Patent 7,432,409, Oct. 7, 2008.
59. COIL 6: 6th International Congress on Ionic Liquids, http://coil6.cjint.kr.
60. ILSEPT, 3rd International Conference on Ionic Liquids in Separation and Purificatin Technology. https://www.elsevier.com/events/conferences/international-conference-on-ionic-liquids-in-separation-and-purification-technology.
61. *Gordon Research Conference*, 2018: https://www.grc.org/programs.aspx?id=17188
62. Kazakov, A.; Magee, J. W.; Chirico, R. D.; Paulechka, E.; Diky, V.; Muzny, C. D.; Kroenlein, K.; Frenkel, M. *NIST Standard Reference Database 147: NIST Ionic Liquids Database - (ILThermo)*, Version 2.0; National Institute of Standards and Technology: Gaithersburg, MD. http://ilthermo.boulder.nist.gov.
63. Dong, Q.; Muzny, C. D.; Kazakov, A.; Diky, V.; Magee, J. W.; Widegren, J. A.; Chirico, R. D.; Marsh, K. N.; Frenkel, M. ILThermo: A Free-Access Web Database for Thermodynamic Properties of Ionic Liquids. *J. Chem. Eng. Data* **2007**, *52* (4), 1151–1159.
64. Marsh, K. N.; Brennecke, J. F.; Chirico, R. D.; Frenkel, M.; Heintz, A.; Magee, J. W.; Peters, C. J.; Rebelo, L. P. N.; Seddon, K. R. Thermodynamic and Thermophysical Properties of the Reference Ionic Liquid: 1-hexyl-3-methylimidazolium bis[(trifluoromethyl)sulfonyl]amide (including mixtures) part 1. Experimental Methods and Results. *Pure Appl. Chem.* **2009**, *81*, 781–790.
65. Chirico, R. D.; Diky, V.; Magee, J. W.; Frenkel, M.; Marsh, K. N. Thermodynamic and Thermophysical Properties of the Reference Ionic Liquid: 1-hexyl-3-methylimidazolium bis[(trifluoromethyl)sulfonyl]amide (including mixtures) part 2. Critical Evaluation and Recommended Property Values. *Pure Appl. Chem.* **2009**, *81*, 791–828.

66. Taylor, B. N.; Kuyatt, C. E. *NIST Technical Note 1297: Guidelines for Evaluation and Expressing the Uncertainity of NIST Measurement Results*; NIST: 1994.
67. Freire, M. G.; Neves, C. M. S. S.; Marrucho, I. S.; Coutinho, J. A. P.; Fernandes, A. M. Hydrolysis of Tetrafluoroborate and Hexafluorophosphate Counter Ions in Imidazolium-Based Ionic Liquids. *J. Phys. Chem. A* **2010**, *114*, 3744–3749.
68. Hough, W. L.; Smiglak, M.; Rodriquez, H.; Swatloski, R. P.; Spear, S. K.; Daly, D. T.; Pernak, J.; Grisel, J. E.; Carliss, R. D.; Soutullo, M. D.; Davis, J. H.; Rogers, R. D. The Third Evolution of Ionic Liquids: Active Pharmaceutical Ingredients. *New J. Chem.* **2007**, *31*, 1429–1436.

Applications

Chapter 2

Translational Research from Academia to Industry: Following the Pathway of George Washington Carver

Oleksandra Zavgorodnya,[1,†] Julia L. Shamshina,[2,3,†] Paula Berton,[2] and Robin D. Rogers[1,2,*]

[1]Department of Chemistry, The University of Alabama, Tuscaloosa, Alabama 35487, United States
[2]Department of Chemistry, McGill University, 801 Sherbrooke Street West, Montreal, Québec H3A 0B8, Canada
[3]525 Solutions, Inc., 720 2nd Street, Tuscaloosa, Alabama 35401, United States
[†]These authors contributed equally to this work.
[*]E-mail: robin.rogers@mcgill.ca

Translation of new sustainable technologies from academia to industry and their commercialization is less based on the technology itself and more so on cost and demonstration of both viability and significant improvements over the current practice. In this chapter, we will discuss various stumbling blocks in the area of translational science, which include substantially different requirements of funding agencies for academia and industry, reluctance of research Universities when it comes to actual technology translation, and difficulties in demonstration of both process and economic viability of new technologies *via* scale-up in an academic environment. The majority of this chapter is dedicated to our own experiences in pursuing the translation of ionic liquid-based technologies for biomass dissolution and the synthesis of materials from Nature's biopolymers.

Introduction

Green Chemistry

Green Chemistry originated at the intersection of synthetic and environmental chemistry, in 1990s, because of increasing attention to problems of chemical pollution and resource depletion. In 1996, EPA's Office of Chemical Safety and Pollution Prevention (*1*) sponsored the Presidential Green Chemistry Challenge Awards (*2*) to both promote and support sustainable development. While many national and international programs at the time focused on removal of pollution consequences and on finding solutions for already apparent environmental issues, Green Chemistry made a distinction between pollution *prevention* at the earliest stages of planning of chemical processes. In 1998, Paul Anastas and John Warner in the book "Green Chemistry: Theory and Practice" (*3*) formulated twelve Green Chemistry principles summarizing the activities of scientific community, industry, and policy makers, altogether directed to reduction or even elimination of toxic and hazardous chemicals in chemical processes.

The field has grown and evolved progressively. Initially, Green Chemistry aimed for the development of greener synthetic pathways and/or greener reaction conditions estimated using the E-factor (*4*). Nowadays, while many industries are still focusing on greener chemical production, people have started to understand that the majority of chemicals and/or materials (from disinfectants to personal care to household items) are acquired based on their *function* rather than a particular chemical structure (*5*). From this standpoint, Green Chemistry has taken a totally new direction - it is more than designing new chemical products and processes that are sustainable, Green Chemistry represents new business opportunities that are sustainable. These business opportunities focus on the current market pain, and encompass sustainable means and methods to fulfill the need not by item substitution, but instead by designing a totally different technology that would allow a function replacement. *Not because these technologies are green, but because they are better.* Yet, somehow these sustainable technologies need to get to the market.

Sustainable Development – Academic or Applied Research?

From the point of new sustainable development, the academic setting is a promising environment, where people are constantly working in the direction of increasing the general body of knowledge through original scientific research. Research findings in different fields, such as chemistry, biology, biophysics, material science, drug discovery and development are published every day in high-impact journals and books. However, a discovery on its own does not mean its implementation into Society. For instance, there are 108,725 hits on SciFinder for the search term 'herbicides', where 32,023 are focused on their preparation (*6*); however on the shelves of stores people find Dicamba (Banvel®), introduced by Velsicol Chemical Corporation in 1967 (*7*) or glyphosate (RoundUp®) marketed by Monsanto in 1974 (*8*). In another example, the National Institutes of Health (NIH) estimated that as many as 90% of funded projects are not getting to the stage of human testing, not to mention commercialization (*9*). Indeed, in many areas of science, translation of exciting new discoveries into products is severely lagging behind the speed of discovery.

There are several causes that can help to explain these phenomena. Part of the reason is that scientists themselves contrapose academia and industry as two completely different worlds, eternally debating the importance of fundamental *vs.* applied research, thus deepening the already existing 'gap' between research done in academia and its translation into marketable products. This is facilitated by substantially different requirements of funding agencies for academia and industry. The main types of available University funding include federal funding to generate fundamental knowledge, funding to support research and development in small University-incubated businesses (Small Business Innovation Research (SBIR), Small Business Technology Transfer (STTR)) (*10*), and industrial funding. The first type of grants for basic fundamental studies is clearly the foundation for later applied science and therefore for potential commercialization efforts; reducing the number of fundamental studies with time will result in few commercialized technologies and products. At the same time, grants for basic studies are not directed to commercialization. This results in a great number of scientific discoveries being published in high-impact journals, but new marketable products or technologies arising from them (even if/when commercialized) may only get to market several decades later. Industry-sponsored research, while valuable for further innovation and is licensed more often than federally sponsored research (*11*), provides only about 5% (some US$3.2 billion (*12*) of U.S. research universities' annual funding. Such separation between funding agencies and inherent conflict between them is the first reason that exciting new findings are barely making it to the market.

The second reason is the mindset of Society. Students in chemistry as they graduate must often choose between academic and industrial careers. Academics are evaluated in their field using the 'academic currency' – number of papers, patents, presentations, impact factor of publishing journals, H-index, etc., and typically not on the number of commercialized technologies, although today patents are becoming more common as a CV-building tool. As a result, original research is viewed as an intellectual competition rather than a pathway towards changing the world. Research universities are reluctant when it comes to actual technology translation, and media are drowning in articles "Translational research *vs.* basic science: comparing apples to upside-down apples" (*13*) or "The basic *vs.* applied research debate" (*14*). The topic is so important that it was recently discussed at the closing panel discussion of the 64th Lindau Nobel Laureate meeting in 2014 (*15*).

Translation of findings from basic science and fundamental research into industrial practice requires a much broader skillset than just a pure knowledge of science, for example, the ability to carry out a complex chemical synthesis to business plan writing, evaluation of resources, comprehension of both fundamental *and* applied research, scale up expertise, and entrepreneurial skills are also needed. Nonetheless, basic and applied research are complementary, with fundamental science discovering exciting new ideas that might be later translated into Society, and applied research posing new questions to be answered. Progress could stall as long as they are viewed as two completely separate worlds. Perhaps we need to more closely examine the Pasteur Quadrant (*16*). In 1872, Louis Pasteur recognized the importance of both to be considered as one in his quote, "*Il n'existe pas de sciences appliquées, mais seulement des applications de la science, liées entre elles comme le fruit à l'arbre qui l'a porté*" (*"There does not exist a category of science to which one can give the name applied science. There are sciences and the applications of science, bound together as the fruit of the tree which bears it."*) (*17*).

Last Piece of the Puzzle? Proving Commercial Viability

Even if we accept the idea that *we*, academics, can ourselves work on the technology translation, we still come to the problem: how will one prove its commercial viability? Therefore, the last, and perhaps the most important point, is that translation of new methods and practices from academy to industry is based on cost and demonstration of both viability and significant improvements over the current practice. Industrial adoption of new methods is difficult: industry has invested millions of dollars to get where they are today, so the new technology needs to be not only better, but so much better to justify new investments. Thus, one of the most difficult moments of moving a technology from an academic concept to a valuable commercialized product, is crossing the line between early innovation and readiness for licensing.

For many technologies in the commercialization stage, scale up is needed to ensure that the process would work in a commercial production environment. The U.S. Department of Energy (DOE), Department of Defense (DOD), Air Force Research Lab (AFRL), Defense Advanced Research Projects Agency (DARPA), Navy, Army, and National Aeronautics and Space Administration (NASA) use metrics called Technology Readiness Levels (or TRL) to estimate technology maturity in order to reduce the risk of investing in immature technologies. Federal agencies and industrial companies prefer licensing technology with a TRL level of not less than 6, that is technology that has a fully functional prototype or scaled-up pilot.

Even though academia is able to secure adequate funding for fundamental research, patenting, and perhaps even product development, typically, academics have no opportunities to conduct the scale-up needed to meet commercialization requirements. Thus, attraction of additional investments that enable scale-up is a difficult task for universities, and the only realistic pathway is to complete the task through manufacturing partners. Outsourcing is another possible pathway, and a few companies are now working as an intermediate link, scaling-up the technologies made in academia. For example, Oleotek, Montreal, Canada (*18*) provides on-site scale-up equipment rental and offers process scale-up services and pilot plant production from laboratory scale, to kilo-lab scale, to pilot plant scale.

Operating inside a University: Is There an 'Academic Business Model'?

Academic settings should be able to use the knowledge they generate and translate it into useful things for Society. Yet, there is again a controversy:

academia operates for a purpose of generating and providing knowledge, while businesses operate to generate profit. So, what should the 'academic business model' be?

Many U.S. universities have now implemented business incubators, helping faculty and students to translate their innovative discoveries into Society by fostering technology startups. The incubators not only provide space, but act as an organizational vehicle, helping in all aspects of technology transfer, such as evaluating the feasibility and marketability of the ideas and securing the patents. Entrepreneurial Centers at the universities provide support in writing business plans, provide business advice, organize targeted business competitions, or 'Launchpads', often offering substantial cash prizes for certain discoveries. They might also facilitate meetings with business experts for help in finding investment.

A substantial piece of the puzzle is nonetheless missing. From the standpoint of Green Chemistry, the incubators should position Green Technology as a priority area for educational and economic advancement and facilitate the transition to environmentally responsible technologies, processes, and materials. Such transition is possible not only through discovery and development of novel high-value products that combine environmental and performance benefits or present entirely different performance, but also through demonstration of process and economic viability via scale-up of new technologies.

Academic scale-up facilities are needed with an access on a weekly/monthly basis, and highly qualified personnel providing technical support during operation to inventors of the technology. This step will bring benefit to the academic side, since students will be involved in an entrepreneurial, multidisciplinary environment that will contribute to their professional development. In addition, scale-up to a pilot plant allows the generation of better techno-economic data to demonstrate the potential of the developed technology, attracting investors that, hopefully, will license the patent and take the technology to industrial scale. Such an academic business model including the formation of interdisciplinary teams to direct R&D efforts towards green technologies ultimately will lead to a 'sustainability revolution.' If this comes to fruition, the economic value will come from new 'products,' perhaps as envisioned by George Washington Carver, who is credited with saving the Alabama economy 100 years ago.

George Washington Carver

George Washington Carver, a professor and later Director of the Agricultural Department at Tuskegee Normal and Industrial School in Tuskegee, Alabama, USA, devoted his work to improving the life of farmers by translating his discoveries into the real world (*19*). One example of his work was the implementation of the crop rotation method and alternative crops grown for improving harvest from heavily planted cotton lands exhausted by cotton farming (*20*). In addition to agricultural innovations, he extensively worked on diversification of farmer's agricultural products by introducing new crops such as sweet potatoes, soybeans, and pecans.

Carver is very well known for introducing and promoting peanuts after devastation of cotton crops caused by the boll weevil (*Anthonomus grandis*) in the early 20th century (*21*). Carver convinced the farmers, who made their living by raising cotton in the mono-culture economy of that time, to grow peanuts instead of cotton, thus converting the entire area to peanut farming. He had also considered which products (or 'market opportunities') were best suited for the poor Alabama region of the time harnessing capacity to diversify the 'product line' for different markets. In search of different products, Carver developed more than 325 possible applications for peanuts including milk, butter, cooking oil, sauce and use of peanut products in cosmetics and medicine (*22*). Not only did Carver invent new products, but he also translated his knowledge to the farmers through teaching them new farming practices and introducing new applications for their crops. He demonstrated direct translation of inventions from academia to their real life applications.

Every new research area, including Green Chemistry, faces similar challenges as Carver did: How to translate new technologies developed during academic research to full scale industrial applications? The next section will explore several examples of technology transfer from academia to industry.

Case Study – Technology Transfer from Academia to Industry

Renewable Polymers as Alternative to Synthetic Plastics

Nature's abundant renewable polymers are attracting significant attention as alternative to synthetic plastics (*23*). Among those, cellulose, which obtained from lignocellulosic biomass and chitin, obtained from the supporting external shell of organisms such as crustaceans, fungi, and insects, are two the most widespread biopolymers on Earth (Figure 1). While lignocellulosic biomass is an excellent renewable source, cellulose and other biopolymers in lignocellulosic biomass have poor solubility in conventional solvents that limits their wide acceptance, although remarkable chemical and mechanical properties of cellulose have resulted in many industrial applications, including textiles, cosmetics, and paper production (*24*, *25*), as well as biofuels. Yet, out of the all the cellulose Nature produces, only < 0.1% is used for product manufacturing annually (*26*). Comparable to the recovery of biopolymers from lignocellulosic biomass, the isolation of chitin from biomass is a chemically intensive process and affects the superior properties of the native biopolymer. Native chitin particularly holds tremendous potential for use in the food industry, medicine, and as separation materials, but challenging to process into advanced materials due to its poor solubility. As a case study in translational research, we will review alternative technologies for dissolution and synthesis of materials from nature-made biopolymers developed by our group.

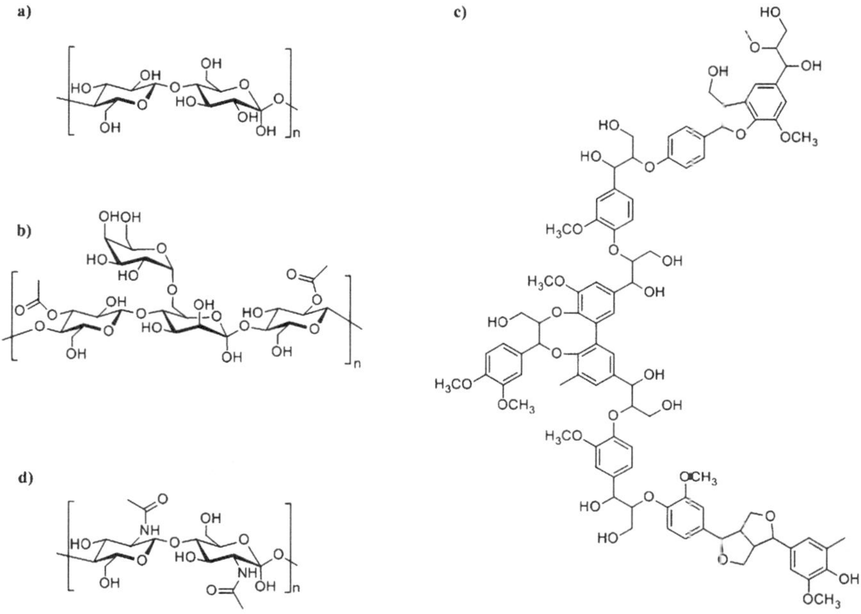

Figure 1. Structure of the biopolymers: a) cellulose, b) representative hemicellulose (galactoglucomannan, major hemicellulose in softwood), c) representative lignin, and d) chitin.

Translation of IL-Based Technologies for Cellulose Dissolution and IL-Assisted Preparation of Functional Materials

In 2002, while at The University of Alabama, we found that cellulose could be dissolved using the ionic liquid (IL, now defined as a salt with a melting point below 100 °C (*27*)) 1-butyl-3-methylimidazolium chloride ($[C_4mim]Cl$) (*28*, *29*). Although ILs were introduced in the late 90s as alternative solvent systems to volatile organic solvents (VOCs), and as electrolytes (*30*), it was the first successful example of complete cellulose dissolution with IL. As compared to industrial process, where cellulose is dissolved by derivatization (*e.g.*, cellophane is produced using a soluble carbon disulfide cellulose derivative (*31–33*), that is upon production of the film is converted back to cellulose by chemical treatment), IL can dissolve cellulose in concentrations sufficient for manufacturing products without the need of derivatization (*28*, *29*). To show versatility of the technology, our group reported preparation of cellulose films, beads, fibers and composites by regenerating cellulose from ILs at the lab scale (*34*, *35*).

The technology has been patented by The University of Alabama who paid a burden of patenting costs. In 2005, BASF obtained exclusive license rights for this technology from The University of Alabama, including dissolution, regeneration, and processing of cellulose (*36*, *37*), before the scale-up of this process was developed.

In the following years, BASF screened a variety of ILs for dissolution of raw cellulosic materials from different biomasses achieving polymer dissolution in amounts from 5 to 25 wt% and launched a product line of cellulose solutions in ILs under the commercial name CELLIONICS™. In collaboration with the Institut für Textilchemie und Chemiefasern (Denkendorf, Germany) and the Thüringisches Institut für Textil- und Kunststoff-Forschung (Rudolstadt, Germany), BASF adapted a dry-wet spinning process for pulling cellulosic fiber from 1-ethyl-3-methylimidazolium acetate ([C_2mim][OAc]) IL at the pilot plant scale (*38*). A distillation process for recycling of the IL after fiber spinning, identified as a major production cost driver, was also developed and optimized. To date, BASF holds about 40 patents related to cellulose dissolution and cellulosic materials production according to SciFinder (search term 'cellulose ILs'), yet, despite great progress achieved in cellulose-based materials synthesis using ILs, this technology has not yet reached wide implementation at a large industrial scale, perhaps because the IL-technology was taken over by the industry at the discovery stage without developing and optimizing scale-up process by the inventors. From industrial point of view, such an early technology takeover required significant financial investments in addition to extra time and efforts, needed to understand and implement the technology at the manufacturing scale, which might be the reason of slow industrialization.

Translation of IL Technology for Separation of Lignocellulosic Biomass Components

Lignocellulosic biomass has three major components: cellulose, hemicellulose, and lignin, in addition to several inorganic and some extractives such as phenolic compounds (*39*). To use cellulose directly from natural feedstocks such as wood, separation of cellulose (pulping) from lignin and hemicellulose is needed. To date, this separation in industry is primarily conducted using Kraft pulping, where the semi-selective chemical degradation of the wood matrix is done by sodium hydroxide/sodium sulfide treatment that introduces significant environmental pollution, in addition to extremely high energy consumption (*40*). Governmental (*41*) and Societal pressure on finding more environmentally friendly pulping methods for cellulose that would at the same time reduce the lignocellulosic waste accumulated every year (*39*, *42*, *43*) spurred the development of alternative technologies based on ILs.

Since ILs were already demonstrated to dissolve cellulose, hemicellulose, and lignin, several academic research groups began to extensively investigate whether lignocellulosic biomass could be itself dissolved and fractionated using ILs (*44–48*). Thus far, the IL [C_2mim][OAc] has been found to be the best solvent for complete dissolution of untreated wood biomass. Moreover, reconstitution of wood dissolved in IL resulted in separation of a free-lignin fraction and cellulose-rich fraction containing only 23.5% of lignin (*44*). However, the cellulose chains in form of macrofibrils incorporated in a matrix of lignin and shorter heteropolysaccharides such as hemicelluloses and pectins that hold these biopolymers together, prevent their complete separation. In 2010, we demonstrated that cleaner separation can be achieved by employing ILs with added catalyst such as polyoxometalates (POMs), combined with oxygen (*49*). Known from 1990, POMs, the metal oxygen clusters of d^0 metal cations (Mo^{VI} or V^V), are used as oxidation catalysts in different applications (*50*). The preliminary results on separation of lignocellulosic biomass by pretreatment with POM catalyst for delignification were excellent and we, in collaboration with 525 Solutions Inc., received U.S. DOE SBIR Phase I funding for further investigation of this technology (*51*). Results of Phase I showed that application of POMs catalyst in [C_2mim][OAc] IL not only decreased biomass dissolution time (from 46 to 15 h), but also resulted in a superior separation of lignin from cellulose with lignin content in cellulose decreasing from 23.5% to as low as 5.4% (*49*).

Based upon our early results, we planned to develop and optimize the scale-up process for this technology, from bench to pilot mini-plant and submitted a U.S. DOE SBIR Phase II application. Here, the overgeneralization-type arguments that ILs are expensive and toxic chemicals prevented us from receiving the Phase II on this project, despite the fact that the term ILs represents a class of compounds defined as salts that melt below 100 °C, and could have almost unlimited variation in chemical composition, price and toxicity. Besides, if 'expensive' IL is the only negative factor to implement the most effective separation technology, we have to consider leveraging its cost through the initial manufacture of high-value products. In such a case, while in the short run the technology would be more expensive than those currently practiced, once in place, the cost will drop down in a long term and, eventually, would become widely acceptable.

What we learned from these experiences is that there are few key components of successful technology transfer from academia to industry that include not only development and optimization of scale-up process, but also development of high-value end products. Furthermore, the successful business plan focused on market pain, and a cost of the technology when scaled is a key for academic groups, who are working on proving scalability. Besides, new technology will not always be expensive, but indeed, the opposite is true.

Are these two case scenarios unique for our group? In our opinion, they are not. Although a lot of research groups in academia perform high quality research generating a lot of Intellectual Property (IP), this IP is not always able to cross the gap between early innovation and technology readiness for industry (*52*).

Translation of ILs Technology for Synthesis of Functional Materials from Chitin: The Beginning

The lesson was learned by the time we were ready to translate our next technology, IL-based chitin extraction from shrimp shell biomass waste (*53*), to an industrially acceptable level. The chitin biopolymer has attracted a lot of attention in applications from metal recovery (*54*, *55*), tissue engineering (*56*, *57*), to drug delivery (*58*) among others, due to its biocompatibility and low toxicity, in addition to availability of functional groups suitable for surface modification.

We developed a chitin extraction process for direct isolation of chitin with high purity and high molecular weight from shrimp shells. The materials prepared from it, appeared to be much stronger compared to those prepared (in cases it was possible) from the commercially available biopolymer (*53*). We have shown IL-based dissolution to form this normally insoluble polymer into multiple useful structures (films, fibers, and complex three-dimensional networks) for functional material preparation (*59–62*).

One of our most interesting applications to date is the development of metal ion specific sorption materials for the selective extraction of metal ions from seawater. Chitin is naturally present in seawater and in search of alternative energy sources, we investigated the possibility to use chitin as a sorbent for uranium (*54*). We have shown that the sorbents in the form of dry jet-wet spun chitin fibers withstood surface modification with uranyl selective groups, resulting in high affinity to uranyl ions; however, for this application high surface area materials were desirable (*63–67*). To increase the surface area of our sorbent materials, we investigated electrospinning (*68*, *69*), a method previously shown to work with cellulose (*70–77*) and in our first effort with chitin (*61*); however, all of these methods used a single needle syringe. To make this technology attractive for industry, we had to show the possibility to scale up materials production using the electrospinning technique. Together with 525 Solutions, Inc., we received a U.S. DOE SBIR Phase I award, where we demonstrated scale up from 3 to 300 mL (100-fold increase) of the electrospinning set-up.

The scale-up was realized through the design and manufacturing of a multi-nozzle stainless steel spinneret equipped with insulating Teflon brackets, which was mounted on the bench-stand frame. Such set-up provided flexibility in adjustment of electrospinning parameters (*e.g.*, working distances) by changing the position of Teflon brackets in the frame. The multi-needle spinneret was built as a pressure fed vessel (*78*), similarly to spinnerets used for electrospinning from VOCs (*79*) and was tested for electrospinning of different chitin solutions. The electrospinning parameters including voltage and distance from the needle to the collector and electrode were adjusted to achieve continuous electrospinning of nanofibers from regenerated chitin solutions (*78*).

However, the scale-up of the electrospinning is not all that needs to be done to make the technology attractive for industrial partners. Indeed, the experience with chitin in the above research has led us to the inescapable conclusion that without a larger and consistent supply of the biopolymer, and supporting technologies to prepare high-value materials, the commercialization of this (and any similar) technology will be quite limited. In this instance, it is truly a case where innovation is stymied by simple lack of supply and scaled economy. Having consistent supplies of chitin (and therefore materials from it), would open the opportunity to develop a diverse product line to various applications, from water purification to medical materials and energy applications.

We are currently working on a U.S. DOE SBIR Phase II (DE-SC0010152), where we are focusing our efforts on designing and implementing continuous chitin extraction at pilot scale, to demonstrate its scale-up viability. Once we have established this critical step, we believe the pathway will be opened for a 'chitin economy' with nearly limitless applications of chitin materials. The ability to consistently produce chitin biopolymer would give us a competitive advantage to diversify the range of possible products and enter several profitable specialized markets, with initial focus on the high-value high-cost products such as medical grade chitin and biomedical chitin products. Building high end markets will help pay for process development and economy of scale and, hopefully, down the road, will result in the revenue stream needed for production of low-value commodity materials.

Future Remarks

Our role as scientists is vital in the development of green transformational technologies and achieving Society's goal of a sustainable world. However, our role as 'academics' does not end with the initial idea or even initial development of novel sustainable processes. Academia needs to take basic research one step further, and be the link between discoveries in research Universities and industries, crossing the line between early innovation and commercial readiness, crossing the so-called Valley of Death (*80*).

Now, of course, our commercialization efforts and thus our experiences and advice are based on a US academic-industrial collaboration model. The challenges of traversing the "Valley of Death" could differ substantially between, for example, Europe and the US, perhaps in part because of a difference in the way technology is transferred from academia to industry. One significant factor in this is the ownership of intellectual property rights (IPR). In the US, as a result of The Bayh-Dole Act of 1980, employment contracts for Universities often state that any IPR resulting from faculty research belongs to the University if/when any university resources are used to create the invention. This leads to less industry-sponsored research activity in US Universities brought about by the fact that industry, in case of IP development, would have to license the IP rights from Universities. Contrarily, in Europe, industry is developing open innovation approaches to R&D, strategically combining in-house and external Universities' resources. A uniform patent policy exists that enables industries to retain IPRs, thus maximizing economic value from their intellectual property.

Here USA can learn from European organizations who have established protocols in place for fostering research collaboration and serve to link basic research with commercialization. With basic discovery and product development being two important links in the overall commercialization process, the translation of original chemical research into Society is only possible through demonstration of process feasibility and its economic viability, which is achieved via scale-up of new technologies. In this regard, academic scale-up facilities are the key for both academicians (to generate techno-economic data and demonstrate the potential of the developed technology) and investors (to reduce the risk of investing in immature technologies). Such scale-up facilities will help attract investors and, hopefully, result in technology licensing and its translation to the industrial scale. From an educational perspective, such technology transfer into Society will benefit future generation of scientists, since students will be involved in an entrepreneurial, multidisciplinary environment.

As academics, we strongly believe that our mission is to help and lead society in the right direction. If global climate change is the *boll weevil* of the next generation, why are we waiting until climate change destroys our world instead of start trying to make a difference now?

Acknowledgments

The authors would like to thank 525 Solutions, Inc. together with the U.S. Department of Energy (DOE) SBIR Office of Science (DE-SC0004198, DE-SC0010152, DE-FG02-13ER90708, DE-FG02-10ER85848) and the U.S. National Science Foundation Small Business Innovation Research (NSF-SBIR) (IIP-1143278) for their support of biomass related research. This research was undertaken, in part, thanks to funding from the Canada Excellence Research Chairs Program.

References

1. EPA US Environmental Protection Agency, Office of Chemical Safety and Pollution Prevention (OCSPP). https://www.epa.gov/aboutepa/about-office-chemical-safety-and-pollution-prevention-ocspp (accessed Nov. 18, 2016).
2. EPA US Environmental Protection Agency, Presidential Green Chemistry Challenge Winners. https://www.epa.gov/greenchemistry/presidential-green-chemistry-challenge-winners (accessed Nov. 18, 2016).
3. Anastas, P. T.; Warner, J. C. *Green Chemistry: Theory and Practice*; Oxford University Press: New York, NY, 1998.
4. Sheldon, R. A. The E-Factor: Fifteen Years On. *Green Chem.* **2007**, *9*, 1273–1283.
5. Bourne, R. A.; Poliakoff, M. Green Chemistry: What is the Way Forward? *Mendeleev Commun.* **2011**, *21*, 235–238.
6. SciFinder search. Search term 'Herbicides,' refining on 'Preparation,' accessed Nov. 18, 2016.
7. *Drinking Water and Health*; Safe Drinking Water Committee, Commission on Life Sciences, Division on Earth and Life Studies, National Research Council; National Academies Press: Washington, DC, 1977; Vol. 1, pp 1–948.
8. Monsanto. History of Monsanto's Glyphosate Herbicides. http://www.monsanto.com/products/documents/glyphosate-background-materials/back_history.pdf (last accessed Nov. 18, 2016.
9. Hobin, J. A.; Galbraith, R. A. Engaging Basic Scientists in Translational Research. *FASEB J.* **2012**, *26*, 2227–2230.
10. SBIR/STTR. https://www.sbir.gov/about/about-sttr (last accessed Nov. 18, 2016.
11. Write, B.; Drivas, K.; Lei, Z.; Merrill, S. A. Technology Transfer: Industry-Funded Academic Inventions Boost Innovation. *Nature* **2014**, *507*, 297–299.

12. National Science Board Science and Engineering Indicators. 2014, 5-13 (NSF, 2014). https://www.nsf.gov/statistics/seind14/content/etc/nsb1401.pdf (last accessed April 24, 2017).
13. Selep, M. Translational Research vs. Basic Science: Comparing Apples to Upside-Down Apples. Nov. 19, 2013. http://blogs.plos.org/thestudentblog/2013/11/19/translational-research-vs-basic-science-comparing-apples-to-upside-down-apples/ (last accessed Nov. 18, 2016).
14. University Affairs Canada, opinion article. The Basic vs. Applied Research Debate. http://www.universityaffairs.ca/opinion/margin-notes/the-basic-vs-applied-research-debate/ (last accessed Nov. 18, 2016).
15. Barré-Sinoussi, F.; Schmidt, B. P. Video – Closing Panel Discussion: Science for the Benefit of Mankind (held on 08-19-2014). Panelists: Barré-Sinoussi, Bassioni, Mgone, Schmidt, Schütte. http://www.mediatheque.lindau-nobel.org/videos/33787/2014-panel-discussion-science-for-the-benefit (last accessed Nov. 18, 2016).
16. Wikipedia. https://en.wikipedia.org/wiki/Pasteur%27s_quadrant (last accessed Nov. 30, 2016).
17. Le Van, M. *The Concise Dictionary of Foreign Quotations*; Routledge: London, 2002; pp 1−332.
18. OLEOTEK offers process scaling-up services as well as pilot plant production from 100 grams (laboratory scale), to tens of kilos (kilo-lab scale), to several tons (pilot plant scale). http://www.oleotek.org/transfert-technologies/info_services_usine-pilote_176_ang.cfm (last accessed Nov. 18, 2016).
19. Kremer, G. *George Washington Carver: in His Own Words*; University of Missouri Press: Columbia, 1991. https://muse.jhu.edu/ (last accessed on Nov. 19, 2016).
20. Epstein, S.; Epstein, B.; Moyers, W. *George Washington Carver: Negro Scientist*; Garrard Press: Champaign, IL, 1960; pp 1−80.
21. Higgs, R. The Boll Weevil, The Cotton Economy, and Black Migration 1910-1930. *Agricultural History* **1976**, *50*, 335–350.
22. Thomas, H. *George Washington Carver*; Van Rees Press: New York, NY, 1958; pp 1−126.
23. Babu, R. P.; O'Connor, K.; Seeram, R. Current Progress on Bio-based Polymers and Their Future Trends. *Prog. Biomater.* **2013**, *2*, 8–16.
24. Pandey, J. K.; Ahn, S. H.; Lee, C. S.; Mohanty, A. K.; Misra, M. Recent Advances in the Application of Natural Fiber Based Composites. *Macromol. Mater. Eng.* **2010**, *295*, 975–989.
25. Edgar, K. J.; Buchanan, C. M.; Debenham, J. S.; Rundquist, P. A.; Seiler, B. D.; Shelton, M. C.; Tindall, D. Advances in Cellulose Ester Performance and Application. *Prog. Polym. Sci.* **2001**, *26*, 1605–1688.
26. Simple and Environmentally Friendly Processing of Cellulose by Means of Ionic Liquids from BASF. http://www.chemicalonline.com/doc/simple-and-environmentally-friendly-processin-0001 (last accessed Nov. 21, 2016).
27. Rogers, R. D.; Seddon, K. R. Ionic liquids - Solvents of the Future? *Science* **2003**, *302*, 792–793.

28. Swatloski, R. P.; Spear, S. K.; Holbrey, J. D.; Rogers, R. D. Dissolution of Cellulose with Ionic Liquids. *J. Am. Chem. Soc.* **2002**, *124*, 4974–4975.
29. Swatloski, R. P.; Holbrey, J. D.; Rogers, R. D. Dissolution and Processing of Cellulose using Ionic Liquids. Int. Pat. WO 03/029329, 2003.
30. Wilkes, J. S. A Short History of Ionic Liquids-from Molten Salts to Neoteric Solvents. *Green Chem.* **2002**, *4*, 73–80.
31. Coss, C. F.; Spruance, D. C. Process of Purifying Viscose. U.S. Patent, 763266, 1904.
32. Lyncke, H. Improvements Relating to the Production or Preparation of Powdered Soluble Alkali-Cellulose Xanthate. GB Patent Office, Pat. No. 190808023, 1908.
33. Hyden, W. L. Manufacture and Properties of Regenerated Cellulose Films. *Ind. Eng. Chem. Res.* **1929**, *21*, 405–410.
34. Turner, M. B.; Spear, S. K.; Holbrey, J. D.; Daly, D. T.; Rogers, R. D. Ionic Liquid-Reconstituted Cellulose Composites as Solid Support Matrices for Biocatalyst Immobilization. *Biomacromolecules* **2005**, *6*, 2497–2502.
35. Turner, M. B.; Spear, S. K.; Holbrey, J. D.; Rogers, R. D. Production of Bioactive Cellulose Films Reconstituted from Ionic Liquids. *Biomacromolecules* **2004**, *5*, 1379–1384.
36. Maase, M.; Massonne, K.; Uerdingen, E.; Vagt, U Solutions of Cellulose in Ionic Liquids. *Aldrich ChemFiles* **2006**, *6.9*, 3 (http://www.sigmaaldrich.com/technical-documents/articles/chemfiles/cellionics.html; accessed Nov. 24, 2016).
37. *UA news*, March 8, 2006; http://uanews.ua.edu/2006/03/basf-ua-formalize-ionic-liquids-partnership/ (accessed Nov. 21, 2016).
38. Hermanutz, F.; Gähr, F.; Uerdingen, E.; Meister, F.; Kosan, B. New Developments in Dissolving and Processing of Cellulose in Ionic Liquids. *Macromol. Symp.* **2008**, *262*, 23–27.
39. Isikgor, F. H.; Becer, C. R. Lignocellulosic Biomass: A Sustainable Platform for the Production of Bio-based Chemicals and Polymers. *Polym. Chem.* **2015**, *6*, 4497–4559.
40. *Profile of the pulp and paper industry*, 2nd ed.; EPA Office of Compliance Sector Notebook EPA/310-R-02-002, U.S. Environmental Protection Agency: Washington, DC, 2002.
41. https://www.whitehouse.gov/energy/securing-american-energy (accessed Nov. 21, 2016).
42. Wang, H.; Block, L. E.; Rogers, R. D. *Catalytic Conversion of Biomass in Ionic Liquids. Catalysis in Ionic Liquids: From Catalyst Synthesis to Application*; RSC Publishing: Cambridge, U.K., 2014; pp 1–19.
43. Upton, B. M.; Kasko, A. M. Strategies for the Conversion of Lignin to High-Value Polymeric Materials: Review and Perspective. *Chem. Rev.* **2016**, *116*, 2275–2306.
44. Sun, N.; Rahman, M.; Qin, Y.; Maxim, M. L.; Rodríguez, H.; Rogers, R. D. Complete Dissolution and Partial Delignification of Wood in the Ionic Liquid 1-ethyl-3-methylimidazolium acetate. *Green. Chem.* **2009**, *11*, 646–655.
45. Fort, D. A.; Remsing, R. C.; Swatloski, R. P.; Moyna, P.; Moyna, G.; Rogers, R. D. Can Ionic Liquids Dissolve Wood? Processing and Analysis

of Lignocellulosic Materials with 1-n-butyl-3-methylimidazolium chloride. *Green Chem.* **2007**, *9*, 63–69.
46. Myllymäki, V.; Aksela, R. Dissolution Method for Lignocellulosic Materials. PCT WO/2005/017001, 2005.
47. Kilpeläinen, I.; Xie, H.; King, A.; Granstrom, M.; Heikkinen, S.; Argyropoulos, D. S. Dissolution of Wood in Ionic Liquids. *J. Agric. Food Chem.* **2007**, *55*, 9142–9148.
48. Lee, S. H.; Doherty, T. V.; Linhardt, R. J.; Dordick, J. S. Ionic Liquid-Mediated Selective Extraction of Lignin from Wood Leading to Enhanced Enzymatic Cellulose Hydrolysis. *Biotechnol. Bioeng.* **2009**, *102*, 1368–1376.
49. Sun, N.; Jiang, X.; Maxim, M. L.; Metlen, A.; Rogers, R. D. Use of Polyoxometalate Catalysts in Ionic Liquids to Enhance the Dissolution and Delignification of Woody Biomass. *ChemSusChem* **2011**, *4*, 65–73.
50. Gaspar, A. R.; Gamelas, J. A. F.; Evtuguin, D. V.; Neto, C. P. Alternatives for Lignocellulosic Pulp Delignification Using Polyoxometalates and Oxygen: A Review. *Green Chem.* **2007**, *9*, 717–730.
51. Gurau, G. Catalytic Fractionation of Biomass in Ionic Liquids. https://www.sbir.gov/sbirsearch/detail/15849 (accessed Nov. 21, 2016).
52. Wapner, J. The Leap to Industry. *Nature* **2016**, *533*, S13–S15.
53. Qin, Y.; Lu, X.; Sun, N.; Rogers, R. D. Dissolution or Extraction of Crustacean Shells Using Ionic Liquids to Obtain High Molecular Weight Purified Chitin and Direct Production of Chitin Films and Fibers. *Green Chem.* **2010**, *12*, 968–971.
54. Barber, P. S.; Kelley, S. P.; Griggs, C. S.; Wallace, S.; Rogers, R. D. Surface Modification of Ionic Liquid-Spun Chitin Fibers for the Extraction of Uranium from Seawater: Seeking the Strength of Chitin and the Chemical Functionality of Chitosan. *Green Chem.* **2014**, *16*, 1828–1836.
55. Schleuter, D.; Günther, A.; Paasch, S.; Ehrlich, H.; Kljajić, Z.; Hanke, T.; Bernhard, G.; Brunner, E. Chitin-based Renewable Materials from Marine Sponges for Uranium Adsorption. *Carbohydr. Polym.* **2013**, *92*, 712–718.
56. Lee, K. Y.; Jeong, L.; Kang, Y. O.; Lee, S. J.; Park, W. H. Electrospinning of Polysaccharides for Regenerative Medicine. *Adv. Drug Delivery Rev.* **2009**, 1020–1032.
57. Singh, N.; Chen, J.; Koziol, K. K.; Hallam, K. R.; Janas, D.; Patil, A. J.; Strachan, A.; Hanley, J. G.; Rahatekar, S. S. Chitin and Carbon Nanotube Composites as Biocompatible Scaffolds for Neuron Growth. *Nanoscale* **2016**, *8*, 8288–8299.
58. Mi, F.-L.; Shyu, S.-S.; Lin, Y.-M.; Wu, Y.-B.; Peng, C.-K.; Tsai, Y.-H. Chitin/PLGA Blend Microspheres as a Biodegradable Drug Delivery System: A New Delivery System for Protein. *Biomaterials* **2003**, *24*, 5023–5036.
59. Shen, X.; Shamshina, J. L.; Berton, P.; Bandomir, J.; Wang, H.; Gurau, G.; Rogers, R. D. Comparison of Hydrogels Prepared with Ionic-Liquid-Isolated vs Commercial Chitin and Cellulose. *ACS Sustainable Chem. Eng.* **2016**, *4*, 471–480.
60. Shamshina, J. L.; Gurau, G.; Block, L. E.; Hansen, L. K.; Dingee, C.; Walters, A.; Rogers, R. D. Chitin – Calcium Alginate Composite Fibers for

Wound Care Dressings Spun From Ionic Liquid Solution. *J. Mater. Chem. B* **2014**, *2*, 3924–3936.

61. Barber, P. S.; Griggs, C. S.; Bonner, J. R.; Rogers, R. D. Electrospinning of Chitin Nanofibers Directly From an Ionic Liquid Extract of Shrimp Shells. *Green. Chem.* **2013**, *15*, 601–607.
62. King, C.; Shamshina, J. L.; Gurau, G.; Berton, P.; Khan, N. F. A. F.; Rogers, R. D. A Platform for More Sustainable Chitin Films from an Ionic Liquid Process. *Green Chem.* **2017**, *19*, 117–126.
63. Min, B.; You, Y.; Kim, J. M.; Lee, S. J.; Park, W. H. Formation of Nanostructured Poly(lactic-co-glycolic acid)/Chitin Matrix and Its Cellular Response to Normal Human Keratinocytes and Fibroblasts. *Carbohydr. Polym.* **2004**, *57*, 285–292.
64. Noh, H. K.; Lee, S. W.; Kim, J. M.; Oh, J. E.; Kim, K. W.; Chung, C. P.; Choi, S. C.; Park, W. H.; Min, B. Electrospinning of Chitin Nanofibers: Degradation Behavior and Cellular Response to Normal Human Keratinocytes and Fibroblasts. *Biomaterials* **2006**, *27*, 3934–3944.
65. Yusof, N. L. B. M.; Wee, A.; Lim, L. Y.; Khor, E. Flexible Chitin Films as Potential Wound Dressing Materials: Wound Model Studies. *J. Biomed. Mater. Res.* **2003**, *66*, 224–232.
66. Liu, P.; Sehaqui, H.; Tingaut, P.; Wichser, A.; Oksman, K.; Mathew, A. P. Cellulose and Chitin Nanomaterials for Capturing Silver Ions (Ag^+) from Water via Surface Adsorption. *Cellulose* **2014**, *21*, 449–461.
67. Huang, Y.; Miao, Y-E.; Liu, T. Electrospun Fibrous Membranes for Efficient Heavy Metal Removal. *J. Appl. Polym. Sci.* **2014**, *131*, 40864.
68. Schiffman, J. D.; Schauer, C. L. A Review: Electrospinning of Biopolymer Nanofibers and Their Applications. *Polym. Rev.* **2008**, *48*, 317–352.
69. Li, D.; Xia, Y. Electrospinning of Nanofibers: Reinventing the Wheel? *Adv. Mater.* **2004**, *16*, 1151–1170.
70. Meli, L.; Miao, J.; Dordick, J. S.; Linhardt, R. J. Electrospinning from Room Temperature Ionic Liquids for Biopolymer Fiber Formation. *Green Chem.* **2010**, *12*, 1883–1892.
71. Viswanathan, G.; Murugesan, S.; Pushparaj, V.; Nalamasu, O.; Ajayan, P. M.; Linhardt, R. J. Preparation of Biopolymer Fibers by Electrospinning from Room Temperature Ionic Liquids. *Biomacromolecules* **2006**, *7*, 415–418.
72. Xu, S.; Zhang, J.; He, A.; Li, J.; Zhang, H.; Han, C. C. Electrospinning of Native Cellulose from Nonvolatile Solvent System. *Polymer* **2008**, *49*, 2911–2917.
73. Quan, S.-L.; Kang, S.-G.; Chin, I.-J. Characterization of Cellulose Fibers Electrospun using Ionic Liquid. *Cellulose* **2010**, *17*, 223–230.
74. Polaskova, M.; Cermaka, R.; Verneyc, V.; Ponizil, P.; Commereuc, S.; Costa Gomes, M. F.; Padua, A. A. H.; Mokrejs, P.; Machovsky, M. Preparation of Microfibers from Wood/Ionic Liquid Solutions. *Carbohydr. Polym.* **2013**, *92*, 214–217.
75. Freire, M. G.; Teles, A. R. R.; Ferreira, R. A. S.; Carlos, L. D.; Lopes-da-Silva, J. A.; Coutinho, J. A. P. Electrospun Nanosized Cellulose Fibers Using Ionic Liquids at Room Temperature. *Green Chem.* **2011**, *13*, 3173–3180.

76. Park, T.-J.; Jung, Y. J.; Choi, S.-W.; Park, H.; Kim, H.; Kim, E.; Lee, S. H.; Kim, J. H. Native Chitosan/Cellulose Composite Fibers from an Ionic Liquid via Electrospinning. *Macromol. Res.* **2011**, *19*, 213–315.
77. Miao, J.; Pangule, R. C.; Paskaleva, E. E.; Hwang, E. E.; Kane, R. S.; Linhardt, R. J.; Dordick, J. S. Lysostaphin-Functionalized Cellulose Fibers with Antistaphylococcal Activity for Wound Healing Applications. *Biomaterials* **2011**, *32*, 9557–9567.
78. Shamshina, J. L.; Zavgorodnya, O.; Bonner, J. R.; Gurau, G.; Di Nardo, T.; Rogers, R. D. 'Practical' Electrospinning of Biopolymers in Ionic Liquids. *ChemSusChem* **2017**, *10*, 106–111.
79. Electrospinning Tong Li Tech, Multi-Needle Spinneret. http://www.electrospinning.com/multi_needle_spinneret.html (accessed Nov. 21, 2016).
80. *Unlocking Our Future: Toward a New National Science Policy*; Committee Print 105-B, Committee on Science, U.S. House of Representatives, One Hundred Fifth Congress, September 1998.

Chapter 3

Current and Future Ionic Liquid Markets

Thomas J. S. Schubert*

IOLITEC Ionic Liquids Technologies GmbH, Salzstrasse 184, D-74076 Heilbronn, Germany
***E-mail: schubert@iolitec.de**

In this book chapter a summary of current and a prediction of future markets of ionic liquids is given. In the first part requirements for creating ionic liquids markets are the focus: Price, availability of information, life-cycle-costing and value-chain-thinking are discussed. Next, important actual examples of using ionic liquids, such as solvents, process chemicals, thermal transport and storage, electrochemical applications, functional fluids and additives, or their use in analytical applications are presented and assessed in terms of actual commercialization, but also in terms of their future commercial success. In the final section, an overview about ionic liquids technologies and their predicted actual technology readiness level is given and an attempt to create an outlook on the general future of ionic liquid related technologies.

1. Introduction

In the early 2000s only a few ionic liquids (ILs) were described in the scientific literature. Nevertheless, a little more than a handful of important review articles and also some books were the starting point of a continuously increasing interest in this class of materials. Because of the nearly unlimited number of potential combinations of ions (some scientists estimated the number to be as high as 10^{18} materials) it seemed at the time that they could be in the position to revolutionize chemistry and many other fields of science and technology as well. Today, in the year 2017, ionic liquids are not new anymore, so it is about time to look back and to summarize which expectations have been met so far.

At the end of 2016 more than 10,000 patent applications and more than 75,000 publications using the concept of ionic liquids were published in the scientific literature (*1*). This clearly indicates that they are not only a lab curiosity anymore and have already received an enormous attention in many different fields, very often also of commercial interest.

2. Requirements for Creating Ionic Liquid Markets

The industry did not wait for ionic liquids. In general all industries are waiting for commercially interesting solutions that will enhance existing or will enable disruptive new technologies. Ionic liquids are often an important or a missing piece within the puzzle.

In the fields of science and engineering the research on ionic liquids related technologies was driven by an increasing knowledge about their properties, delivered by universities and research institutes. Until today this information was used to tune and to tailor ionic liquids towards first examples of successful commercialization. But what will happen within the next 5, 10, or 25 years? Which requirements besides technical performance have to be fulfilled so that ionic liquids can enter more markets?

2.1. Price

If the technical performance of an ionic liquid is fully demonstrated, it is trivial that at a certain point also the price determines the final commercial success of a product. It is sometimes neglected by academic research groups that at the end somebody has to pay for a given material or process.

"Life-Cycle-Costing"-Thinking (LCC)

In industry such questions can be rationalized for processes by taking a look at the "Life-Cycle-Costing" (LCC): In some cases comparable high investment costs for ionic liquids may be compensated by:

- lower operating costs (e.g. reduced consumption of energy, reduced service frequencies),
- lower disposal costs (e.g. by recycling and reuse, or by alternative use).

If for a new process the reduction in operating and/or disposal costs is higher than additional acquisition costs, e.g., for the "first fill" with an ionic liquid, it should automatically lead to the conclusion to invest in such a process. This may become evident from Figure 1, where the additional acquisition costs of the new process B* has to be lower than the sum of the additional operating and disposal costs, in order for the new process to make sense from a financial point of view.

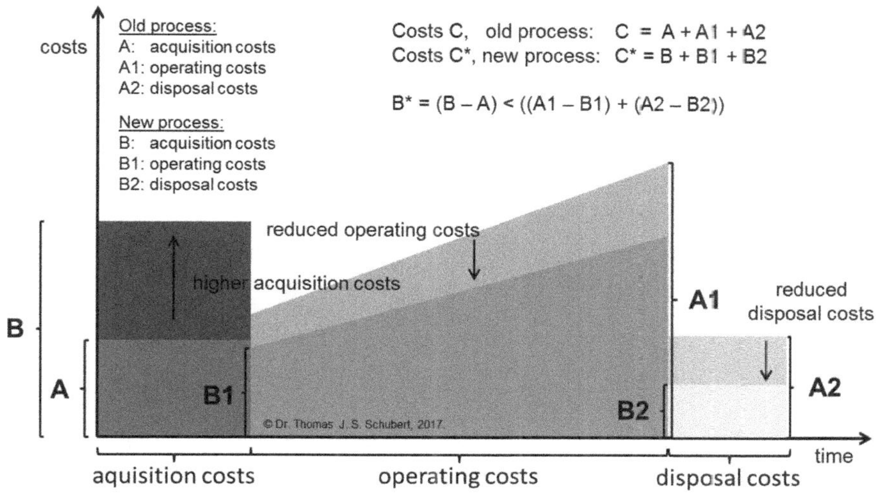

Figure 1. Life-Cycle-Costing (LCC) to compare process costs.

However, in the past these costs advantages for potential processes were maybe not worked out completely from most ionic liquid producing companies. As a consequence, sometimes a rational basis for decision makers was missing. If this gap of information is closed for each technical feasible process, it is very probable that we'll see more ionic liquid-based processes running in the future.

Added-Value

But to lower costs should never be the only argument that determines the success of technologies: In fact one should argue also with the additional value provided by innovation. If we take for example a battery that uses an ionic liquid as electrolyte, providing an additional value is, e.g., a longer lifetime, safety, and/or a faster charging. If such a battery is more expensive at the point of its market entry, the simple question is, if somebody is willing to pay a corresponding higher price for such an additional value or not. In this case it is not helpful to compare costs, instead is more likely to work out the benefits of the additional value. But what does it mean for ionic liquids?

Value Chains

Let's have a look at the example of an innovative battery having a couple of advantageous properties, resulting from using an ionic liquid based electrolyte: The battery itself will not be produced by an ionic liquid producing company, but instead by a battery manufacturer. As a consequence, it is the battery manufacturer that has to convince their customers about the innovative character of the novel product. This does mean that the introduction of an ionic liquid as an innovative electrolyte depends on the final success of a system at the end of the value chain. In our case a system like a battery may be *e.g.,* part of an electrically driven car or a smartphone, and thus depends also on the final success of the corresponding concepts.

At this point a good marketing strategy is to explain the higher costs that have to be paid for the additional value. However, it is worth noting that some materials maybe more expensive, but rarely by one or more magnitudes. In this case even the best marketing strategies can fail. Figure 2 demonstrates price levels of different chemicals.

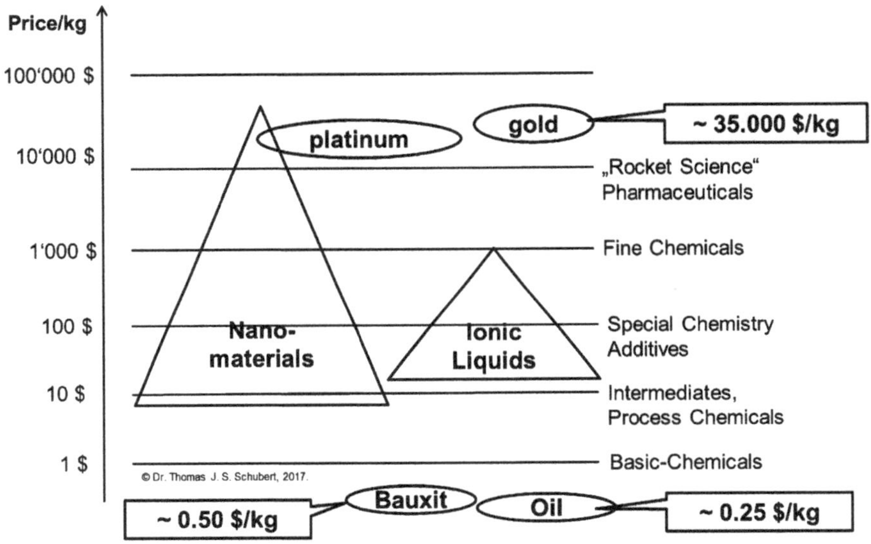

Figure 2. Price levels of selected groups of chemicals.

2.2. Frame Conditions

Frame conditions may also have an influence on the successful implementation of technology into the market. In terms of materials driven R&D important frame conditions are:

- climate change & global warming,
- e-mobility,
- regulatory issues.

For many ionic liquid-based technologies the reduction of CO_2-emissions or the use of CO_2 as a raw material is a driving force. In Asia, North America and Europe numerous research projects have received funding to develop novel technologies in this field. Furthermore, the rise of battery-powered cars, sparked especially by TESLA, led at least in Europe to many activities to develop novel types and concepts for batteries, initiated by the car manufacturers.

Another frame condition, which is often the starting point of R&D-activities, are regulatory issues: Perfluorooctysulfonates (PFOS) and Perfluoroctylsulfonic acids (PFOA) are known to be persistent substances, showing a very poor biodegradability combined with a tendency of bioaccumulation. As a consequence, use and distribution of these materials is prohibited. Nevertheless, there are some exceptions, where their use is still allowed, *e.g.,* as surfactants in chromium plating baths to avoid the formation of chromium containing fogs. As soon as there will be other, less problematic substances available, PFOS has to be replaced instantaneously by law.

2.3. Pressure To Innovate

Frame conditions and prices (the price might be also interpreted as a frame condition!) are creating within companies the pressure to innovate. Furthermore, if one company has already developed and introduced an innovative product into the market, their competitors feel often themselves challenged to start developments of a similar, sometimes slightly enhanced product ("me-too-products").

As soon as there are more reported successful product developments based on ionic liquids technologies, then there will be more products which will enter the market and this will inspire R&D in similar fields of applications.

2.4. Data, Data, Data!

The worldwide academic research created the basis for all activities using the concept of ionic liquids and it is of course a still ongoing process. Today the available information in the scientific literature about ionic liquids is already enormous. Information about their properties and in particular reliable experimental data are of fundamental relevance to set scientists and engineers in the position to design interdisciplinary R&D innovative processes, devices etc. In this context, it is worth stressing that reliable data are of great importance, but reliability should not be interpreted that the data has to be generated using ultrapure ionic liquids. It does mean instead that the purity, and also the content of impurities, has to be characterized sufficiently. It is an important difference if an ionic liquid with a purity of 99% has a content of 1% residual water or chloride.

In summary, property data for ionic liquids used as solvents or process chemicals already exists and will be the key to future and numerous ionic liquid based innovations.

2.5. The Simpler, the Cheaper, the Better?

From the academic or fundamental research point of view, it is important to demonstrate what is possible, even if economics are neglected. In the following it is the task of market oriented research to transfer results into marketable products. This sometimes means to simplify structures or also to work with lower purities. In other words at a certain point this question is important, can a desired effect be sufficiently achieved using a lower purity material, which also maybe significantly cheaper. This does mean in other words, the complexity of a structure and the purity of an ionic liquid should only be as high as necessary!

If this thinking is a bit more established in applied R&D, the overall goal to enter as many potential markets as possible, will definitely be achieved much more often in the future.

3. Current and Future Markets for Ionic Liquids

Without any intention to be complete, it has been attempted in the following sections to give an overview of many fields, the current situation and a prediction about future markets for a couple of ionic liquids technologies and applications.

3.1. Solvents

The use of ionic liquids as polar, non-volatile solvents was one of the earliest intentions. This may also become evident by the first company, which was founded based on ionic liquid technology, "Solvent Innovation". In the early 2000s most activities were focusing on organic chemistry. At that time ionic liquids were often linked with the labels "green solvents" and "green chemistry", which was in many cases not helpful, because early studies on toxicity indicated that some ionic liquids were toxic and/or non-biodegradable. However it was not helpful to give a complete class of materials that combines by design organic with inorganic chemistry to generate thousands of compounds with the label "green": Some ionic liquids are toxic, some can be eaten, and others may be used as pharmaceuticals or explosives.

Nevertheless, the negligible vapor pressure in combination with interesting ability to dissolve substances led to numerous ideas to use them as solvents. In particular in the 2000s nearly every named organic reaction was tested to determine if there was a better yield or an easier work-up using an ionic liquid as the solvent. However, to the best of our knowledge it was not reported that a relevant organic or pharmaceutical product is or has being synthesized in a commercial process using an ionic liquid. This can be verified easily by looking at which ionic liquids are being registered for use in any large scale process. Concerning synthesis protocols on the lab scale, the situation might be different, but this is of course difficult to estimate or verify.

The reasons, why there has been so far no real breakthrough, are surely in many cases the high costs compared to common solvents. On the other hand, if there's a real increase in the yield and a decrease in costs then ionic liquids will find their way into high priced products and niche applications.

Dissolution of Biopolymers

So far the most prominent example is the dissolution of biopolymers, introduced by Rogers *et al.* in 2002 (*2*). In particular the dissolution of cellulose and lignocellulose was intensively studied in the following years and is still today the subject of numerous R&D activities, such as biomass to liquids. Many of those studies and activities have reached the level of demonstration. The general trend for materials from renewable feedstocks will also give support to R&D-activities based on ionic liquid technology, making it quite probable that the first processes could have a commercial success within the next few years.

In addition, the dissolution of biopolymers followed by the generation of ionic liquid-based composites from biopolymers will be an interesting field in the future (*3*).

High-Temperature-Solvent for the Sabatier-Process

Another example for an ionic liquid is the use as a high-temperature-solvent in the catalytic transformation of CO_2 with H_2 to generate methane (CH_4), known as *Sabatier-process*. The efficiency of this process increases with its temperature, thus a sufficient reaction rate is reached above 250°C. Furthermore, the choice of a suitable reaction medium is also important for the exchange of generated heat during the continuous reaction. Within a project called "SEE", an ionic liquid with a high thermal stability was selected, which was used slightly below its temperature of degradation. This technology has so far reached the level of demonstration and suffers currently from a slightly degradation of the ionic liquid in contact with the used catalysts (Figure 3) (*4*).

Inorganic Synthesis

In terms of using ionic liquids as solvents the most underestimated field is probably the synthesis of inorganic, in particular of nano-scaled materials. Many inorganic salts can be dissolved within ionic liquids under polar aprotic conditions. Furthermore, ionic liquids are in some cases like a solvent consisting of pure ligands. This is making them to be ideal media for the size control of nanoparticles (*5*). Dependent from the anion of the used ionic liquid, a structure directing effect can be observed (*6*). As a consequence, numerous protocols are being described to synthesize or to modify nano-materials (*7*).

Synthesis of Polymers

In principle for the synthesis of polymers the same is true as for inorganic synthesis. In this context, next to their good properties as solvents ionic liquids can sometimes also be useful as process additives, e.g., for emulsion polymerizations. This field is also from an academic or scientific point of view at comparable early stage, but with a large potential and a positive outlook (*8*).

Figure 3. Gas-bubble-reactor for running the Sabatier-process.

Stabilization of Proteins

1) Ionic liquids can be designed to stabilize proteins as additives, but also if used as neat solvents. In suitable ionic liquids the denaturation starts at higher temperature if compared with aqueous environments. This effect can be used for enzymatic catalyzed reactions to run them at higher temperatures, leading to higher turnover frequencies and thus to faster chemical transformation, which was demonstrated in numerous publications (*9*). Though much knowledge, also with some very promising results, has been generated between 2000 to 2010, a real breakthrough has not occurred, leading to a decreasing interest in this field of research over the past few years (*10*).

It is worth noting, that the stabilizing effect can be used for some applications in the field of analytics.

3.2. Process Chemicals

Process chemicals are typically used to enhance industrial processes. Instead of solvents they are providing typically an additional value towards a process, such as an increased reaction rate (catalysts), more safety, replacement of hazardous or toxic chemicals, an easier work-up etc. Over the past years a couple of processes using ionic liquids were realized. One should keep in mind that there may also be others, which are not taken into account, because no information was shared with the public. Until today there are three active major chemical processes applying ionic liquid technology, but it is very probable that others will follow in the future.

3.2.1. BASIL™-Process

The BASIL™-Process (Biphasic Acid Scavenging utilizing Ionic Liquids), first published in 2002, was the first industrial example using ionic liquids in larger quantities (*11*). A very interesting point is that the concept of low melting compounds was applied to improve a process. It uses 1-methylimidazole instead of triethylamine as a base to scavenge the acid HCl. The major advantage is that triethylamine forms a viscously slurry, while 1-methylimidazole forms a low melting liquid with HCl. This has obvious advantages in the work-up. After regeneration, the 1-methylimidazole can be reused in the process.

3.2.2. Chevron's Replacement of HF as Alkylation Catalyst by an Ionic Liquid

The oil company Chevron developed a process using an ionic liquid instead of HF as alkylation catalyst. It was reported that this technology, which is also licensed to the company Honeywell, could have a big impact on how refining industry carries out alkylation. In order to replace dangerous HF, this technology will find its way into novel sites. In 2020 a major site in Salt Lake City will start the production. It will be the largest-scale chemical synthesis using ionic liquids (*12*).

3.2.3. Petronas' Mercury Removal

In many regions of the world the production of hydrocarbons is quite challenging, because the feedstock of natural gas contains mercury and mercury compounds. In cooperation with the Queen's University of Belfast the oil company Petronas developed a mercury removal technology using suitable ionic liquids. By applying this technology, the content of mercury could be lowered to < 0.01 ppb. This technology, marketed under the label HycaPure™, is already deployed at Petronas' facility in Bintulu, Malaysia.

3.2.4. Electrodeposition, Electropolishing and Recycling of Metals

Electrodeposition

Next to the use of ionic liquids as solvents, electrodeposition of metals is one of the earliest investigated fields of ionic liquids research. Today many known metals can be deposited purely or as alloys. Of particular industrial interest are metals which cannot be deposited from water. Because of its great technical relevance and its availability aluminum is of great importance and also tantalum has received a lot of interest over the past two decades.

Deposits of aluminum are of interest in order to lower weight, because of its mechanically stable protection against corrosion, which can be enhanced by additional passivation, but also for decorative reasons. As a consequence, many major industries such as automotive, aerospace and electronics are interested in this technology.

The introduction of this technology is today close to market entry. In North America and the European Union there are several projects running, which have reached the technology readiness levels (TRL) of 6 or 7. As often asked, one question is the cost of the process. In this regard it is worth to note that it has been demonstrated successfully that ionic liquids can be regenerated and reused in deposition baths without loss of performance.

Electropolishing

The use of ionic liquids as electrolyte for electropolishing is less known, though there has been some activity. In electropolishing processes the metal is the anode of the electrochemical cell. By this process roughness is removed by electrochemical forced dissolution of the metal, leading to smooth and shiny surfaces. As for electrodeposition, aluminum is specifically of interest. The feasibility is already shown, but further development is necessary to implement such processes (*13*).

Recycling of Metals

The recycling of metals is already important today and will be even more important in the future. In focus are expensive materials, such as noble metals, but also rare earth metals. e.g., the export restrictions from 2010 to 2015 of rare earth metals from China hyped the activities to develop processes for recycling of such metals. Based on the very good selectivity of specifically designed ionic liquids, a couple of recycling processes were brought at least to a level of demonstration. These can use extraction techniques, but also processes based on electrodeposition were also suggested (*14*).

Following the megatrend of renewable resources, it is very probable that ionic liquids will participate on increased activities in this field in the future.

Supported Ionic Liquids Phase (SILP)

A very fundamental concept in ionic liquid related research is the "supported ionic liquids phase"-technology (SILP-technology). In principle it combines the advantages of heterogeneous with those of homogenous catalysis in order to prevent leaching of expensive catalysts by immobilization. To achieve the goal of generating maximum surfaces, a porous solid is modified by dispersing a thin film of an ionic liquid on its surface. The low vapor pressure allows a nearly permanent coating of the substrate. By choosing different combination of anions and cations the solubility, reactivity and also the coordination properties of the specific surface can be tuned towards a specific application (*15*).

The situation concerning the commercialization of this technology is not clear, but it has at least reached a level of demonstration in an operational environment, corresponding with a TRL of 7.

3.3. Thermal Transport and Storage

In the early 2000s it was considered to use ionic liquids for the transport as well as storage and transformation of heat. As in many other applications it is their low vapor pressure that makes the difference if ionic liquids are compared with other materials. One of the starting points described by *Rogers* in 2001, for the ionic liquid 1-methyl-3-octyl-imidazolium tetrafluoroborate which was reported to have a very high thermal stability of 480°C (*16*). Though this value had to be corrected later towards much lower values, this publication generated a lot of interest.

3.3.1. Thermal Fluids

Thermal fluids are used in numerous industrial processes, in particular in chemical industries, as well as in solar thermal applications to absorb and to transport heat. Since ionic liquids have generally low vapor pressures combined with sufficient to good heat capacities they were suggested to be interesting candidates for using them as thermal fluids. Many other properties are in the same range compared to state of the art materials, except viscosity, which is typically higher for ILs.

Well-chosen materials have rarely a long term thermal stability above 275°C, if biodegradability is taken into account (*17*). Though it is possible by design to synthesize e.g., highly fluorinated ionic liquids which can achieve higher stabilities, those compounds suffer typically from a poor biodegradability and also the potential to accumulate in the environment.

The low vapor pressure of ionic liquids is the main advantage, because it offers engineers the possibility to design and construct equipment which operates at much lower pressures, leading to lower costs. Furthermore, ionic liquid based fluids can also be used to transport heat under vacuum conditions.

Solar Thermal Applications

Based on *Rogers*' publication, in 2003 and the following years it was suggested to use ionic liquids to replace eutectic mixtures or polychlorinated biphenyls (PCBs) in solar thermal power stations. These eutectic mixtures suffer from becoming solids below about 250°C. Organic PCBs are today already banned because of their persistency, but they still had an operating exception in the early 2000s.

At low temperatures ionic liquids have the clear advantage of a lower solidification point if compared with molten salts, but, as mentioned above, at temperatures above 275°C, even the best candidates start to decompose. As a consequence, ionic liquids are not currently a real alternative to inorganic molten salts.

In principle the same reasons prevented their use in domestic solar thermal systems, which are used to generate warm water. In this context, the driving force to apply ionic liquids was low vapor pressure in order to design systems with low operating pressures, and the suggestion to identify fluids which might be more stable than conventional water-propylene-glycol-mixtures. The latter requirement was important, because during summertime, when less warm water is required within households, thermal fluids based on organic moieties had the tendency to degrade and form dark brown to black tars, reducing step by step the efficiency of such solar collectors. Some feasibility studies clearly demonstrated that ionic liquids degraded under harsh conditions in a similar way as water-propylene-glycol-mixtures and thus did not represent a real advantage.

The Future

The design of organic based thermal fluids suffers in most cases from the stability of the weakest covalent bond within the molecule. The fact that in ionic liquids coulomb interaction may provide an additional contribution to their thermal stability, leads not to significantly higher degradation temperatures, if compared with conventional thermal fluids or also with similar non-ionic species.

This does mean that the uniqueness of ionic liquids is limited to their low vapor pressure, leading to the conclusion, that they will find their place more in niche applications, in particular if heat under vacuum conditions has to be transported.

3.3.2. Phase Changing Materials

Phase changing materials (PCMs) are substances with a high heat of fusion, which enables them to store or to release large amounts of energy. Heat is absorbed when the material melts and is released when it solidifies.

Today's state of the art material is paraffin, which allows tuning the phase-transition-temperature by choosing the right chain-length. Paraffin has the fundamental problem to be a flammable material. Keeping in mind that the use of PCMs is in particular interesting for using it for active thermal insulation combined with the function of buffering heat within houses, its flammability is a clear disadvantage.

In this context, ionic liquids initially provided as a class of materials some interesting properties, e.g., non-flammability, but also melting points in nearly every temperature range combined with sufficient to good heat capacities. But their weak point is for many of them the difference in melting and solidification point. While melting points are typically sharp transitions, many ionic liquids show the tendency for supercooling, meaning that their solidification point is lower than their melting point. The hysteresis between melting and solidification is a clear disadvantage for most known applications of PCMs, in particular for those in which heat should be buffered.

Though the use of ionic liquids was investigated by academic and industrial research groups so far no materials have become commercially available (*18*).

3.3.3. Sorption Cooling

Air conditioning, but also industrial cooling units for processes are consuming large amounts of electricity, which is typically needed to generate the mechanical work for the compression, which is followed by the decompression, which is generating the chill via the Joule-Thomson-effect. Absorption refrigeration cycles are an alternative concept, though they have been known for more than hundred years. Instead of the principle of vapor compressors, it can use low-quality energy sources, e.g., solar-energy, waste heat or district heating systems, to provide the energy needed to regenerate the absorption medium. Common working pairs are lithium bromide-water or water-NH_3. Since absorption refrigerators can be constructed to be more silent than vapor-compressors, they are already widely used for hotel fridges. The water-NH_3-working pair chillers are used in technical applications, where temperatures of -70°C can be achieved.

Ionic liquids provide a couple of interesting properties as (ab-)sorption medium, such as being liquid, even at lower temperatures and concentrations, a large degassing width, but also a low corrosion compared to Lithium bromide (*19*). Disadvantages are their comparable high viscosity and their high molecular weight.

Because of the big market potential there was a large interest from chemical companies as well as from manufacturers of chillers to explore benefits of ionic liquids. This led to patents from different companies and a prototype shown in Figure 4. The technology has reached the level of demonstration, but until now has not been commercialized. The principle of solar cooling and also the better use of multiple waste heat sources may lead to wider use of sorption cooling in the future which could include the use of ionic liquids.

Figure 4. Prototype of a Sorption Cooling Device, operated with an ionic liquid.

3.4. Electrochemical Applications: Batteries, Supercaps, DSSCs and Fuel Cells

The main driving force for increased research activities in the field of electrochemistry for storage of electricity are e-mobility as well as the reduction of CO_2-emissions in view of a climate change. In this context, many of the actual research and developments are focusing on generation and even more storage of electricity from renewable sources, since it is for obvious reasons one of the biggest bottlenecks. In many of those developments ionic liquids are of interest, because as electrolytes they are believed to supply interesting novel and unique combinations of properties.

3.4.1. Batteries

Developments of batteries using ionic liquids as electrolytes ran already through a couple of learning iterations. It is of course not as simple as to expect a higher energy density by simply replacing conventional electrolytes with ionic liquids. Batteries in terms of their electrodes, electrolytes, etc. are high-end compositions. All components and materials are tuned very carefully to each other, especially at the solid electrolyte interfaces (SEI).

The motivation to use ionic liquids is given by their non-flammability as shown in Figure 5, but also, generally spoken, to interact specifically with other materials and components in view of higher energy densities. It has to be stressed that ionic liquids should be involved from the beginning in the development of novel battery concepts, because it is in most cases not possible to integrate them at a later stage of the development.

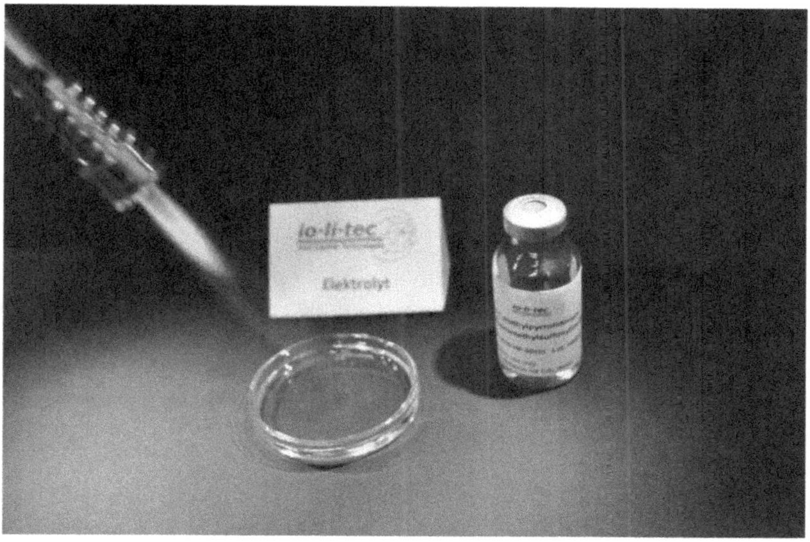

Figure 5. Non-flammability is one of the major advantages of ionic liquids if used as electrolyte.

Currently worldwide activities in the field of battery R&D involving ionic liquids are enormous (*20*), thus, it is not possible to give a detailed overview within this book chapter. In the following only a rough overview about the current situation is given.

Metal-Air-Batteries

Metal-air-batteries are in terms of costs a potentially interesting alternative. The focus is on cheap metals, which have a good outlook in their supply such as zinc, aluminum, or silicon. So far, most proof-of-concepts involving metal-air batteries are primary batteries, e.g., zinc-air-batteries are available for years to power hearing aid devices.

The challenge is to develop the fundamentals for secondary batteries. Thus, battery concepts have to be new and disruptive. As a consequence, many of those new concepts are involving ionic liquids. It has to be noticed that for each type of metal another specific set of properties is important.

This field today is at the stage of fundamental research, but it is very dynamic. First reported results are very promising, so it is probable that one or more concepts will be commercial (*21*).

Lithium-Ion-Batteries

The outlook for lithium as a raw material is not as good as for other metals, but in terms of reliability, and energy density lithium-ion-batteries are still the benchmark. So far it has not been reported that ionic liquids are providing any advantage in lithium-ion-batteries as electrolytes. The major problem is their incompatibility with the used electrodes, but also their high viscosity. Thus, research involving ionic liquid based electrolytes will presumably concentrate on other types of batteries.

Lithium-Batteries

Lithium-batteries are primary batteries. In terms of their energy density are interesting alternatives, but the use of pure lithium-metal within the battery has prevented intensification of R&D-activities. In this context, ionic liquids may change the game, since they are providing a couple of benefits, especially under the aspect of safety issues. The main advantage of ionic liquids is to prevent dendritic growth of the lithium-metal, which can lead to short circuits followed by dangerous breakdowns of the battery (*22*).

The lithium-battery may have a renaissance, as soon as it is more commonly accepted that safety is guaranteed for this type of battery.

Lithium-Sulfur-Batteries

Though this concept has been known for nearly 60 years, lithium-sulfur-batteries are rarely present in the scientific literature over the past decades, mainly because of poor cyclability. Because of their high theoretical energy density (2.6 kWh/kg) they are nevertheless an interesting concept.

Recently the research on this type of battery was stimulated by numerous novel electrode materials and also by ionic liquids (*23*). In terms of cyclability there close to those values of lithium-ion-batteries. Thus, there is a good chance that ionic liquids will be part of a breakthrough for this technology.

Redox-Flow-Batteries

Redox-flow-batteries are hybrids between fuel cells and rechargeable batteries. The energy is stored in two different chemical components, which are dissolved in liquids and can be stored outside the cell in different tanks, which is one of their main advantages, since their energy capacity is just limited by the volume of the tanks. Both parts of the cell are separated by a membrane, where the ion exchange occurs (*24*).

Some concepts of redox-flow-batteries are using zinc-bromine. To reduce the self-discharge and to reduce the vapor-pressure within the system, salts and ionic liquids are being used in order to complex the bromine by reaction to the tribromide-anion. The major advantage of using bromide containing ionic liquids is that they form a liquid material even at lower temperatures, which is important, since the liquids have to remain pumpable (*25*).

For some applications redox-flow-batteries will be interesting concepts that have a good chance to be commercialized. This is in particular the case for local applications, but also if the necessary space for the tanks is not too limited. If higher energy densities can be achieved, the door may also be opened for other fields, eventually for e-mobility, where the major advantage could be a fast charging by exchanging discharged versus charged electrolyte.

The Future of Ionic Liquids in Batteries

Driven by a couple of megatrends, such as CO_2-reduction and e-mobility, numerous R&D-efforts are focusing on the development of batteries. In many concepts ionic liquids play a role as novel types of electrolytes. Though it is still not clear which type of batteries will be the next big thing, it seems quite probable that at least in a few ionic liquids will play a significant role.

3.4.2. Fuel Cells

Ionic liquids are particular interesting for proton exchange membrane fuel cells, also known as polymer electrolyte membrane fuel cells (PEM-FCs). In this context, they were suggested to be used as an electrolyte for wetting the proton exchange membrane, which is typically NAFION™, enabling the operation at temperatures greater than 100°C, which increases the overall coefficient of performance. Other activities are focusing on polymerizable ionic liquids to generate alternatives to the current state-of-the-art, NAFION™ (*26*).

Though today a couple of patent applications involving ionic liquids exist, it seems that there have been no real breakthroughs. Initiated by the success of TESLA, the development of fuel cells has become less important – at least for the moment. However, the main advantage of fuel cells is that a fast refueling process instead of long lasting recharging is possible. How intense future efforts in this field will be is strongly depending on fundamental decisions, such as a hydrogen-based economy, but also on the success of e-mobility in the near future. As mentioned above, ionic liquids can be interesting candidates as electrolytes or polymers.

3.4.3. Dye Sensitized Solar Cells

A dye sensitized solar cell (DSSCs) is a photo-electrochemical system belonging to the group of thin film solar cells. It uses a photo-sensitized anode and an electrolyte to form a semiconductor (*27*). In the development of DSSCs ionic liquids were involved since the mid-1990s. An interesting fact is that a couple of ionic liquids, especially anions, were developed as electrolytes for DSSCs. One of the most prominent is the bis(trifluoromethylsulfonyl)amide (known as BTA, TFSI, TFSA, or NTf_2) (*28*), but also iodides and thiocyantes are connected to DSSC R&D. As often the case, it was their low vapor pressure combined with their electrical conductivity that made them the electrolytes of choice.

The Future

In terms of their commercialization DSSCs have not passed the level of demonstration yet. In nearly all demonstration projects electrolytes based on ionic liquids were used. However, the rise of alternative technologies such as perovskite-solar cells and organic photovoltaics (OPV) with similar or better efficiencies at lower production costs prevented their final market introduction and success. This leads to the assumption that DSSCs may not become a commercial success and therefore not lead to a future ionic liquid market.

3.4.4. Electrochemical Double Layer Capacitors (EDLCs, Supercapacitors)

Electrochemical double layer capacitors (EDLCs, also called supercapacitors or ultracaps) are today in terms of power densities the best and in terms of recharging the fastest storage device for storing electricity. On the other hand they are having much lower energy densities when compared with lithium-ion-batteries. As a consequence, one target of actual research in this field is to combine the advantages of EDCLs and batteries to generate fast charging systems with high energy densities.

In this context, ionic liquids are used in numerous novel concepts to realize such systems. Because of their non-flammability, but also because of their electrochemical stability they are interesting candidates for this technology (*29*). A few developments are already available in the market or close to market introduction. An indicator for the dynamic character in this field is that a couple of start-up companies were founded to market this interesting technology. The worldwide activities and promising results are strong indications that supercaps based on ionic liquids will lead to new types of energy storage devices. In particular future developments may close the gap between available power density optimized supercaps and energy density optimized lithium-ion-batteries.

3.5. Functional Fluids & Additives

The use of ionic liquids as functional fluids and additives is underestimated and underrepresented in publications and conference contributions. An explanation might be that related R&D-activities are typically located more at companies and/or applied research institutes. As a consequence, results are typically not published, due to reasons of nondisclosure.

However, functional fluids and in particular additives are generally high-priced, high-performance chemicals. If an ionic liquid is providing a real benefit in these two fields, the criterion of their price is less important as for high-volume applications such as solvents.

Within this subchapter a few examples are provided. Actually this field is very dynamic and it is quite probable that soon there will be numerous examples of successful commercializations.

3.5.1. Lubricants

Lubricants play an important role in many applications, in which parts of machines, engines etc. are in motion, in order to reduce friction between to surfaces. Other roles of lubricants are to transmit forces, to transport particles, or to cool surfaces.

In this context, ionic liquids can provide a couple of interesting novel properties or unique combination of properties and attracted thus from the mid-2000s an increasing interest to use them as novel types of base oils, but later also as additives for different types of base oils.

Ionic liquids as Base Oils for Lubricants

Many ionic liquids are in principle interesting as lubricants by themselves, because of their tendency to interact with metallic surfaces to reduce friction and to protect against wear (*30*). However, the broad use of suitable ionic liquids as base oils has not reported. A successful introduction into the market suffers not from their performance, but mainly from their comparable high production costs.

Compared to base oils such as mineral oil, ionic liquids are too expensive. Compared to synthetic base oils, ionic liquids have at least some small to large niche applications, e.g., vacuum lubrication.

Ionic liquids as Additives for Base Oils

In lubrication technology ionic liquids can be used as base oils, but they are also interesting additives. Many of the beneficial properties, particularly the surface-active ones, can be transferred if just a few percent of an ionic liquid are added to the base oil. Properties that can be influenced are friction, wear, conductivity and the behavior under extreme pressure (EP-additives) as shown in Figure 6 (*31*).

In terms of commercialization this is one of the most promising future markets for ionic liquids. A comparable high level of the prices is often accepted in this field, in particular if no other technical solutions are available.

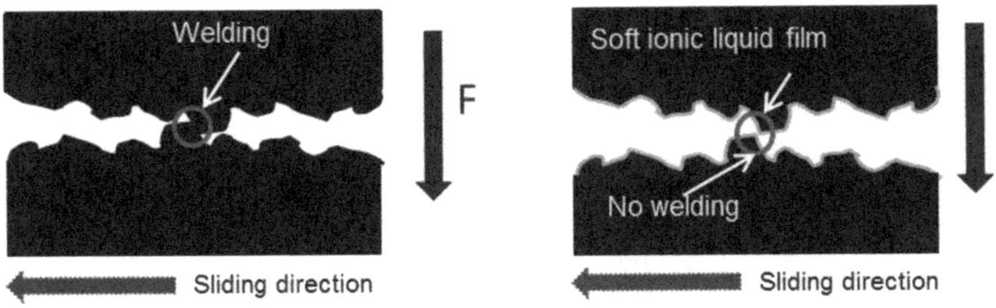

Figure 6. Ionic liquids as EP-additives. Left: Without, right: with EP-additive.

3.5.2. Hydraulic Fluids

Hydraulic fluids are used to transmit mechanical forces. Classical fluids can form ultra-small bubbles, which are responsible for mechanical damage and corrosion. This effect is known as cavitation.

A material that possesses a negligible vapor pressure typically has a reduced to no tendency for cavitation. Hence, ionic liquids seem to be ideal candidates for use as hydraulic fluids (*32*).

Ionic liquid Piston Compressor

The ionic liquid piston compressor is a device for the densification of gases, developed by the German company Linde (*33*). Ionic liquids seem to be ideal fluids for this purpose, because of the low tendency for cavitation and their often low tendency to dissolve gases. By applying an ionic liquid for the densification of hydrogen it was possible to reduce the number of moving parts within the compressor significantly. The hydrogen compressor is already introduced into the market, and it is highly probable that other types of gas compressors will follow.

3.5.3. Antistatic & Conductivity Additives

Many modern polymeric compounds are insulators and share the problem of electrostatic charging. For many reasons electrostatic charging should be avoided, e.g., in furniture, because it attracts dust, or the soles of safety shoes used in EX-areas where electrostatic discharge may cause ignition of flammable materials. To avoid this charging, a slight electrical conductivity is necessary, which can be achieved by adding antistatic additives. Such additives are interesting for many polymeric compounds such as rubbers, plastics, glues, pigments, etc.

In this context, one of the first examples (in 2004) of using ionic liquids as antistatic agents was to increase the conductivity of an aqueous based cleaning fluid, which was used in an industrial cleaning machine. The principle of this machine was to spray small water droplets onto the top of very tiny filaments. Sitting at the top of them these droplets are in the position to absorb small dust particles by adhesive forces. Due to the oval construction of the brush, these droplets including the dust load are ejected by centrifugal forces as shown in Figure 7.

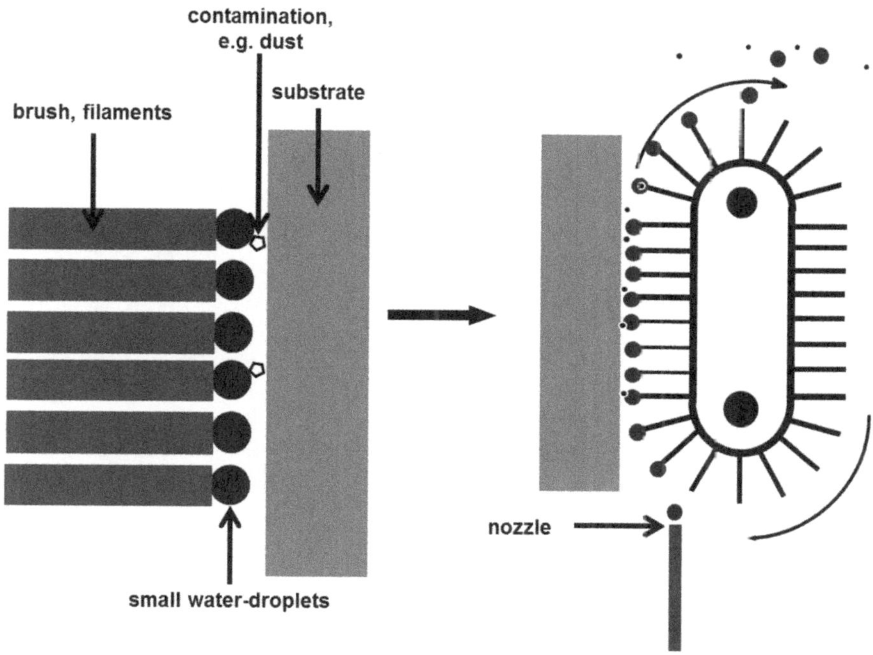

Figure 7. Scheme of the Surface-Cleaning-Device, using a cleaning fluid with an ionic liquid-based antistatic-additive.

If the nylon-filaments are in contact with other insulators it leads very quickly to electrostatic charging. To avoid this, sodium chloride was added, which led in time to an encrustation of the spray-nozzle. To prevent this, an ionic liquid was added instead of NaCl, which avoided both, the electrostatic charging and the encrustation of the nozzle, because the ionic liquid was liquid at the operating temperature.

Only a few examples like this one have been publicized, but it is known that ionic liquids have been applied successfully in a couple of other similar commercial applications. It is very probable that numerous challenges and problems will be solved by similar approaches.

3.5.4. Dispersing Agents for Nano-Scaled Materials

The use of ionic liquids as reaction medium or solvents for the synthesis of nano-scaled materials was described above. In addition, it is also possible to apply the same or similar principle for the preparation of stable dispersions of nano-particles (*34*). This is in particular interesting for those types of nano-scaled materials, which are manufactured e.g., by cheaper gas-phase-reaction techniques, where no sufficient dispersing agents for a chosen solvent are available. This development is just at the beginning and a few commercial applications are already in the market as shown in Figure 8. Thus, it is very probable that numerous examples will follow in the near future.

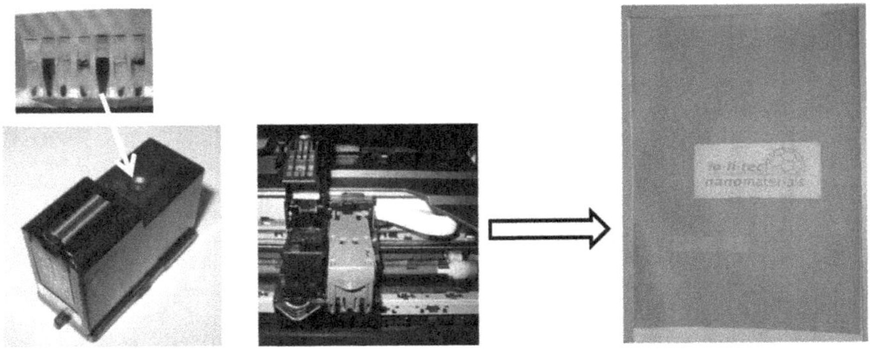

Figure 8. Printable nanomaterials – dispersed by ionic liquids.

3.6. Analytical Applications & Reagents

The use of ionic liquids in analytical applications and as reagents is now a commercial success due to fewer regulatory issues and that higher prices corresponding to lower production volumes are typically accepted. As a consequence, the barriers for a successful market entry are much lower compared to other fields.

3.6.1. Gas-Chromatography

GC Headspace

The GC headspace analysis is today a very common analytical method. Since many volatile materials can be dissolved in suitable ionic liquids, they can be vaporized at elevated temperatures to bring them into the headspace of the vial, while ionic liquids do not evaporate due to their negligible vapor pressure (*35*).

So far the use of ionic liquids as solvents for GC headspace has not become a standard method, though the advantages are quite obvious. The situation may be different in a few years, in particular if manufacturers of gas chromatographs can be convinced to recommend this method and to supply examples of beneficial use.

Materials for GC-Columns

Ionic liquids as GC stationary phase columns represent another example of an already successfully commercialized technology. In particular the extensive work by *Armstrong*, protected by numerous patent applications, led to novel groups of ionic liquids (e.g., dicationic structures) and finally to innovative stationary phases (*36*). Those have a couple of advantages, such as being more stable against moisture/oxygen and can be operated at higher temperatures. Furthermore, by rational design the selectivity can be tuned towards specific analytes, resulting in better peak shapes.

The development is still an ongoing process, so there will be further innovations in this field in the next few years.

Inverse Gas-Chromatography

The classic inverse gas-chromatography is a method to investigate the characteristic of solids. Due to the fact that ionic liquids have a low vapor pressure, they can be used to coat solids. If the retention time of the same analyte for two equal solids each coated with a different ionic liquid is determined, it is a measure for the activity of an ionic liquid towards this analyte. In other words it is possible to identify active materials by this safe method.

Though this method is very elegant, it seems that there are until today no examples of its commercialization. However, it is very useful for R&D-purposes to identify materials for separation techniques without time and material consuming equilibrium measurements.

3.6.2. Mass Spectrometry

Matrix-assisted-laser-desorption-ionization-time-of-flight mass spectrometry (MALDI-TOF-MS) is a mild method to analyze biologically and medically relevant species, such as enzymes, antibodies, etc. The matrix is typically a solid organic material that absorbs energy of the used LASER-wavelength very well. Today a toolbox of organic compounds does already exist, since different wavelengths, solvent-environments etc. are needed to prepare samples. Typical molecules are e.g., 3,5-dimethoxy-4-hydroxycinnamic acid or α-cyano-4-hydroxy-cinnamic acid. After the energy uptake by the matrix, ablated parts of matrix-molecules together with the analytes, protonated by small amounts of added acids (e.g., TFA), are ionized and accelerated within the electric field.

The properties of chosen ionic liquids to dispense bioorganic molecules, proteins, and biopolymers together with the possibility to design them to absorb LASER-light is making them interesting candidates as alternative matrix materials - with the advantage of being liquid. The fact that a droplet instead of a crystal can be positioned at the sample target, has a couple of advantages, e.g., a faster sample preparation, no need to wait for the co-crystallization of matrix and analyte, but also a homogenous solution instead of so called "hot spots" of the analyte, corresponding with a high concentration (*37*).

A wider use of ionic liquids as matrix-materials has so far been prevented by intellectual property rights (IPR) but wider commercial availability may be realized in a few years as the IPR run out.

3.6.3. Karl-Fischer-Titration

Karl-Fischer-Titration (KFT) is a quantitative method for the determination of water contents within a sample. If a material is not fully soluble within the used solvent, it is not guaranteed that all water is available for the titration process. The broad and tunable properties of ILs, which are in particular interesting for the dissolution of cellulose and other biopolymers, can be applied to yield homogenous solutions, which can be titrated directly.

Though the proof-of-concept of this method was already developed in 2004, it was so far not introduced into the market. Nevertheless, as soon as there are some standard operating procedures available, it will surely simplify the sample preparation and will thus generate a market. Since methanol is used as the common solvent in KFT, there are some tendencies to reduce its use if there were alternatives. This may be another driving force to work on methods using ionic liquids as a solvent or at least as a solvent additive.

3.6.4. Stabilization of Biomolecules

Stabilization of Biomolecules

Biomolecules and in particular enzymes are stabilized by covering their surfaces with a spherical layer of weakly coordinating cation and anion, making them more resistant against denaturation by temperature. This stabilizing effect can be applied to store enzymes for a longer time, but also to use them at elevated temperatures (see also 3.1, "Solvents").

Crystallization of Proteins

To determine the molecular structure of proteins by x-ray crystallography, it is necessary to generate crystals. Since proteins are very complex structures, the crystallization using the so called sitting or hanging drop methods are time consuming processes and need some experience.

The stabilizing effect of ionic liquids on proteins can be used to crystallize proteins. In particular by applying the sitting drop method to proteins dissolved in water-ionic-liquid-solutions it is possible to achieve in some cases crystals of good quality, which can be analyzed by XRD.

The proof-of-concept has evolved since 2004 (*38*), but until today ionic liquids as solvents or additives for protein crystallization have not become a commercial success.

3.6.5. Electron Microscopy

Over the past few years the number of publications using ionic liquids in electron microscopy, scanning electron microscopy (SEM) and transmission electron microscopy (TEM) has increased every year. Both techniques are operated under vacuum conditions, so the samples have to be dry solids. Ionic liquids as non-volatile conductive materials are providing important requirements for the sample preparation.

SEM

A key publication in this context was the work of *Torimoto et al.*, who discovered that coating an insulating sample with an ionic liquid provided the necessary conductivity as coating with metal or carbon, which is important to observe a clear picture via scanning electron microscopy (SEM) (*39*). This technique is widely used in SEM.

TEM

The same group used an ionic liquid to suspend a phosphatidylcholine liposome and to visualize it by transmission electron microscopy (TEM) (*40*).

The Future

The application of ionic liquids in sample preparation for electron microscopy has the potential to become an accepted standard procedure, which is clearly indicated by numerous publications in this field. It is simplifying preparation techniques and enlarges also the scope of substrates that can be examined by SEM or TEM.

3.7. Other Applications

Active Pharmaceutical Ingredients

The question about the toxicity of ionic liquids received increasing attention with their increasing popularity in the scientific literature. In comparison to carbon allotropes, where toxicity can only be rarely influenced, ionic liquids had a chance to be “benign by design” (*41*).

Toxicity and pharmaceutical activity often go hand in hand. "Solely the dose determines that a thing is not a poison", is a famous quote by Paracelsus, and it does of course also work very well for ionic liquids. Just their structure and concentration determines if an ionic liquid can be eaten, if it is a poison, or if it is maybe an active pharmaceutical ingredient (API). In pharmaceutical chemistry it is a well-known principle to make a salt from an API, based on an organic molecule, to enhance its disposability. These salts often suffer from the problem of polymorphism, meaning that one salt may have different crystal structures and therefore different activities. This can be avoided if an API is delivered as a liquid salt that avoids its crystallization.

A complete new approach is to combine two APIs with different activity, while one can be in the form of cation, while the other can be in the form of an anion, leading to so called "combination salts". It was demonstrated that numerous types of activity can be combined in one molecule, e.g., antibacterial combined with non-steroidal anti-inflammatory activity (*42*). Using this principle may also lead to novel strategies for the design of pharmaceuticals.

In terms of potential markets maybe one of the most interesting and important fields could be the design of novel types of antibiotics: The "strategy" e.g., of methicillin-resistant *Staphylococcus aureus* (MRSA) is to protect themselves against antibiotics by the formation of stable biofilms, avoiding that the active species get into contact with bacteria. Since some ionic species are known to act as anti-biofilm agents, e.g., 1-alkylquinolinium bromide, or 1-alkyl-3-methylimidazolium chloride, it would be interesting to combine them or similar structures with an antibiotic anion.

Active Ingredients for Crop Sciences

The same principles valid for APIs can be transferred to active ingredients for crop sciences.

Composite Materials

Earlier within this chapter a few examples concerning the use of ionic liquids as additives were described. By definition additives are substances being added to one or more other main-components in low concentrations to enhance one or more properties. In material sciences there were already a couple of publications on composite materials, combining ionic liquids as a main-component with other materials. Striking examples were "ionogels" (ionic liquids with silica) (*43*) and "bucky gels" (ionic liquids with CNTs) (*44*), but the combination with the broad variety of novel materials available from nano-sciences should inspire scientists to many novel high-tech-materials.

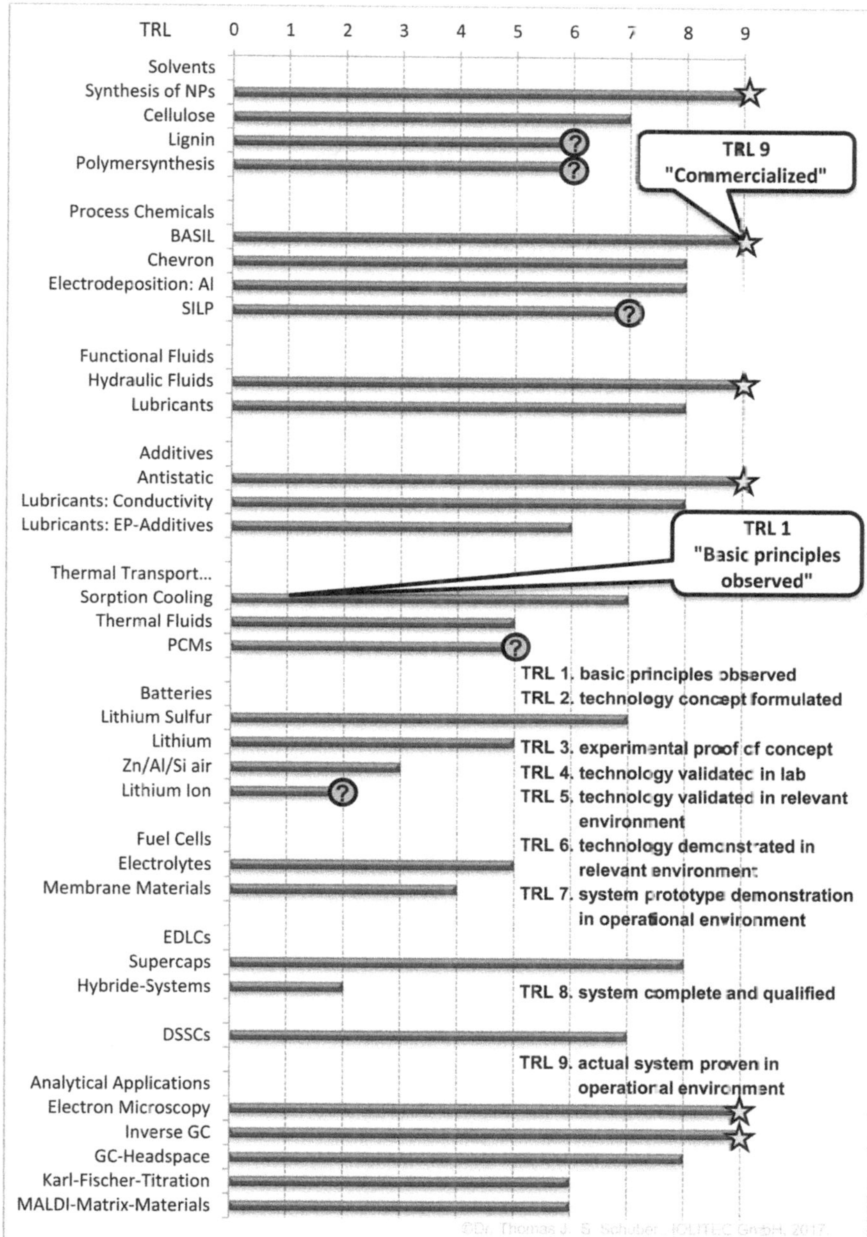

Figure 9. Technology readiness levels (TRL) of ionic liquid-based technologies.

4. Summary and Outlook

Summary

Over the past two decades ionic liquids have become very popular in the scientific literature. This scientific success has not directly lead to an economic success, however a couple of technologies have been successfully commercialized. Numerous fields were overestimated in terms of their market potential and only a few have met expectations. The potential of other fields to be commercialized were completely underestimated, because of the fact that corresponding R&D was not published and happened in corporate labs.

The research on ionic liquids also follows trends. The main activities for today's ionic liquid related applied research are different from the ones at the beginning of the century. In the future there will be other challenges, where ionic liquids may also contribute interesting aspects.

Figure 9 visualizes and compares technology readiness levels for many applications of ionic liquids. Please note that this is not intended to be complete.

Outlook

Scientists and engineers have done an excellent job and within about 20 years a huge amount of information is now available about ionic liquids. Numerous materials have been synthesized and many of them are more or less characterized. A lot of their properties are today understood in a sufficient way. Nevertheless, many challenges still remain. Ionic liquids should not only be seen as a new class of materials, but they should also be interpreted as a concept to think different about chemistry. The core of this concept is the extension of coulomb interactions to all fields of chemistry.

It is very likely that other new concepts and applications using ionic liquids will be invented in the future. There will also be some applications, which are today not covered by any category in this chapter. So it quite feasible that an update of this chapter in 10 or 20 years will presumably have new ideas, concepts, applications, and commercially successful products.

The job of manufacturers of ionic liquids is to summarize available knowledge and to translate it into marketable products and/or processes. In some cases they have to find technical solutions to lower the costs or to find answers for questions such as "how pure an ionic liquid should be for my application?" This is all part of the innovation process, which is different for every single application.

In Figure 9 several potential markets for ionic liquid technologies is provided. It is not very probable that ionic liquids will enter all of these potential markets, but because of the many fields it is probable that they will reach significant market volume in the future!

References

1. Source: SciFinder™.
2. Swatloski, R. P.; Spear, S. K.; Holbrey, J. D.; Rogers, R. D. *J. Am. Chem. Soc.* **2002**, *124*, 4974–4975.
3. Haverhals, L. M.; Reichert, W. M.; De Long, H. C.; Trulove, P. C. *Macromol. Mater. Eng.* **2010**, *295* (5), 425–530.
4. Sahin, H.; Wimberg, J.; Schubert, T. J. S. Presented at 5th Congress on Ionic Liquids (COIL 5), Portugal (Algarve), 21–25 April 2013; 2nd International Conference on Materials for Energy (EnMat II), Karlsruhe, 12–16 May 2013.
5. Antonietti, M.; Smarsly, D. B.; Zhou, Y. *Angew. Chem.* **2004**, *116*, 5096–5100.
6. Beyersdorff, T. F.; Janiak, C.; Klingele, M.; Redel, E.; Schubert, T.; Klingele, M. H. Univ Freiburg Albert-Ludwigs. Patents WO2009040107-A2, DE102007045878-A1, DE102007045878-B4, WO2009040107-A3.
7. Torimoto, T.; Tsuda, T.; Okazaki, K.-I.; Kuwabata, S. *Adv. Mater.* **2010**, *22*, 1196–1221.
8. Winterton, N. *J. Mater. Chem.* **2006**, *16*, 4281–4293.
9. Beck, M.; Neise, C.; Ahrenberg, M.; Schick, C.; Kragl, U.; Kessler, O. *Int. J. Microstruct. Mater. Prop.* **2016**, *11*, 359–372.
10. Mutschler, J.; Rausis, T.; Bourgeois, J.-M.; Bastian, C.; Zufferey, D.; Mohrenza, I. V.; Fischer, F. *Green Chem.* **2009**, *11*, 1793–1800.
11. BASF SE. Method for the Separation of Acids from Chemical Reaction Mixtures by Means of Ionic Fluids. World Patent WO/2003/062251, Jul. 31, 2003.
12. McCoy, M. *Chem. Eng. News* **2016**, *94*, 16.
13. Ban, A.; Rademacher, M.; ReichardtT. *Elektropolieren in ionischen Flüssigkeiten und nicht waessrigen Elektrolyten*; Forschungsvorhaben AiF, FKZ: 2011.
14. Vannder Hoogerstraete, T.; Wellens, S.; Verachtert, K.; Binnemans, K. *Green Chem.* **2013**, *15*, 919–927.
15. Fehrmann, R., Riisager, A., Haumann, M., Eds. *Supported Ionic Liquids – Fundamentals and Applications*; Wiley-VCH Verlag GmbH & Co. KGaA: Weinheim, Germany, 2014.
16. Wu, B.; Reddy, R. G.; Rogers, R. D. *Proceedings of Solar Forum 2001, Solar Energy: The Power to Choose*; April 21–25, 2001, Washington, DC.
17. Sahin, H.; Wimberg,J.; Schubert, T. J. S. *Storage of Energy from Renewable Sources Within the Natural Gas Net – Electrolysis of Water and Synthesis of Gas Components* Final Report; IOLITEC Ionic Liquids Technologies GmbH: 2014 (funded by the BMBF, Germany, FKZ: 033 RC 1010A-G).
18. Klingele, M.; Schubert, T. J. S. *Development of Micro-PCM-Emulsions and Ionic Liquids for Heat Storage Applications*; Final Report; IOLITEC Ionic Liquids Technologies GmbH: 2008 (funded by the BMWi, Germany, FKZ: 0327384A-C).
19. Constantinescu, D.; Schaber, K.-H.; Agel, F.; Klingele, M. H.; Schubert, T. J. S. *J. Chem. Eng. Data* **2007**, *52*, 1280–1285. Römich, C.; Merkel, N. C.;

Valbonesi, A.; Schaber, K.-H.; Sauer, S.; Schubert, T. J. S. *J. Chem. Eng. Data* **2012**, *57*, 2258–2264.
20. Osada, I.; de Vries, H.; Scrosati, B.; Passerini, S. *Angew. Chem., Int. Ed.* **2016**, *55*, 500–513.
21. Drillet, J.-F.; Endres, F.; Reichardt, H.-U. *Batteriesystem Zink/Luft , Handbuch Elektromobilität*; EW Verlag: 2013; pp 145−155. Endres, F.; Abbott, A. P.; MacFarlane, D. R. *Electrodeposition from Ionic Liquids*; Wiley-VCH: 2008. Liu, Z.; Zein El Abedin, S.; Endres, F. *Electrochim. Acta* **2013**, *89*, 635–643. Al-Salman, R.; Zein El Abedin, S.; Endres, F. *Phys. Chem. Chem. Phys.* **2008**, *10*, 4650–4657.
22. Seki, S.; Kobayashi, Y.; Miyashiro, H.; Ohno, Y.; Usami, A.; Mita, Y.; Watanabe, M.; Terada, N. *Chem. Commun.* **2006**, 544–545. Garcia, B.; Lavallee, S.; Perron, G.; Michot, C.; Armand, M. *Electrochim. Acta* **2004**, *49*, 4583–4588.
23. Song, M.-K.; Zhang, Y.; Cairns, E. J. A Long-Life, High-Rate Lithium/Sulfur Cell: A Multifaceted Approach to Enhancing Cell Performance. *Nano Lett.* **2013**, *13*, 5891–5899.
24. Alotto, P.; Guarnieri, M.; Moro, F. *Renewable Sustainable Energy Rev.* **2014**, *29*, 325–335.
25. Noack, J.; Fischer, P.; Tübke, J.; Pinkwart, K. Fraunhofer Gesellschaft zur Förderung der angewandten Forschung e.V. Method for Storing Electrical Energy in Ionic Liquids. WO2010094657, Aug. 26, 2010.
26. Hübner, G.; Huth, A. Volkswagen AG. Ionic liquid, Useful e.g., as an Electrolyte, for the Manufacture of Fuel Cell Membrane, Comprises a Polymer or Copolymer Derived From Allyl- or Vinyl-Monomer and Containing Ammonium-Cation and an Anion. DE102006054951. June 29, 2008.
27. Hagfeldt, A.; Graetzel, M. *Acc. Chem. Res.* **2000**, *33*, 269–277. Wang, P.; Zakeeruddin, S. M.; Moser, J.-E.; Graetzel, M. *J. Phys. Chem. B* **2003**, *107*, 13280–13285.
28. Graetzel, M.; Bonhôte, P.; Dias, A.-P. Asulab SA. Liquid Hydrophobic Salts, Their Preparation and Their Use in Electrochemistry. EP0718288, June 26, 1996.
29. Brandt, A.; Pohlmann, S.; Varzi, A.; Balducci, A.; Passerini, S. *MRS Bull.* **2013**, *38*, 554–559.
30. Predel, T.; Pohrer, B.; Schlücker, E. *Chem. Eng. Technol.* **2010**, *33*, 132–136.
31. Schubert, T. J. S. Ionic Liquids and Ionic Liquids Mediated Dispersions of Nanomaterials as High Performance Additives for Lubricants. Conference Contribution, OILDOC Conference, Rosenheim, Germany, 2017.
32. Boesmann, A.; Schubert, T. J. S. IoLiTec Ionic Liquids Technologies GmbH. Liquid Pressure Transmitting Medium, Useful in Hydraulic Component or Machines, Comprises Ionic Liquid as Pressure Transmitting Medium. DE102004033021, February 2, 2006.
33. Maas, H.-J.; Meilinger, M.; Neuendorf, S.; Schoedel, N. Linde AG. Method for Compressing a Gaseous, Oxygen-Containing Medium, Comprises Condensing the Medium by Means of a Liquid Piston Containing Ionic

Liquid. DE102006014335, October 4, 2007. Adler, B.; Adler, R.; Mayer, H. Linde AG. PISTON-FREE COMPRESSOR. WO2008025432, March 6, 2008.
34. Schubert, T. J. S. The Influence of Ionic Liquidson the Synthesis of Nanostructured Materials. Conference Contribution at the EUCHEM Conference on Molten Salts and Ionic Liquids, Kopenhagen, 2008.
35. Koch, P.; Kuesters, E. Novartis AG. Analytical Process. EP1529205, May 11, 2005.
36. Anderson, J. L.; Armstrong, D. W. *Anal. Chem.* **2005**, *77*, 6453–6462.
37. Stahl, B.; Boehm, G.; Mank, M. Nutricia NV. MALDI-MATRIX. EP1549958, July 6, 2005.
38. Boesmann, A.; Schubert, T. J. S. IoLiTec Ionic Liquids Technologies GmbH. Crystallization of Polymers (e.g., Polyolefin, Polyamide and Polyurethane) and Biopolymers (e.g., Starch, Chitin and Nucleotide) Comprises the Utilization of an Ionic Fluid. DE102004027196, Dec. 22, 2005.
39. Kuwabata, S.; Kongkanand, A.; Oyamatsu, D.; Torimoto, T. *Chem. Lett.* **2006**, *35*, 600.
40. Kuwabata, S.; Torimoto, T.; Nakazawa, E. *Spectroscopy* **2009**, *44*, 61.
41. Stolte, S.; Arning, J.; Bottin-Weber, U.; Mueller, A.; Pitner, W.-R.; Welz-Biermann, U.; Jastorff, B.; Ranke, J. *Green Chem.* **2007**, *9*, 760–767.
42. Stoimenovski, J.; MacFarlane, D. R.; Bica, K.; Rogers, R. D. *Pharm. Res*. **2010**, *27*, 521–526. Kumar, V.; Malhotra, S. V. In *Ionic Liquid Applications: Pharmaceuticals, Therapeutics, and Biotechnology*; American Chemical Society: Washington, DC, 2010; pp 1– 12.
43. Wang, P.; Zakeeruddin, S. M.; Comte, P.; Exnar, I.; Gratzel, M. *J. Am. Chem. Soc.* **2003**, *125*, 1166. Ueno, K.; Hata, K.; Katakabe, T.; Kondoh, M.; Watanabe, M. *J. Phys. Chem. B* **2008**, *112*, 9013. Ueno, K.; Inaba, A.; Kondoh, M.; Watanabe, M. *Langmuir* **2008**, *24*, 5253. Ueno, K.; Imaizumi, S.; Hata, K.; Watanabe, M. *Langmuir* **2009**, *25*, 825.
44. Fukushima, T.; Kosaka, A.; Ishimura, Y.; Yamamoto, T.; Takigawa, T.; Ishii, N.; Aida, T. *Science* **2003**, *300*, 2072.

Materials

Chapter 4

Photopolymerization of Alkyl- and Ether-Functionalized Coordinated Ionic Liquid Monomers

John W. Whitley, Michael T. Burnette, Shellby C. Benefield, and Jason E. Bara*

Department of Chemical & Biological Engineering, University of Alabama, Tuscaloosa, Alabama 35487-0203, United States
***E-mail: jbara@eng.ua.edu**

Coordinated ionic liquid (IL) monomers prepared from small organic monomers and lithium bistriflimide ($LiTf_2N$) provide a versatile and facile approach to building polymer + inorganic salt hybrid materials. These reactive coordinated IL species are advantageous as they exhibit improvements in photopolymerization kinetics and monomer conversion compared to the bulk monomer, which may enable expedited fabrication of 2-D and 3-D objects with little or no residual unreacted small organic species. In the present study, the investigation of the photopolymerization behaviors of coordinated ILs were extended to systems comprising"long-chain" alkyl- and ether-functionalized acrylates with $LiTf_2N$. An analysis of the polymerization kinetics using real time attenuated total reflectance Fourier transform infrared spectroscopy (ATR-FTIR) revealed improvements in reaction behavior for all salt-containing mixtures. However, an inspection of the FTIR spectra suggested different mechanisms for these improvements depending on monomer side chain length and polarity. This work furthers the understanding of the photopolymerization behaviors or coordinated IL monomers and expands the range of monomers known to be usable in this manner.

Introduction

Among the different methods of radical polymerization, those initiated by light (i.e. photopolymerizations) offer unique advantages (*1–3*). In addition to the ability to exert greater control over reaction progress, photopolymerizations performed under ambient conditions decrease the risk of side reactions that can occur at elevated temperatures (*1–3*). The desirable features of photopolymerization reactions have been traditionally used for the preparations of coatings and adhesives, especially in dentistry. Photopolymerization also finds great utility in applications such as 3-D printing, photolithography and medical procedures (*2–5*). However, as photopolymerization reactions are often performed in bulk conditions (i.e., where only monomer and initiator are present), they are often subject to mass transfer limitations at moderate conversions that can result in large fractions of unreacted monomer in the product (*2, 5–10*).

"Structured" media offer a means of reducing the problems associated with bulk polymerization by decreasing the effects of reactant diffusion on polymerization kinetics (*11–13*). Crystalline and microporous solids along with liquid crystals (LCs) have been extensively studied as media in radical polymerization, with improved polymerization kinetics observed with increasing ordering of the LC phase (*12–20*). More recently, it has been proposed that ionic liquids (ILs) may also affect polymerization behavior because of their organizational properties. Investigations of the use of ILs as media in radical polymerizations have demonstrated the ability of these materials to increase the rate of polymer propagation (R_P) while decreasing the rate of termination (R_T) (*21–23*). Although the relatively high polarities and viscosities of ILs are often identified with these effects, the "dual" ionic and non-polar natures of many IL cations may contribute to improvements in polymerization behavior (*24–29*). Both computational and spectroscopic studies have demonstrated the propensity of ILs to aggregate into ionic/polar and non-polar domains, a property with implications for their use as solvents in the preparation of organic and inorganic materials (*30–33*). These effects may also extend to polymer synthesis, as observed in a study performed by Thurect and co-workers. In that work, the polymerization of styrene in imidazolium-based ILs in which the observed chain transfer behavior and polymer molecular weight suggested organization/confinement of reactive species (e.g. monomers, macroradicals) within IL nanodomains (*28*). This influence was confirmed via rotating frame nuclear Overhauser effect spectroscopy (ROESY) wherein aggregation of methyl methacrylate (MMA) at the domain interfaces of an IL solvent was observed, demonstrating the ability of IL nanostructure to affect monomer organization (*29*).

While classical ILs such as 1-butyl-3-methylimidazolium bistriflimide ($[C_4mim][Tf_2N]$) are composed of molecular ions, "coordinated" ILs based on ionic coordination complexes can also be synthesized from a variety of coordinating ligands and metal cations (*34–41*). In the case where the ligand(s) is polymerizable (e.g., acrylate), then a coordinated IL monomer is formed. In a previous work, we prepared and photopolymerized coordinated IL monomers from 1-vinylimidazole (VIm) and lithium bistriflimide ($LiTf_2N$) (*42*), a alkali metal salt with a bulky, weakly coordinating anion commonly used in the preparation of classical ILs such as [C_4mim][Tf_2N]. In comparison to neat VIm and solutions containing uncoordinated monomer, coordinated ILs displayed improved polymerization behavior, attaining full monomer conversion within the time of irradiation (*42*). Moreover, $LiTf_2N$ could be removed from the resulting poly(vinylimidazole) product using an appropriate solvent (*42*). We later extended this polymerization technique to a series of commercially important (meth)acrylic monomers, observing similar improvements in photopolymerization kinetics as well as increases in the molecular weights of the polymer products (*43*). More recently, coordinated IL monomers prepared from a variety of alkali and alkaline earth metal bistriflimide ($M^{n+}(Tf_2N^-)_n$) salts were photopolymerized to investigate the influence of coordinating metal cation on the polymerization behavior of these materials (*44*).

As suggested by our previous work, differences in the reaction kinetics of the monomers may due, in part, to variation in the aggregation of the coordinated ILs into polar and non-polar domains. The extent of this behavior has been shown to be dependent on the hydrophobicity of side chains present in the ILs. Among the first reports of these differences was reported by Hayes and co-workers in the examination of the organizational heterogeneity of two protic ILs, ethylammonium nitrate ([$EtNH_3$][NO_3]) and ethanolammonium nitrate ([(HOEt)NH_3][NO_3]) (*45*). Utilizing empirical potential structure refinement (EPSR) in the analysis of neutron diffraction spectra, these ILs were found to have distinct intermolecular structures, with the alkyl-functionalized IL displaying a greater degree of organization in its non-polar domains than the hydroxyl-functionalized IL. Further investigations of the organizational characteristics of aprotic ILs containing alkyl and ether groups have provided similar results, with X-ray scattering and molecular dynamics (MD) experiments revealing side chain related differences in nanoscale aggregation that were attributed to differences in chain conformation as well as disruption of polar domains by ether groups (*46–48*). To our knowledge, the effects of chain polarity on the polymerization kinetics of coordinated IL monomers have not been investigated.

Experimental

To examine the effects of side chain polarity on the polymerization kinetics of these materials, six acrylate monomers containing large ($\geq$ 4 atoms) alkyl or ether groups were combined with $LiTf_2N$ under gentle heating at monomer:salt ratios of 1:1, 2:1, 3:1, 4:1, and 10:1, followed by addition of the photoinitiator 2-hydroxy-2-methylpropiophenone (1 wt. %). The mixing of the components was performed in open air at ambient conditions. No change in sample mass due to moisture absorption or vaporization of monomer was observed during the mixing process. Structures of the monomers and used in the study and $LiTf_2N$ are presented in Figure 1. Samples were photopolymerized under UV light (~1 W, ~2.6 W/cm^2, 250-400 nm filter) and conversion profiles were obtained using similar methods to those used in previous studies, with "real-time" attenuated total reflectance Fourier transform infrared spectroscopy (ATR-FTIR) being used to monitor the normalized area of the vibrational band corresponding to the acrylate vinyl wagging deformation (*42–44, 49*). FTIR spectra were also used to examine the influence of electronic and aggregation related effects of polymerization behavior as reflected in infrared band position.

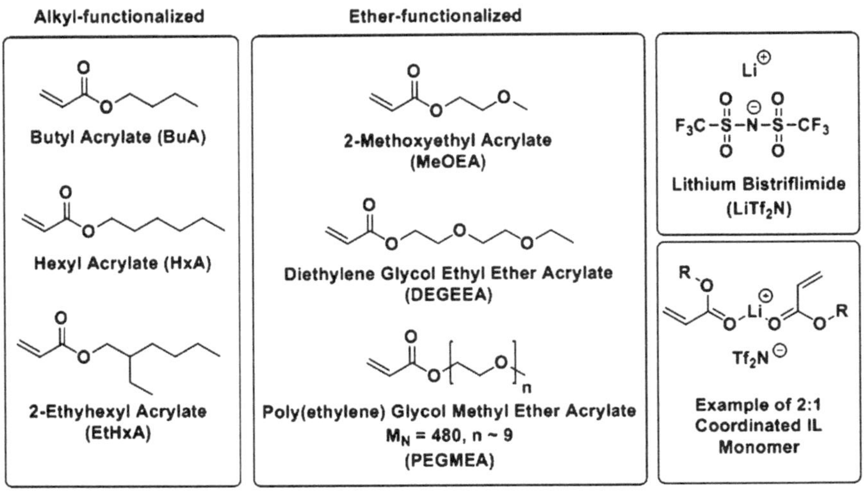

Figure 1. Molecular structures of alkyl- (left) and ether- (center) functionalized monomers examined; $LiTf_2N$ (right, top); generic structure of 2:1 coordinated IL monomer (right, bottom).

Results and Discussion

Systems containing $LiTf_2N$ displayed rapid polymerization kinetics, with all monomer:salt combinations achieving complete, or nearly complete, conversion of acrylate groups within 20 s. These similarities in behavior, however, mask differences in the effects of $LiTf_2N$ on reaction kinetics. In contrast to the coordinated ILs and mixtures, neat monomer samples displayed a relatively large degree of variation in polymerization behavior, as displayed in Figure 2 and summarized in Table 1. Although all samples attained ~100% conversion of acrylate groups to polymer within 6 min (Figure 2), monomers containing non-polar, alkyl side groups clearly polymerized slower than their ether-functionalized counterparts. The conversion at 20 s of butyl acrylate (BuA), for instance, is ~50 % less than that of 2-methoxyethyl acrylate (MeOA), despite the similarities in the monomer size (i.e., same number of atoms in side chain).

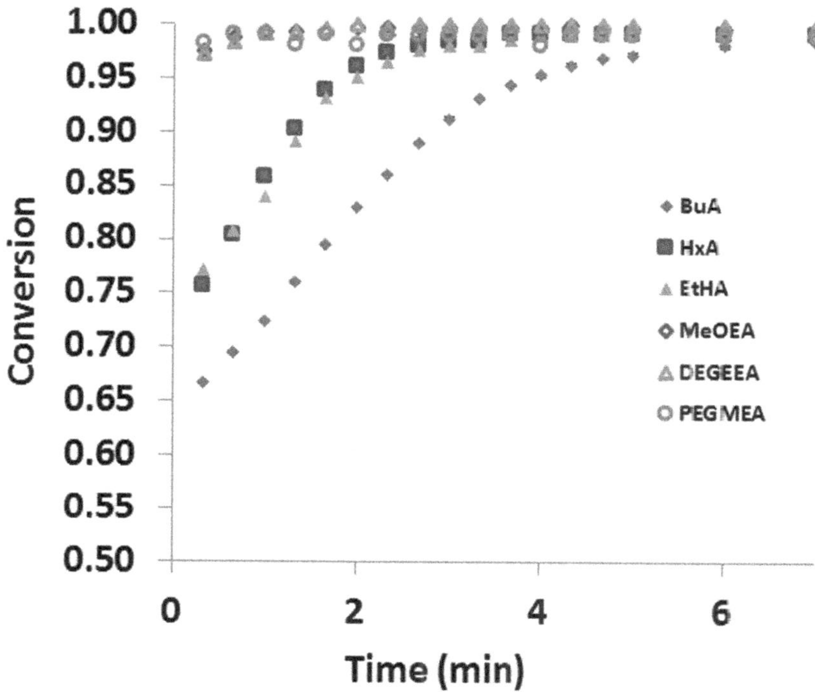

Figure 2. Conversion with time for neat monomers without addition of $LiTf_2N$.

Table 1. Improvement in conversion at 20 s of monomer-$LiTf_2N$ mixtures[a] relative to neat monomer samples.

Monomer	*Increase in Conversion at 20 s (%)*
BuA	49.85
HxA	32.14
EtHxA	29.73
MeOEA	2.62
DEGEEA	2.92
PEGMEA	1.91

[a] Since all salt-containing solutions displayed ~100% monomer conversion at 20 s, the % increase in conversion was found using a conversion of 1.00 for solutions containing $LiTf_2N$

Differences in coordination were also evident for the two monomer types, particularly for those with larger side groups. As with our prior work, extent of monomer coordination was determined by examining splitting patterns of the ~1720 cm^{-1} carbonyl (C=O) stretching band, as bands corresponding to coordinated C=O groups appear at lower wavenumbers than uncoordinated functionalities (*43*, *44*). Systems of both alkyl acrylates and MeOEA displayed coordination numbers of 2, in agreement with short chain (meth)acrylate monomers (*43*). Different behaviors were observed, however, for two larger ether-functionalized acrylates, DEGDEEA and PEGMEA. In contrast to other monomers examined, a carbonyl coordination number of 1 was observed for DEGDEEA, while PEGMEA-$LiTf_2N$ systems contained uncoordinated carbonyl groups at a monomer:salt ratio of 1:1. The different coordination behaviors displayed by the two types of monomers are illustrated in Figures 3a,b showing the C=O stretching bands of systems containing HxA and PEGMEA. It should be noted that the spectra of several 2:1 monomer:salt ILs suggest the presence of small amounts of uncoordinated monomer. This behavior has been observed in similar mixtures and is likely a result of the relatively weak alkali metal cation - ligand interactions occurring in these systems (*43*, *50*). For these coordinated ILs, however, uncoordinated C=O bands appear as small shoulders of coordinated peaks and only become prominent at larger monomer:salt ratios. Figure 4 presents a simplistic depiction of the differences in Li^+ coordination between BuA and MeOEA to illustrate the possible coordination sites in alkyl-functionalized and ether-functionalized acrylate monomers with $LiTf_2N$.

Differences in the polymerization kinetics of the neat monomers examined reveal variation in the effects of the addition of salt on reaction behavior. As was shown in Table 1, the presence of $LiTf_2N$ results in dramatic improvements in the polymerization behaviors of monomers containing long alkyl chains, with an average increase of ~37 % in conversion at 20 s for alkyl functionalized monomers as opposed to ~2.5 % improvement for compounds with ether side groups. It is likely that these differences are the result of variation in electronic and structuring effects of the salt on the different classes of monomer. Previous studies have shown that the addition of Li^+ salts to vinyl monomers can both improve reaction kinetics and increase polymer molecular weight, effects attributed to a combination of electronic and electrostatic factors (*42–44*, *51–55*). Pedron, et al. examined the polymerization of ether-functionalized methacrylates in the presence of lithium triflate (LiTfO), where increases in conversion were spectroscopically observed and FTIR spectra revealed blueshifting of the vinyl C-H overtone band in the presence of the salt (*51*). As similar effects were taken to be indicative of increased reactivity of the vinyl group with respect to radical addition reactions, the observed increases in conversion were attributed to vinyl group electronic redistributions following monomer coordination to the Li^+ cation (*51*, *56–58*). Similar effects were suggested by both ^1H and ^{13}C NMR spectroscopic analysis in studies of the reaction behavior of (meth)acrylamides in the presence of $LiTf_2N$ (*54*, *55*). In addition to catalytic effects, work by Hermosilla and co-workers using both electron paramagnetic resonance spectroscopy (EPRS) and computational methods in the study of methyl methacrylate (MMA) - LiTfO systems has suggested that decreases in R_T occurring as a result of increased electrostatic repulsion between coordinated polymer chain ends also contributes to the observed increases in rate of polymerization (*52*).

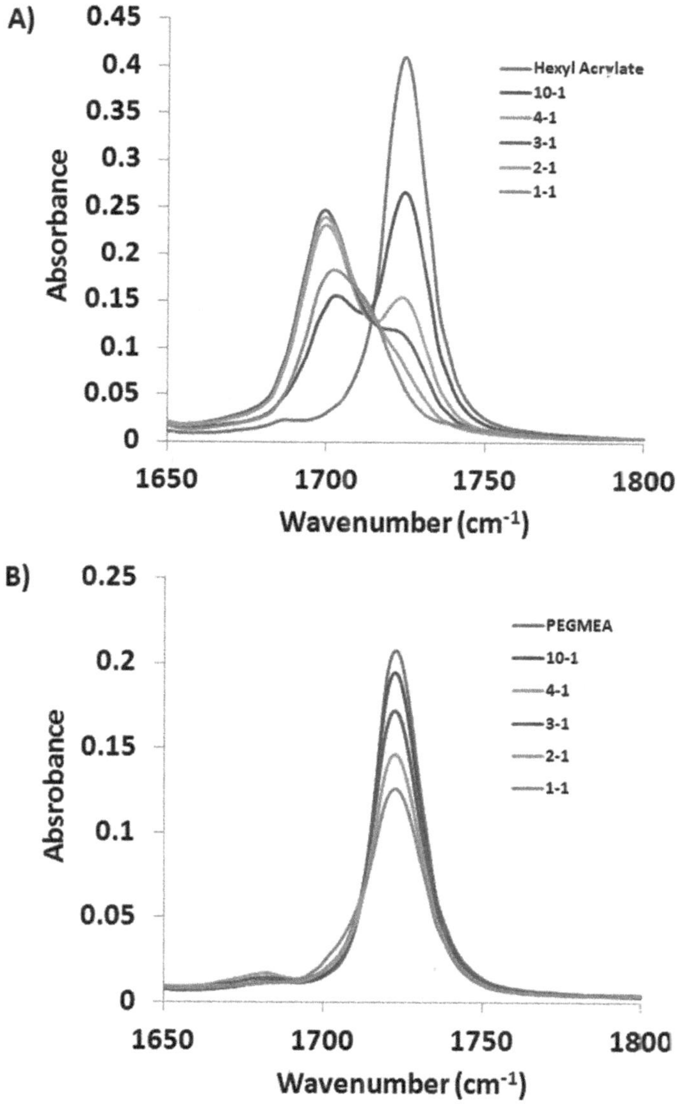

Figure 3. FTIR bands for C=O stretching deformation for coordinated IL systems containing $LiTf_2N$ with a) HxA and b) PEGMEA

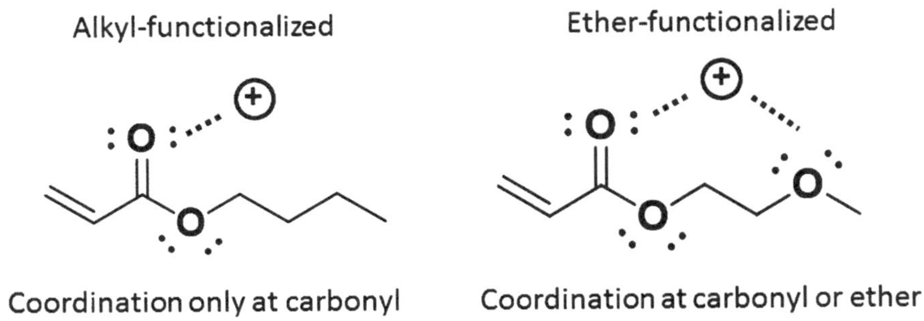

Figure 4. Illustration of coordination sites in alkyl-functionalized acrylate monomers (left) and ether-functionalized monomers (right).

Similar spectroscopic phenomena were found for the systems examined here. As displayed in Figure 5, the position of the vinyl C-H wagging infrared band shifts to higher wavenumbers for nearly all of the monomers examined. With the exception of PEGMEA, all bands were shifted by at least 5 cm^{-1}. Interestingly, changes in the positions of the alkyl-functionalized monomers were greater than those of those with ether side groups, with an average difference in blueshifting of ~2.7 cm^{-1} between the two acrylate groups. These differences may result in part from differences in coordination behavior between the classes of monomer. In contrast to those with alkyl groups, ether-functionalized monomers are capable of interacting with Li^+ at both their C=O groups as well as through the ethers in their side chains (Figures 3, 4). Competition for coordination between the two sites may lead to decreased interaction between the C=O group and the Li^+ cation, resulting in decreased blueshifting of the vinyl band. Indeed, for many of the monomers, the rate of blueshifting with salt content appeared to correspond to the coordination state of the mixture, with increases in band position for alkyl acrylates decreasing as the monomer:salt ratio decreased beyond the coordination number of the monomers (i.e., $n = 2$). The limited C=O coordination of ether-functionalized monomers suggests that the relatively low degree of blueshifting observed in their FTIR spectra may be due to competition between multiple coordination sites, contributing to differences in reaction behavior between alkyl- and ether-functionalized coordinated IL monomers.

Differences in coordination behavior may also contribute to differences in the extent of aggregation between the two monomer classes. As discussed previously, ILs are capable of organizing into polar and non-polar domains, a feature that has important implications in the use of these materials as synthesis media (*30–33*). Moreover, conventional ILs composed of molecular ions, this behavior is partially dependent on side chain polarity (*45–48*). This also appears to be true for coordinated ILs, as evidenced by differences in the FTIR spectra of alkyl and ether acrylates with $LiTf_2N$. Previous work has shown that increased ion clustering in systems containing $LiTf_2N$ is associated with blueshifting of the anion S-N stretching band at ~740 cm^{-1} (*59*). As shown in Figure 5, the spectra of many of the ILs and solutions examined in our study display shifts of this band to higher wavenumbers with increasing salt content. However, the extent of these shifts and their relationship to chain length depends on the polarity of the monomer side group. For alkyl-functionalized monomers, increases in $LiTf_2N$ content corresponded to relatively large blueshifts in the position of the anion S-N stretching band, with shifts ≥ 5.82 cm^{-1} observed for all monomers in this group examined. However, blueshifts in band position for ether-functionalized monomers were comparatively small. In addition, the extent of shifting decreased as the size of the side group increased. These observations suggest that, like conventional ILs, these systems display side chain polarity dependent differences in aggregation.

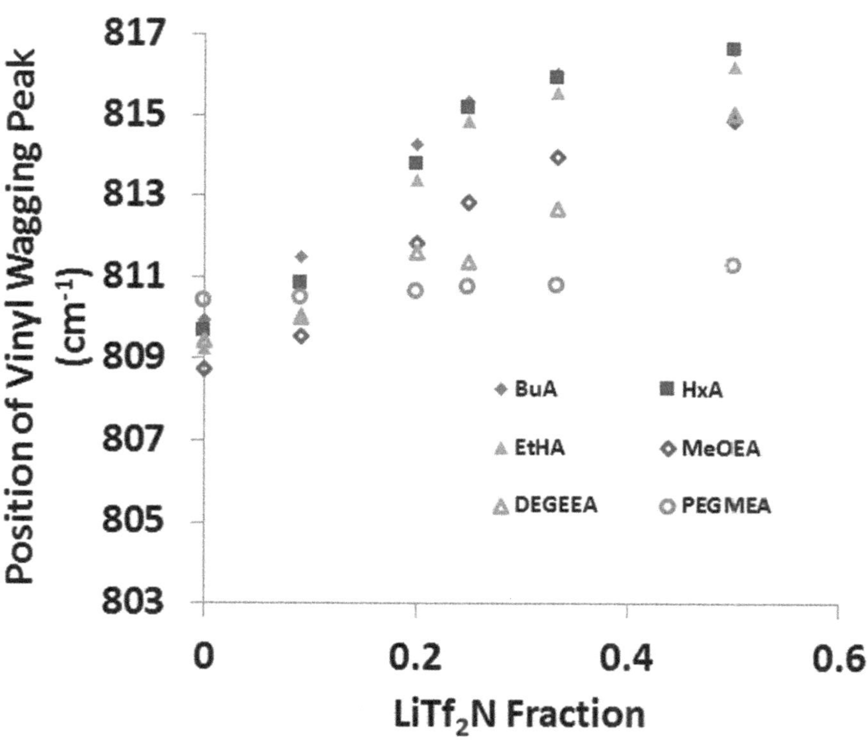

Figure 5. Plot displaying position of band corresponding to vinyl wagging deformation (+/- 0.013 cm^{-1}) with salt content for acrylate monomers.

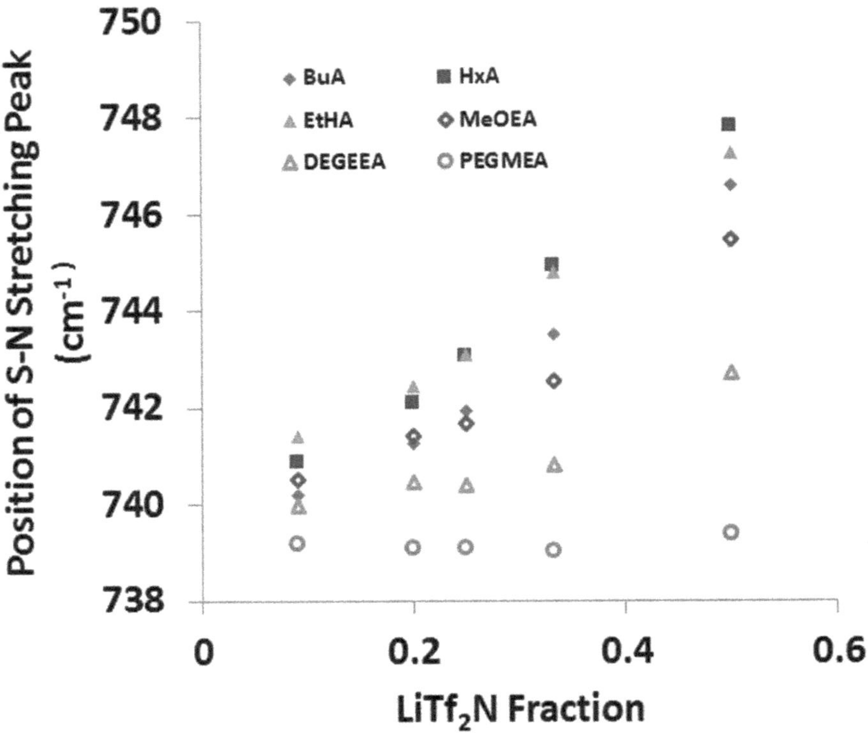

Figure 6. Position of anion S-N stretching band (+/- 0.018 cm^{-1}) relative to fraction of $LiTf_2N$ for each acrylate monomer

Conclusions

Recent studies have demonstrated the ability of coordinated IL containing polar vinyl monomers to improve both reaction kinetics and polymer product properties relative to uncoordinated monomer samples (*42–44*). These effects have been explained as a combination of favorable electronic redistribution resulting from monomer coordination and IL aggregation behavior (*42–44*). However, despite the probable influence of IL organization on reaction behavior, the effects of monomer side chain polarity and size on polymerization kinetics of these systems have not been investigated. In this study, coordinated ILs and more dilute salt containing solutions were prepared with $LiTf_2N$ and six acrylate monomers containing alkyl and ether side chains of varying length. Although all salt containing mixtures displayed similar improvements in reaction behavior, FTIR analysis revealed monomer size and polarity dependent differences in the causes of this behavior. Thus, we have shown that the concept of a coordinated IL monomer can be extended to acrylates with large side chains, with the presence of $LiTf_2N$ improving overall monomer conversion, especially for alkyl-functionalized species. The ability to select from many types of organic species in the formulation of coordinated ILs with $LiTf_2N$ enables access to potentally vast arrays of polymer-inorganic composites that can be formed via photopolymerization. Polymers produced from coordinated IL monomers may ultimately find use in 3D-printing as a means of controlling properties such as density, hardness, tensile strength, conductivity or even color if a transition metal cation is used instead of Li^+. Our future work will explore the physical properties of these polymer-inorganic composites produced via the photopolymerization of coordinated IL monomers using molds and/or 3-D printing.

Acknowledgments

Acknowledgment is made to the Donors of the American Chemical Society Petroleum Research Fund for partial support of this research.

References

1. O'Brien, A. K.; Bowman, C. N. Modeling Thermal and Optical Effects on Photopolymerization Systems. *Macromolecules* **2003**, *36*, 7777–7782.
2. Bowman, C. N.; Kloxin, C. J. Toward an Enhanced Understanding and Implementation of Photopolymerization Reactions. *AIChE J.* **2008**, *54*, 2775–2795.
3. Yagci, Y.; Jockusch, S.; Turro, N. J. Photoinitiated Polymerization: Advances, Challenges, and Opportunities. *Macromolecules* **2010**, *43*, 6245–6260.
4. Torgersen, J.; Qin, X.-H.; Li, Z.; Ovsianikov, A.; Liska, R.; Stampfl, J. Hydrogels for Two-Photon Polymerization: A Toolbox for Mimicking the Extracellular Matrix. *Adv. Funct. Mater.* **2013**, *23* (36), 4542–4554.
5. Nesvadba, P. Radical Polymerization in Industry. In *Encyclopedia of Radicals in Chemistry, Biology and Materials*; Chatgilialoglu, C., Studer, A., Eds.; John Wiley & Sons Ltd.: 2012; Vol. 4, pp 1701−1736.
6. Brazel, C. S.; Rosen, S. L. *Fundamental Principles of Polymeric Materials*, 3rd ed.; Wiley & Sons Inc.: Hoboken, NJ, 2012.
7. Goodner, M. D.; Lee, H. R.; Bowman, C. N. Method for Determining the Kinetic Parameters in Diffusion-Controlled Free-Radical Homopolymerizations. *Ind. Eng. Chem. Res.* **1997**, *36*, 1247–1252.
8. Anseth, K. S.; Kline, L. M.; Walker, T. A.; Anderson, K. J.; Bowman, C. N. Reaction Kinetics and Volume Relaxation during Polymerizations of Multiethylene Glycol Dimethacrylates. *Macromolecules* **1995**, *28*, 2491–2499.
9. Berchtold, K. A.; Lovestead, T. M.; Bowman, C. N. Coupling Chain Length Dependent and Reaction Diffusion Controlled Termination in the Free Radical Polymerization of Multivinyl (Meth)acrylates. *Macromolecules* **2002**, *35*, 7968–7975.
10. Dietz, J. E.; Peppas, N. A. Reaction Kinetics and Chemical Changes During Polymerization of Multifunctional (Meth)acrylates for the Production of Highly Crosslinked Polymers used in Information Storage Systems. *Polymer* **1997**, *38*, 3767–3781.

11. Tajima, K.; Aida, T. Controlled Polymerizations with Constrained Geometries. *Chem. Commun.* **2000**, 2399–2412.
12. O'Brien, D. F.; Armitage, B.; Benedicto, A.; Bennett, D. E.; Lamparski, H. G.; Lee, Y.-S.; Srisiri, W.; Sisson, T. M. Polymerization of Preformed Self-Organized Assemblies. *Acc. Chem. Res.* **1998**, *31*, 861–868.
13. *GROMACS*. www.gromacs.org.
14. Allen, M. P.; Tildesley, D. J. *Computer Simulation of Liquids*; Clarendon Press: Oxford, 1987.
15. Rankin, R. B.; Liu, J. C.; Kulkarni, A. D.; Johnson, J. K. Adsorption and Diffusion of Light Gases in ZIF-68 and ZIF-70: A Simulation Study. *J. Phys. Chem. C* **2009**, *113*, 16906–16914.
16. *CHARMM*. http://www.charmm.org/.
17. Guymon, C. A.; Bowman, C. N. Kinetic Analysis of Polymerization Rate Acceleration During the Formation of Polymer/Smectic Liquid Crystal Composites. *Macromolecules* **1997**, *30*, 5271–5278.
18. Lester, C. L.; Guymon, C. A. Phase Behavior and Polymerization Kinetics of a Semifluorinated Lyotropic Liquid Crystal. *Macromolecules* **2000**, *33*, 5448–5454.
19. Lester, C. L.; Colson, C. D.; Guymon, C. A. Photopolymerization Kinetics and Structure Development of Templated Lyotropic Liquid Crystalline Systems. *Macromolecules* **2001**, *34*, 4430–4438.
20. An, W.; Turner, C. H. Structural, Electronic, and Magnetic Features of Platinum Alloy Strings Templated on a Boron-Doped Carbon Nanotube. *Phys. Rev. B* **2010**, *81*, 205433.
21. Lu, J.; Yan, F.; Texter, J. Advanced Applications of Ionic Liquids in Polymer Science. *Prog. Polym. Sci.* **2009**, *34*, 431–448.
22. McGuinness, D. S.; Cavell, K. J.; Yates, B. F.; Skelton, B. W.; White, A. H. Oxidative Addition of the Imidazolium Cation to Zerovalent Ni, Pd, and Pt: A Combined Density Functional and Experimental Study. *J. Am. Chem. Soc.* **2001**, *123*, 8317–8328.
23. Andrzejewska, E. Photopolymerization in ionic liquids. In *Basics and Applications of Photopolymerization Reactions*; Fouassier, J. P., Allonas, X., Eds.; Research Signpost: Trivandrum, India, 2010; Vol. 2, pp 245–257.
24. Woecht, I.; Schmidt-Naake, G.; Beuermann, S.; Buback, M.; Garcia, N. Propagation Kinetics of Free-Radical Polymerizations in Ionic Liquids. *J. Polym. Sci., Part A: Polym. Chem.* **2008**, *46*, 1460–1469.
25. Frenkel, D.; Smit, B. *Understanding Molecular SImulation: From Algorithms to Applications*, 2nd ed.; Academic Press: San Diego, 2002.
26. Fukumoto, K.; Yoshizawa, M.; Ohno, H. Room Temperature Ionic Liquids from 20 Natural Amino Acids. *J. Am. Chem. Soc.* **2005**, *127*, 2398–2399.
27. Yazaki, S.; Funahashi, M.; Kagimoto, J.; Ohno, H.; Kato, T. Nanostructured Liquid Crystals Combining Ionic and Electronic Functions. *J. Am. Chem. Soc.* **2010**, *132*, 7702–7708.
28. Kato, T.; Mizoshita, N.; Kishimoto, K. Functional Liquid-crystalline Assemblies: Self-Organized Soft Materials. *Angew. Chem., Int. Ed.* **2006**, *45*, 38–68.

29. Puttick, S.; Davis, A. L.; Butler, K.; Lambert, L.; El Harfi, J.; Irvine, D. J.; Whittaker, A. K.; Thurecht, K. J.; Licence, P. NMR as a Probe of Nanostructured Domains in Ionic Liquids: Does Domain Segregation Explain Increased Performance of Free Radical Oolymerization? *Chem. Sci.* **2011**, *2*, 1810–1816.
30. Canongia, L. J. N. A.; Padua, A. A. H. Nanostructural Organization in Ionic Liquids. *J. Phys. Chem. B* **2006**, *110*, 3330–3335.
31. Russina, O.; Triolo, A. New Experimental Evidence Supporting the Mesoscopic Segregation Model in Room Temperature Ionic Liquids. *Faraday Discuss.* **2012**, *154*, 97–109(Ionic Liquids).
32. Ohno, H.; Fukumoto, K. Amino Acid Ionic Liquids. *Acc. Chem. Res.* **2007**, *40*, 1122–1129.
33. Weber, C. C.; Masters, A. F.; Maschmeyer, T. Pseudo-Encapsulation - Nanodomains for Enhanced Reactivity in Ionic Liquids. *Angew. Chem., Int. Ed.* **2012**, *51*, 11483–11486.
34. Pratt, H. D., III; Rose, A. J.; Staiger, C. L.; Ingersoll, D.; Anderson, T. M. Synthesis and Characterization of Ionic Liquids Containing Copper, Manganese, or Zinc Coordination Cations. *Dalton Trans.* **2011**, *40*, 11396–11401.
35. Pratt, H. D.; Leonard, J. C.; Steele, L. A. M.; Staiger, C. L.; Anderson, T. M. Copper Ionic Liquids: Examining the Role of the Anion in Determining Physical and Electrochemical Properties. *Inorg. Chim. Acta* **2013**, *396*, 78–83.
36. Brooks, N. R.; Schaltin, S.; Van, H. K.; Van, M. L.; Binnemans, K.; Fransaer, J. Copper(I)-Containing Ionic Liquids for High-Rate Electrodeposition. *Chem. Eur. J.* **2011**, *17*, 5054–5059.
37. Zhang, P.; Gong, Y.; Lv, Y.; Guo, Y.; Wang, Y.; Wang, C.; Li, H. Ionic Liquids with Metal Chelate Anions. *Chem. Commun.* **2012**, *48*, 2334–2336.
38. Huang, J.-F.; Luo, H.; Dai, S. A New Strategy for Synthesis of Novel Classes of Room-Temperature Ionic Liquids Based on Complexation Reaction of Cations. *J. Electrochem. Soc.* **2006**, *153*, J9–J13.
39. Gao, Y.; Twamley, B.; Shreeve, J. M. The First (Ferrocenylmethyl)-imidazolium and (Ferrocenylmethyl)triazolium Room Temperature Ionic Liquids. *Inorg. Chem.* **2004**, *43*, 3406–3412.
40. Gardas, R. L.; Coutinho, J. A. P. A Group Contribution Method for Viscosity Estimation of Ionic Liquids. *Fluid Phase Equilibr.* **2008**, *266*, 195–201.
41. Katritzky, A. R.; Millet, G. H.; Noor, H. M.; Yates, F. S. Heterocycles in Organic Synthesis 11. Reactions of Heteroaromatic N-Oxides with Pyridine and Diazoles. *J. Org. Chem.* **1978**, *43*, 3957–3960.
42. Hu, Y. F.; Liu, Z. C.; Xu, C. M.; Zhang, X. M. The Molecular Characteristics Dominating the Solubility of Gases in Ionic Liquids. *Chem. Soc. Rev.* **2011**, *40*, 3802–3823.
43. Whitley, J. W.; Adams, I. A.; Terrill, K. L.; Hayward, S. S.; Burnette, M. T.; Bara, J. E. Photopolymerization of Coordinated Ionic Liquid Monomers: Realizing the Benefits of Structured Media Using Only Common Reagents. *J. Polym. Sci., Polym. Chem.* **2016**, *54*, 2004–2014.

44. Radziszewski, B. Ueber Einige Neue Glyoxaline. *Ber. Dtsch. Chem. Ges.* **1883**, *16*, 747–749.
45. Brandrup, J.; Immergut, E. H.; Grulke, E. A.; Abe, A.; Bloch, D. R. *Polymer Handbook*, 4th ed.; John Wiley & Sons: New York, 1999.
46. Tosoni, M.; Laschat, S.; Baro, A. Synthesis of Novel chiral ionic liquids and their phase behavior in mixtures with smectic and nematic liquid crystals. *Helv. Chim. Acta* **2004**, *87* (11), 2742–2749.
47. Gasparini, J. P.; Gassend, R.; Maire, J. C.; Elguero, J. Study on Oragnosilyll Azole Series 1. Action of Alkyl-Halides, Acid-Chlorides and Halogenated Ketones. *J. Organomet. Chem.* **1980**, *188*, 141–150.
48. Buchel, K.-H.; Falbe, J. F. A Process for Preparing an n-Alkyl or n-Alkenyl Imidazole. Australian Patent AU1967021134, 1968.
49. Socrates, G. *Infrared and Raman Characteristic Group Frequencies: Tables and Charts*, 3rd ed.; John Wiley & Sons: West Sussex, U.K., 2001.
50. Hamilton, J. R.; Abedini, A.; Zhang, Z.; Whitley, J. W.; Bara, J. E.; Turner, C. H. Enhancing the Pre-Polymerization Coordination of 1-Vinylimidazole. *Chem. Eng. Sci.* **2015**, *138*, 646–654.
51. Pedron, S.; Guzman, J.; Garcia, N. Polymerization Kinetics of Ethylene Oxide Methacrylates in Ionic Media. *Macromol. Chem. Phys.* **2011**, *212*, 860–869.
52. Hermosilla, L.; Calle, P.; Tiemblo, P.; Garcia, N.; Garrido, L.; Guzman, J. Polymerization of Methyl Methacrylate with Lithium Triflate. A Kinetic and Structural Study. *Macromolecules* **2013**, *46*, 5445–5454.
53. Noble, B. B.; Smith, L. M.; Coote, M. L. The Effect of $LiNTf_2$ on the Propagation Rate Coefficient of Methyl Methacrylate. *Polym. Chem.* **2014**, *5*, 4974–4983.
54. Hirano, T.; Saito, T.; Kurano, Y.; Miwa, Y.; Oshimura, M.; Ute, K. Dual Rrole for Alkali Metal Cations in Enhancing the Low-Temperature Radical Polymerization of N,N-Dimethylacrylamide. *Polym. Chem.* **2015**, *6*, 2054–2064.
55. Hirano, T.; Segata, T.; Hashimoto, J.; Miwa, Y.; Oshimura, M.; Ute, K. Syndiotactic- and Heterotactic-Specific Radical Polymerization of N-*n*-Propylmethacrylamide Complexed with Alkali Metal Ions. *Polym. Chem.* **2015**, *6*, 4927–4939.
56. Siegmann, R.; Jelicic, A.; Beuermann, S. Propagation and Termination Kinetics of PEGylated Methacrylate Radical Polymerizations. *Macromol. Chem. Phys.* **2010**, *211*, 546–562.
57. Beuermann, S.; Nelke, D. The Influence of Hydrogen Bonding on the Propagation Rate Coefficient in Free-Radical Polymerizations of Hydroxypropyl Methacrylate. *Macromol. Chem. Phys.* **2003**, *204*, 460–470.
58. Beuermann, S. Impact of Hydrogen Bonding on Propagation Kinetics in Butyl Methacrylate Radical Polymerizations. *Macromolecules* **2004**, *37*, 1037–1041.
59. Abbrent, S.; Lindgren, J.; Tegenfeldt, J.; Wendsjo, A. Gel Electrolytes Prepared from Oligo(Ethylene Glycol) Dimethacrylate: Glass Ttransition, Conductivity and Li^+-Coordination. *Electrochim. Acta* **1998**, *43*, 1185–1191.

Chapter 5

Self-Assembly of Block Copolymers in Ionic Liquids

Ru Xie,[1] Carlos R. López-Barrón,[2] and Norman J. Wagner[1,*]

[1]Center for Neutron Science, Department of Chemical and Biomolecular Engineering, University of Delaware, 150 Academy St., Newark, Delaware 19716, United States
[2]Advanced Characterization Department, ExxonMobil Chemical Company, 5200 Bayway Drive, Baytown, Texas 77520, United States
[*]E-mail: wagnernj@udel.edu

Self-assembly of amphiphilic block copolymers can impart desired discrete or continuous nanostructures, such as micelles utilized as drug delivery vehicles in aqueous solvents, or cross-linked micelles for stretchable electronics fabrications in ionic liquids. These appealing applications have motivated significant research efforts to understand the nano- and microstructure as well as the structure-property relationships underlying such self-assembled systems, with the ultimate goal being effective formulation. To take full advantage of the bottom-up self-assembly approach, a comprehensive understanding of the factors that govern the self-assembly behavior of dilute, concentrated and functionalized system of non-ionic block copolymers self-assembly in ionic liquids, as well as robust characterization methods for quantifying the microstructure and properties relationship must be reviewed. For each system, the most significant challenges are presented and discussed. In addition, current and potential applications of block copolymers/ionic liquid system are also discussed. This provides a measure of the current state of understanding, which motivates a brief look forward towards future directions in this field.

Introduction

Well-defined nanostructured materials are of considerable interest due to their emergent properties that can be exploited in various applications (*1*). As a result, the development of methodologies for their formulation and fabrication motivates research into their nanostructure and its relationship to the desired properties. One of the most universal techniques to prepare those materials with tunable and controllable morphologies is by self-asssembly of amphiphic block copolymers (ABCs) in selected self-assembly media. Self-assembled nanostructures in aqueous solutions have found applications in emerging nanotechnologies, including drug delivery vehicles and aqueous nanoreactors. However, self-assembly in ionic liquids (ILs) has been mainly utilized in material science for engineering novel, hierarchically structured solid or soft-solid materials in ion-conducting electrolytes or wearable electronics applications. In addition to those technological applications, ABC/IL systems also provide important fundamental scientific significance. Prior to the use of IL as a self-assembly media, ABC aqueous micellar solutions has been a commonly used model system for studying molecular interactions between non-ionic surfactants and molecular solvents, as well as thermodynamics and kinetics involved in the micellization process. Furthermore, with the addition of charged ions, ABC wormlike micellar nanocolloids create another model system for studying non-linear flow phenomena and flow instabilities such as shear banding. Using ILs as self-assembly media not only opens up a new line of investigation where self-assembly is not determined in part by Coulombic interactions, but also opens a new model paradigm for studying micellization and the associated physiochemical properties of microphase transitions. As will be demonstrated in this chapter, there has been a significant amount of recent research exploring the microstructure and possible emergent properties of self-assembled ABC/IL systems is warranted. Accordingly, this chapter provides an up-to-date literature and patent review on the field of ABCs self-assembly in ILs, which identifies some key findings that should be of interest to practitioners and researchers alike working with block co-polymers and non-ionic surfactants in ILs. This chapter starts with a brief review on the differences and similarities between the self-assembly of ABCs in ILs versus water environment, which is followed by a detailed review on the self-assembled microstructure-property relationship of representative ABC/IL micellar systems under static and flow conditions. We conclude with a brief discussion of emerging and potential applications as well as future directions in this field. For the purpose of this review, only non-ionic ABCs will be included. We do not seek to comprehensively survey all existing ABC/IL systems, but instead to provide some perspectives on the current understanding, characterization techniques, challenges and opportunities that could be realized to assist in better formulation and in the development of new applications. While space limitations restrict how much that can be included, we refer the reader to recent, relevant reviews of the structure of ILs (*2*, *3*), ILs as self-assembly media (*4*, *5*), polymers in ILs (*6*) and IL polymer electrolytes (*7*, *8*).

Current Understanding on Self-Assembly of Amphiphilic Block Copolymers in Ionic Liquids versus Other Self-Assembly Media

In conventional material design, due to the limited amount of known self-assembly media (*5*), tailoring a self-assembly system to meet specific design properties often requires modifying the components, compositions and architecture of ABCs. This subject has been extensively reviewed (*1*) and is now a textbook topic (*9–11*). With the identification of ILs to be capable of promoting ABCs self-assembly, the number of self-assembly systems has increased dramatically. Currently, one molecular solvent (water), three protic ILs (PILs, mainly with alkylammonium cations) and nine aprotic ILs (AILs, mainly with imidazolium cations) have been experimentally demonstrated to support ABCs self-assembly, restricted only by the ABCs solubility. However, it should be possible to select or create new ILs with different cation and anion combinations as self-assembly media if favorable molecular interactions between the IL and the block copolymer, which governs the self-assembly and resultant phase behavior, can be identified.

Properties of Self-Assembly Media

Aside from high ionic conductivity and electrochemical stability, ILs have other remarkable properties that make them candidates for self-assembly media, including negligible vapor pressure, good chemical and thermal stability, and wide liquid temperature ranges (*12*). These vastly different physicochemical properties between water and ILs is a direct consequence of very different molecular structures. Atkin (*13*, *14*) and Hayes (*2*, *15–17*) *et al.* have carried out systematic studies and have reviewed the literature to reveal that ILs are highly structured solvents. Further, a recent review shows that the microstructures of ILs can be modeled over a broad range of length scales, ranging from supramolecular (ion pairs, H-bonded networks, ion and clusters) to self-assembled (micelle-like, mesoscopic) solvent structures (*2*). This difference between ILs and classical molecular liquids manifests in many ILs as a long-range disordered sponge (L_3 phase), where the ions form a network of polar and non-polar domains due to electrostatic and solvophobic clustering of like molecular groups (*2*). This order is on the nanometer length scale and as such, may not directly influence the self-assembly of block copolymers which have mesophase structures typically an order of magnitude larger. However, this topic remains relatively unexplored. What may be important, however, is that this propensity of ILs to self-organize is due to their amphiphilic structure when not dissociated, such that ionic liquids can act as co-surfactants in molecular mesophase formation. This can lead to unique features, such as the stabilization of water nanoclusters in ILs (*18*) or even the formation of spontaneous vesicles with a single surfactant (*19*).

Thermodynamics and Kinetics of Micellization in ILs

The solution self-assembly of block copolymers may be limited by thermodynamics or kinetics. Thermodynamic considerations include the development of a phase diagram, which includes the critical micellization concentration (CMC) and critical gelation concentration (CGC), as well as the progression of equilibrium mesophases. In addition to the specific chemical composition, temperature and pressure, the polymer molecular mass and polydispersity will dictate the length scale(s) of the mesophase. Concepts traditional to surfactant self-assembly can also be valuable, such as packing parameter. Kinetic considerations arise when considering mixing and possible nonequilibrium phases, as well as the sometimes slow equilibration of mesophases.

From a thermodynamics perspectives, ABC solution self-assembly in solution is primarily thought to be an entropically driven process, whereby the solvent molecules surrounding the solvophobic blocks acquire a lower entropy H-bonded network configuration through self-assembly than when they are surrounded by other water molecules. The strength of this entropic driving force for a given IL is characterized through the solvent's Gordon parameter, which is a measure of cohesive energy density of the solvent. A higher Gordon parameter implies a stronger driving force for phase separation. In the mixture this translates into, a hydrophobic effect (*20*) in aqueous micellar solutions, or a solvophobic effect for ILs (*21*).

Aside from this largely entropic solvophobic interaction, the chemical nature of the block copolymer and the enthalpic interactions with the IL play a critical role. This is the basis of the classic Flory χ parameter that is foundational to polymer thermodynamics (e.g., the Flory-Huggins constitutive model) and is a determining parameter in the solution and melt thermodynamics of block copolymers more broadly.

Additional factors that dictate the solution self-assembly include specific block copolymer properties (e.g., molecular mass, polydispersity, relative block lengths, primary chain configuration, chain branching topology) (*9*), Such molecular properties are often coarse-grained into concepts such as the packing parameter, whereby the specific self-assembled topology is determined by a free energy balance that includes the chain stretching in the core, the inter-coronal interactions, and the excess core-corona interfacial energy (*20*, *22*, *23*). Among the various molecular considerations, the specific role of H-bonding in self-assembly is not fully understood. Hayes *et al.* report that most of the PILs form 3-D H-bond networks which are different from that in water, and are confined within the polar domain between adjacent proton donor and acceptor sites of the IL's nanostructure (*2*). Furthermore, Hayes *et al.* argued that these H-bonds in ILs are accommodated between ions, but do not control self-assembly nor are they the principle driving force for structure formation (*2*). However, these H-bonds are important for applications as they control the IL solvent properties and proton transfer (*2*, *24*).

All of these factors play a role in self-assembly, which is inherently a cooperative process that requires the simultaneous participation of numerous amphiphilic and solvents molecules and ions. As an example, the relatively simple case of surfactant micellization in water is characterized by a CMC and a critical micellization temperature (CMT). The hydrophobic effect mandates a lower micelle size limit as assemblies of low aggregation number of amphiphilic molecules cannot eliminate the unfavorable solvophobic/solvent interface, meanwhile repulsive interactions between the charged head-groups of amphiphilic molecules impose an upper size limit and restrict the micelle growth. Due to the inherent microstructure and ionic nature of ILs, as well as their inherent amphiphilic nature when they are not dissociated, the self-assembly of ABCs in ILs is inherently more complex than in aqueous solution as will be demonstrated in more detail in the next section.

The dynamics of self-assembled block copolymers, which include micellar exchange relaxation (termed as "relaxation kinetics" for non-equilibrium micelles and "equilibrium dynamics" for equilibrium micelles), and unimer to micelle formation(termed as "micellization kinetics") are often very slow in ABC assemblies in aqueous system (*25–29*). Because of this extremely slow dynamics, the self-assembled structure are highly dependent on assembly pathway, and the "final" morphology is often a kinetically-trapped or long-lived metastable state (*28*). A number of experimental (*26*, *27*, *30–32*), theoretical (*23*, *33–35*), and computational (*36*, *37*) research has provided valuable insights on kinetics of ABC aqueous micelles. Generally speaking, two mechanisms are found to be key for stimulating the changes in micelle size and structures, namely, single chain exchange and micelle fusion/fission. Although the role of those two mechanisms in polymeric micelle dynamics in still a subject of research, it is generally accepted that single chain exchange mechanism is dominant for micelle microstructures near equilibrium (*26*, *30*, *33*), while the fusion/fission mechanism dominates micelle growth far from equilibrium (*32*). Specific measurements of such kinetics for ABCs in ILs are discussed in the next section.

Microstructure-Property Relationships for Self-Assembled Block Copolymer/Ionic Liquid Systems

Self-assembly of ABCs in ILs not only provides new sources of complex colloidal model systems for study, but also constitutes a platform for combining the outstanding physicochemical properties of ILs with a number of favorable attributes offered by ABCs contributing to the field of soft materials science. In particular, ABCs self-assemble into well-defined nanostructures, which can be engineered to provide a durable and stretchable mechanical scaffold for templating ILs into the self-assembled ion-conducting channels. Fundamental understanding of the structure and thermodynamics of ABCs in ILs is critical for developing microstructure-property relationship in predictable design of hierarchically structured, functional soft materials.

Dilute Solutions of Block Copolymers in Ionic Liquids

Investigation of the ABCs self-assembled phase behavior in ILs typically begins with the study of micellization in dilute solution. The thermal stability and low volatility of PILs and AILs open up new opportunities to investigate the self-assembly of ABCs over a wide range of temperatures. We examine micellization for each class of IL as follows.

The earliest reports of ILs capable of supporting self-assembly were for PILs in 1980s, mainly belonging to the family including alkylammoium nitrate (*38*, *39*). The use of alkylammonium nitrate as a self-assembly medium has often been considered with reference to water, due to the evidence of similar H-bonding structures found in EAN as revealed by Evans *et al.* (*38*) Self-assembly in aqueous systems are often used to guide investigations in alkylammoium nitrate, as they are thought to have similar driving forces for micellization (*4*). As briefly mentioned in the previous section, the driving force for micellization in alkylammonium nitrate is due to the favorable solvent entropy gain from shielding the solvophobic blocks out of the polymer, which include unfavorable organization in the surrounding PIL (*20*, *38*). However, one of the main quantitative differences between alkylammonium nitrate and water is that the hydrocarbons in the ABCs are more soluble in alkylammonium nitrate, which results in a weaker solvophobic effect of the ILs as compared to the equivalent hydrophobic effect of water (*40*). This quantitative difference leads to higher CMCs in alkylammonium nitrate, as the greater solubility in alkylammonium nitrate requires more unimers or longer ABC chains to impose similar aggregation behavior to that in water (*41*, *42*). To have a better look at the behavior of CMC in water versus ILs, CMC measurements for currently reported ABCs in PILs and AILs are listed in Table 1, along with the corresponding CMCs in water and other relevant self-assembly properties if available. Nevertheless, micellization in both water and alkylammonium has been shown to be a spontaneous process, with a clear discontinuity in solution properties below and above CMC, and also followed closed association micellization model (*20*), reflecting the single step unimers to micelles equilibrium (*4*, *38*). However, micellization of alkyammonium nitrate in water causes an increase in partial molar volume, while no change in partial molar volume was observed in EAN, suggesting micellization process does not significantly change EAN's solvent structure (*4*, *43*).

Table 1. Micellization of block copolymers in ionic liquids compared to aqueous solvents.

Line	*Block copolymers*	*Formula*	*M_n (g/mol)*	*HLB[a] (44)*
1	Pluronic F127	EO_{100}–PO_{65}–EO_{100}	12600	22
		EO_{106}–PO_{70}–EO_{106}	13388	
		EO_{106}–PO_{70}–EO_{106}	11500	
2	Pluronic F68	EO_{79}–PO_{30}–EO_{79}	8400	29

Continued on next page.

Table 1. (Continued). Micellization of block copolymers in ionic liquids compared to aqueous solvents.

Line	*Block copolymers*	*Formula*	*M_n (g/mol)*	*HLB[a] (44)*
3	Pluronic P123	$EO_{19}-PO_{69}-EO_{19}$	5750	8
		$EO_{20}-PO_{70}-EO_{20}$ $EO_{20}-PO_{70}-EO_{20}$	5820	
		$EO_{20}-PO_{70}-EO_{20}$	6710	
		$EO_{20}-PO_{70}-EO_{20}$		
		$EO_{20}-PO_{70}-EO_{20}$	7000	
4	Pluronic P105	$EO_{37}-PO_{56}-EO_{37}$	6500	15
		$EO_{37}-PO_{56}-EO_{37}$	6500	
5	Pluronic P85	$EO_{26}-PO_{40}-EO_{26}$	4600	16
		$EO_{25}-PO_{40}-EO_{25}$	4500	
		$EO_{26}-PO_{40}-EO_{26}$	4600	
6	Pluronic P65	$EO_{18}-PO_{29}-EO_{18}$	3400	
		$EO_{20}-PO_{30}-EO_{20}$	3500	
		$EO_{19}-PO_{30}-EO_{19}$	3400	
7	Pluronic L121	$EO_{5}-PO_{68}-EO_{5}$ $EO_{5}-PO_{70}-EO_{5}$	4400 3720	1
		$EO_{5}-PO_{68}-EO_{5}$	4400	

Continued on next page.

Table 1. (Continued). Micellization of block copolymers in ionic liquids compared to aqueous solvents.

Line	*Block copolymers*	*Formula*	*M_n (g/mol)*	*HLB[a] (44)*
8	Pluronic L81	EO_6–PO_{39}–EO_6 EO_3–PO_{43}–EO_3	2750 2750	2
9	Pluronic L64	EO_{13}–PO_{30}–EO_{13} EO_{13}–PO_{30}–EO_{13}	2900 2900	15
10	Pluronic L61	EO_3–PO_{30}–EO_3	2000	3
11	PEGE-PEO	EGE_{109}-EO_{54} EGE_{113}-EO_{115} EGE_{104}-EO_{178}	13600 16700 18600	
12	PGPrE-PEO	$GPrE_{98}$-EO_{260}	21400	
13	PS-PEO	PS-PEO PS-PEO PS-PEO PS-PEO (20-13) PS-PEO (20-13) PS-PEO (20-8) PS-PEO (20-5) PS-PEO (1-3)	28700 14100 8500 20000 20000 20000 20000	
14	Micelle shuttle	PB-PEO (9-20) PB-PEO (9-10) PB-PEO (9-17) PB-PEO (9-4)	30200 19100 16400 13200	
15	PS-PMMA	PS-PMMA (3-13) PS-PMMA (7-8) PS-PMMA (11-4)	3000 7000 11000	

Continued on next page.

Table 1. (Continued). Micellization of block copolymers in ionic liquids compared to aqueous solvents.

Line	*Block copolymers*	*Formula*	*M_n (g/mol)*	*HLB[a] (44)*
16	Micelle shuttle	PB-PEO (9-21) PB-PEO (9-10) PB-PEO (9-17) PB-PEO (9-4)	30200 19100 16400 13200	
17	LCST	PnBMA-PEO (12-9)		
18	LCST	PBnMA-PMMA (2.2-19) PBnMA-PMMA (7.3-19) PBnMA-PMMA (28-41)		
19	LCST	PS-PMMA (1.7-19)		
20	UCMT	PEO-b-P(AzoMA-r-NIPAm)	3500	
21	UCMT/LCMT	PEO-PNIPAm		
22		PS-PMMA-PS	13050 11130 9760	

Line	*Solvents*	*Temp(°C)*	*CMC (wt%)*	*c (wt%)[b]*
1	H_2O H_2O d_3EAN	20 25 30 35 40 15 25 42 25 31	4[c] 0.7[c] 0.1[c] 0.025[c] 0.008[c] 0.7[d] 0.005[d] 5 1.8	
2	H_2O [BMIM][PF_4] [BMIM][PF_6]	30 25	~10 ~30	

Continued on next page.

Table 1. (Continued). Micellization of block copolymers in ionic liquids compared to aqueous solvents.

Line	*Solvents*	*Temp(°C)*	*CMC (wt%)*	*c (wt%)*[b]
3	H_2O	20	0.18[c]	
		25	0.03[c]	
		30	0.005[c]	
		35	0.001[c]	
	H_2O	25	0.004[d]	
	H_2O			
				1
				3
				5
				10
	d_3EAN	15	5[e]	20
		25	0.1[e]	
	EAN			
				3
				5
				10
	[BMIM][PF_6]	25		20
		25		1
		25		5
		25		10
		25		15
		25		20
		25		30
		35		20
		45		20
		55		20
		65		20
	[BMIM][FAP]			20
				0~15
4	H_2O	20	2.2[c]	
		25	0.3[c]	
		30	0.025[c]	
		35	0.005[c]	
		40	0.001[c]	
	d_3EAN			
5	H_2O	25	4[c]	
		30	0.9[c]	
		35	0.2[c]	
		40	0.05[c]	
		45	0.014[c]	
	D_2O	≤15		
		~25		Low
		60-70		
	d_3EAN			

Continued on next page.

Table 1. (Continued). Micellization of block copolymers in ionic liquids compared to aqueous solvents.

Line	*Solvents*	*Temp(°C)*	*CMC (wt%)*	*c (wt%)^b*
6	H_2O	30	4^c	
		35	1^c	
		40	0.35^c	
		45	0.1^c	
		50	0.04^c	
	H_2O	25	8^d	
		30	3.2^d	
		32		
		40	0.15^d	
	d_3EAN	25		
		63		
		63		1, 10
				1
				10
7	H_2O	37	0.001mM^f	
	d_3EAN	25		1
				5
	d_3EAN	25		1
		63		1, 10
8	H_2O	37		
	d_3EAN	25		1
		63		1
		63		10
9	H_2O	30	1.5^c	
		35	0.4^c	
		40	0.1^c	
		45	0.02^c	
	[BMIM][BF_4]	24.85	0.000221^d mol/L	
10	[BMIM][BF_4]	24.85	0.0000375^d mol/L	
		34.85	0.0000335^d mol/L	
		44.85	0.0000303^d mol/L	
	[BMIM][BF_6]	24.85	0.0181^d mol/L	
		34.85	0.0122^d mol/L	
		44.85	0.0084^d mol/L	
11	D_2O	25		
	EAN	25		
	PAN	45		
		70		
		100		
	D_2O	25		
	EAN	25		
	PAN	45		
		70		
		100		
	D_2O	25		
	PAN	70		
		100		

Continued on next page.

Table 1. (Continued). Micellization of block copolymers in ionic liquids compared to aqueous solvents.

Line	*Solvents*	*Temp(°C)*	*CMC (wt%)*	*c (wt%)*[b]
12	D_2O	25		
	EAN	25		
	PAN	10		
		25		
		70		
		100		
13	H_2O		1.0[g]	
			3.2[g], 1.6[h]	
			2.8[g], 2.8[h]	
	[BMIM][BF_4]		Check SI	
	[EMIM][TFSA]		0.13[h]	
			0.12[h]	
			0.088[h]	
			~0.5	
	[BMIM][BF_4]			
14	H_2O			
	[BMIM][PF_6]			
	[BMIM][PF_6]			
	[BMIM][PF_6]			
	H_2O			
	[BMIM][PF_6]			
15	[EMIM][TFSA]		0.40[h]	
			0.14[h]	
			0.078[h]	
16	H_2O			
	[EMIM][TFSI]	25		
		40		
		60		
		80		
		100		
	H_2O	25		
	[EMIM][TFSI]	40		
		60		
		80		
		100		
	H_2O			
	[EMIM][TFSI]			
	H_2O			
	[EMIM][TFSI]			

Continued on next page.

Table 1. (Continued). Micellization of block copolymers in ionic liquids compared to aqueous solvents.

Line	*Solvents*	*Temp(°C)*	*CMC (wt%)*	*c (wt%)*[b]
17	[EMIM][TFSA]/ [BMIM][TFSA] blends 67.5/32.5 wt% 75/25 wt% 87/13 wt% 100/0 wt%	30 60 85 100		
18	[EMIM][NFf_2]	25 130 25 130 25 130		
19	Acetone [EMIM][NFf_2]	25 25 130		
20	[BMIM][PF_6]	60		
21	[EMIM][BF_4]/ [BMIM][BF_4] blends 70/30 wt% 85/15 wt% 100/0 wt%	80 80 80		
22	[BMIM][PF_6]			

Line	*Self-assembled morphology*	*Morphology geometry parameters (nm)*[i]
1	Spherical micelle Core-shell sphere	R=5.7[j] R=7.2[e]
2		

Continued on next page.

Table 1. (Continued). Micellization of block copolymers in ionic liquids compared to aqueous solvents.

Line	*Self-assembled morphology*	*Morphology geometry parameters (nm)*[i]
3		
	Spherical micelle	R=5.77[j]
	Core-shell sphere	R_c =5.16, T= 3.58[k]
	Core-shell sphere	R_c =5.15, T= 3.44[k]
	Core-shell sphere	R_c =5.21, T= 3.15[k]
	Core-shell sphere	R_c =5.34, T= 2.17[k]
	Core-shell sphere	R_c =6.0, T= 2.07[k]
	Core-shell sphere	
	Core-shell sphere	R=7.4[e]
	Core-shell sphere	R_c =5.1, T= 2.17[k]
	Core-shell sphere	R_c =5.24, T= 2.10[k]
	Core-shell sphere	R_c =5.36, T= 2.65[k]
	Core-shell sphere	R_c =5.23, T= 2.55[k]
	Core-shell sphere	R_c =6.00, T= 2.86[k]
	Core-shell sphere	R_c =5.69, T= 3.10[k]
	Core-shell sphere	R_c =5.58, T= 2.85[k]
	Core-shell sphere	R_c =5.68, T= 2.54[k]
	Core-shell sphere	R_c =5.50, T= 2.53[k]
	Core-shell sphere	R_c =5.10, T= 2.32[k]
	Core-shell sphere	R_c =5.47, T= 2.44[k]
	Core-shell sphere	R_c =5.65, T= 2.48[k]
	Core-shell sphere	R_c =5.49, T= 2.87[k]
	Core-shell sphere	R_c =5.57, T= 3.10[k]
	Core-shell sphere	R_c =5.37, T= 3.56[k]
	No evidence of micelles or any self-assembled nanostructures[k]	
4		
5		
	Gaussian coil	R_G= 1.7[e]
	Spherical micelle	R_c=4~5[e]
	Prolate ellipsoid	

Continued on next page.

Table 1. (Continued). Micellization of block copolymers in ionic liquids compared to aqueous solvents.

Line	*Self-assembled morphology*	*Morphology geometry parameters (nm)*[i]
6		
	Gaussian coil	R_G= 1.5 ±0.2, T=3.2[e]
	Spherical micelle	R_G= 1.5[e]
7		
	Vesicles	R_c=30.3, T_1 = 8.9[e]
	Vesicles	R_c=13.2, T_1= 20.7, T_2= 77.4[e]
	Lamellar vesicles	
	Lamellar stacks	d~20, T_{PPO}=4.6[e]
8		R_H= 1.5[l]
	Gaussian coil	R_G= 1.4[e]
	Lamellar stacks	d~18[e]
	Polydisperse vesicle	R=120, T=4.3[e]
9		
10		
11	Disk	R_c > 300[e]
	Disk	R_c > 300[e]
	Disk	R_c > 300, T= 11[e]
	Disk	R_c > 300, T= 11[e]
	Disk	R_c > 300, T= 13[e]
	Core shell sphere	R_c = 13, T =7[e]
	Disk	R_c > 300[e]
	Sphere	R_c = 10[e]
	Sphere	R_c = 13[e]
	Sphere	R_c = 14[e]
	Core shell sphere	R_c = 9, T =10[e]
	Sphere	R_c = 12[e]
	Sphere	R_c = 13[e]
12	Core shell sphere	R_c = 10, T =8[e]
	Core shell sphere	R_c = 12, T =7[e]
	Sphere	R_c = 9[e]
	Sphere	R_c = 12[e]
	Sphere	R_c = 9[e]
	Sphere	R_c = 9[e]

Continued on next page.

Table 1. (Continued). Micellization of block copolymers in ionic liquids compared to aqueous solvents.

Line	*Self-assembled morphology*	*Morphology geometry parameters (nm)*[i]
13	Core-shell micelle	R_H= 22[m] R_H= 14[m] R_H= 10[m]
14	Spherical micelles Spherical micelles Spherical micelles Spherical micelles Wormlike micelle Wormlike micelle Bilayered vescile Wormlike micelle Bilayered vescile	R_c = 22.2 R_c = 15.2[n], R_H= 46[l] R_c = 18.6[n], R_H= 36[l] R_c = 20.7[n], R_H= 54[l] R_c = 12.0[n] R_c = 15.8 R_c = 8.8 R_c = 14.5[n], R_H= 97[l] R_c = 12.0[n]
15	Core-shell micelle	
16		R_H= 74[l] R_H= 54[l] R_H= 55[l] R_H= 56[l] R_H= 57[l] R_H= 55[l] R_H= 61[l] R_H= 47[l] R_H= 46[l] R_H= 47[l] R_H= 50[l] R_H= 48[l] R_H= 95[l] R_H= 60/61[l] R_H= 103[l] R_H= 71/72[l]
17		R_H= 10[l] R_H= 11[l] R_H= 9[l] R_H= 7[l]
18		R_H= 4.4[l] R_H= 2.5/126[l] R_H= 4.2[l] R_H= 92[l] R_H= 6.9[l] R_H= 61[l]
19		R_H=3.5/59[l] R_H= 25[l] R_H= 27[l]

Continued on next page.

Table 1. (Continued). Micellization of block copolymers in ionic liquids compared to aqueous solvents.

Line	*Self-assembled morphology*	*Morphology geometry parameters (nm)*[i]
20		R_H= 100[l] (under dark) R_H= 125[l] (under UV light irradiation)
21		R_H= 13[l] R_H= 14[l] R_H= 7[l]
22	Spherical micelles Spherical micelles/vesicles	R_c = 20~30[o] ,R = 22.5[o] R < 50[o] R < 50[o]

Line	*Block copolymers*	*N_{agg}(Temp)*	*Ref*
1	Pluronic F127	22[e], 37[j] 9.8[e]	*(45)* *(46)* *(47)*
2	Pluronic F68		*(48)* *(49)*
3	Pluronic P123	86[j], 93[e] 85.6[e] 85.1[e] 88.1[e] 94.7[e] 134.6[e] 164[e] 82.6[k] 89.9[k] 95.9[k] 89.1[k]	*(45)* *(46)* *(50)* *(47)* *(50)* *(51)*

Continued on next page.

Table 1. (Continued). Micellization of block copolymers in ionic liquids compared to aqueous solvents.

Line	*Block copolymers*	N_{agg}*(Temp)*	*Ref*
			(51)
4	Pluronic P105		*(45)* *(52)*
5	Pluronic P85	37(20), 78(40)[e]	*(45)* *(53)* *(52)*
6	Pluronic P65	1[j] 4[j] 21[j] 2[j] 17[j]	*(45)* *(46)* *(52)*
7	Pluronic L121		*(44)* *(47)* *(52)*
8	Pluronic L81		*(54)* *(52)*

Continued on next page.

Table 1. (Continued). Micellization of block copolymers in ionic liquids compared to aqueous solvents.

Line	*Block copolymers*	N_{agg}*(Temp)*	*Ref*
9	Pluronic L64		(*45*) (*49*)
10	Pluronic L61		(*49*)
11	PEGE-PEO	450 200 510 679 170 420 510	(*55*)
12	PGPrE-PEO	210 350 190 230 150 30	(*55*)
13	PS-PEO		(*56*) (*31*) (*57*)
14	Micelle shuttle		(*58, 59*) (*60*) (*60*) (*60*) (*58, 59*) (*58, 59*) (*60*) (*60*)
15	PS-PMMA		(*31*)
16	Micelle shuttle		(*61*)
17	LCST		(*62*)
18	LCST		(*63*)

Continued on next page.

Table 1. (Continued). Micellization of block copolymers in ionic liquids compared to aqueous solvents.

Line	*Block copolymers*	N_{agg}*(Temp)*	*Ref*
19	LCST		(*63*)
20	UCMT		(*64*)
21	UCMT/LCMT		(*65*)
22			(*66*)

[a] HLB (hydrophilic-lipophilic balance) values of the copolymers were determined by manufacturer. [b] c is block copolymer concentration. [c] The CMC values were measured using dye solubilization method. [d] The CMC values were measured using small angle neutron scattering (SANS). [e] The CMC values were measured using small angle neutron scattering (SANS). [f] The CMC values were measured using pyrene probe. [g] The CMC values were measured using excitation spectra. [h] The CMC values were measured using I3/I1 plots. [i] The morphology geometry parameters are: R is overall micelle radius, R_c is micelle core radius, T is overall corona thickness, R_H is the hydrodynamic radius, R_G is the radius of gyration, d is the lamellar repeat spacing, T_1 is shell thickness of the 1st shell, T_2 is the shell thickness of 2nd shell, T_{PPO} is the thickness of PPO layer, and N_{agg} is aggregation number. [j] The CMC values were measured using static light scattering (SLS). [k] The CMC values were measured using small angle X-ray scattering (SAXS). [l] The CMC values were measured using dynamic light scattering (DLS). [m] The CMC values were measured using quasi-elastic light scattering (QELS). [n] The CMC values were measured using cryogenic transmission electron microscopy (Cryo-TEM). [o] The CMC values were measured using transmission electron microscopy (TEM).

Pluronic is one of the most studied commercially available non-ionic block copolymers in water and ILs. The aforementioned comparable characteristics of self-assembly in water versus in PIL can be further elaborated through the example of Pluronic F127 self-assembly in water and partially deuterated ethylammonium nitrate (d_3EAN). As shown in Figure 1, both Pluronic F127/water and Pluronic F127/d_3EAN systems have similar phase behaviors originating from the similar self-assembly driving force, where spherical micelles from at low concentration and those micelles further self-assemble into a cubic and crystal phases at elevated concentration and temperature (*46*, *47*). As tabulated in Table 1, the micelle sizes in both systems are of the same order of magnitude, but CMC values and aggregation numbers are 2 orders of magnitude larger in d_3EAN than in water. A significant lowering of ΔG_m^0 in Table 2 indicates the lower solvophobicity of the solvophophobic blocks of the polymer in d_3EAN as compared to water. Interestingly, similar results were obtained when comparing the self-assembly of Pluronic F68 in water versus AIL (*49*). This was attributed to the much a weaker driving force in d_3EAN. Utilizing small angle neutron scattering (SANS) and combining with mass balance, López-Barrón *et al.* proved that there are less than 10 vol% of d_3EAN in the spherical micelle core formed by Pluronic F127 (*68*). Compared to self-assembly in water (*69*), this extremely low degree of solvent penetration in the micelle core verified the weaker segregation strength in d_3EAN.

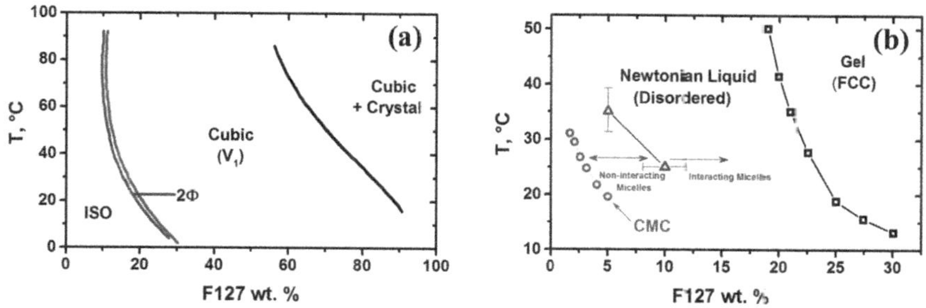

Figure 1. (a) Phase diagram of Pluronic F127 in water. The figure was adapted from (46). Reprinted with permission from (46). Copyright @ 1994, American Chemical Society. (b) Phase diagram of Pluronic F127 in d_3EAN. The figure was reproduced from (47). Reprinted with permission from (47). Copyright @2014, American Chemical Society.

Table 2. Thermodynamics of self-assembly of Pluronic triblock copolymers in ionic liquids compared to aqueous solvents.

Line	*Block copolymers*	*Formula*	*Solvent*
1	Pluronic F127	EO_{106}–PO_{70}–EO_{106} EO_{100}–PO_{65}–EO_{100}	H_2O H_2O
2	Pluronic F108		H_2O
3	Pluronic P123	EO_{20}–PO_{70}–EO_{20} EO_{19}–PO_{69}–EO_{19} EO_{20}–PO_{70}–EO_{20}	H_2O H_2O d_3EAN
4	Pluronic P105		H_2O
5	Pluronic P85		H_2O
6	Pluronic L121	EO_5–PO_{70}–EO_5 EO_5–PO_{70}–EO_5	H_2O d_3EAN
7	Pluronic L64	EO_{13}–PO_{30}–EO_{13} EO_{13}–PO_{30}–EO_{13}	H_2O [BMIM][BF_4]
8	Pluronic L61	EO_3–PO_{30}–EO_3	[BMIM][BF_4] [BMIM][BF_6]

Line	*Temp(°C)*	*ΔG_m^0 (kJ/mol)*	*ΔH_m^0 (kJ/mol)*
1	23.1 CMT at 1wt%	 -27.5	485[a] 253[b]
2	CMT at 1wt%	-28.4	266[b]

Continued on next page.

Table 2. (Continued). Thermodynamics of self-assembly of Pluronic triblock copolymers in ionic liquids compared to aqueous solvents.

Line	*Temp(°C)*	*ΔG_m^0 (kJ/mol)*	*ΔH_m^0 (kJ/mol)*
3	13.7 CMT at 1wt% 22.85 23.85 24.85	-24.9	351[a] 329[b] 58.9[c] 74.6[c] 65.2[c]
4	CMT at 1wt%	-25.6	331[b]
5	CMT at 1wt%	-25.5	229[b]
6	17 18 20 22 25		419[a] 32.52[c] 38.86[c] 65.03[c] 74.10[c]
7	24.85	-24.5	230[b] 21.33
8	24.85 34.85 44.85 24.85 34.85 44.85	-29.35 -30.61 -31.86 -13.84 -15.31 -16.80	7.67 8.19 8.74 28.34 30.27 32.27

Line	*ΔS_m^0 (kJ/mol)*	*$-T\Delta S_m^0$ (kJ/mol)*	*Ref*
1	0.944		*(46)* *(45)*
2	0.975		*(45)*
3	1.223		*(46)* *(45)* *(67)*
4	1.212		*(45)*
5	0.842		*(45)*
6			*(46)* *(67)*
7	0.835		*(45)* *(49)*
8	0.835	-37.02 -38.8 -40.6 -42.17 -45.58 -49.06	*(49)*

[a] The ΔH_m^0 value was measured by DLS. [b] The ΔH_m^0 value was measured by dye solubilization method. [c] The ΔH_m^0 value was measured by ITC.

To interrogate the effect of relative solvophilic to solvophobic block size on the self-assembled microstructure, the microstructure and rheological properties of micellar solutions of a homologous series Pluronics with similar PPO block, F127, P123 and L121, in d_3EAN were studied (*47*). In analogy to aqueous solutions, the larger PEO/PPO molar ratios promotes the formation of spherical micelles, whereas small PEO/PPO ratios favor the formation of less curved microstructures, such as vesicles, wormlike micelles (WLMs), and lamellar phases. The CMCs for Pluronics F127 and P123 show LCST-like CMC behavior, but at much higher concentrations than observed in water. These micelles develop strong intermicellar interactions upon increasing the Pluronic concentration, and further increase in concentration or temperature results in an order–disorder transition to a supramolecular cubic microstructure (FCC for F127 and BCC for P123) with a gel-like rheological behavior, as shown in their corresponding phase diagrams in Figure 1(b) and Figure 2(a). Utilized SANS, ultra-small angle neutron scattering (USANS) and confocal microscopy, it was found that Pluronic L121 exhibits a much richer phase diagram, as shown in Figure 2(b)-(g).

Micellization in Aprotic Ionic Liquids

The use of AILs as self-assembly media is relatively more recent, whose history could be dated to 2003 (*70*). To investigate the microstructure of amphiphilic block copolymer self-assembly in AILs, Lodge and co-workers conducted a comprehensive study on a series of carefully selected diblock copolymers/AIL model systems, including PB-PEO/[BMIM][PF_6] (*60*), PS-PEO/[EMIM][TFSA] (*31*), and PS-PMMA/[EMIM][TFSA] (*31*). The PB-PEO/[BMIM][PF_6] model system formed self-assembled nanostructures consisting of micelles: insoluble dense PB cores surrounded by well-solvated PEO coronas. Combining direct visualization of micelles via cryo-TEM (Figure 3) with DLS measurements, Lodge *et al.* revealed that decreasing the solvophilic PEO molecular weight (MW) resulted in universal nonergodic micellar morphology, ranging from spherical micelles, WLMs, to bilayered vesicles (*60*). This nonergodicity effect is attributed to strong amphiphilicity and relatively low solvent compatibility of block copolymers. The observed structural changes resulting from the variation of copolymer composition qualitatively resembles that observed in water (*25*, *59*). With this prior knowledge in mind, the effect of block copolymer MW and composition on CMC were examined on two well-defined and systematically prepared model systems: PS-PEO (varying solvophilic PEO MW, maintaining solvophobic PS MW) and PS-PMMA (varying solvophobic PS MW, maintaining overall MW) in [EMIM][TFSA]. Literature suggests a strong CMC dependence on solvophobic block MW at lower MW, but a weaker dependence at higher MW. However, it was found that both systems shows a very weak CMC dependence on solvophobic molecular weight. The authors hypothesized that this discrepancy results from the collapse of unimer solvophobic blocks and potential role of kinetic limitations.

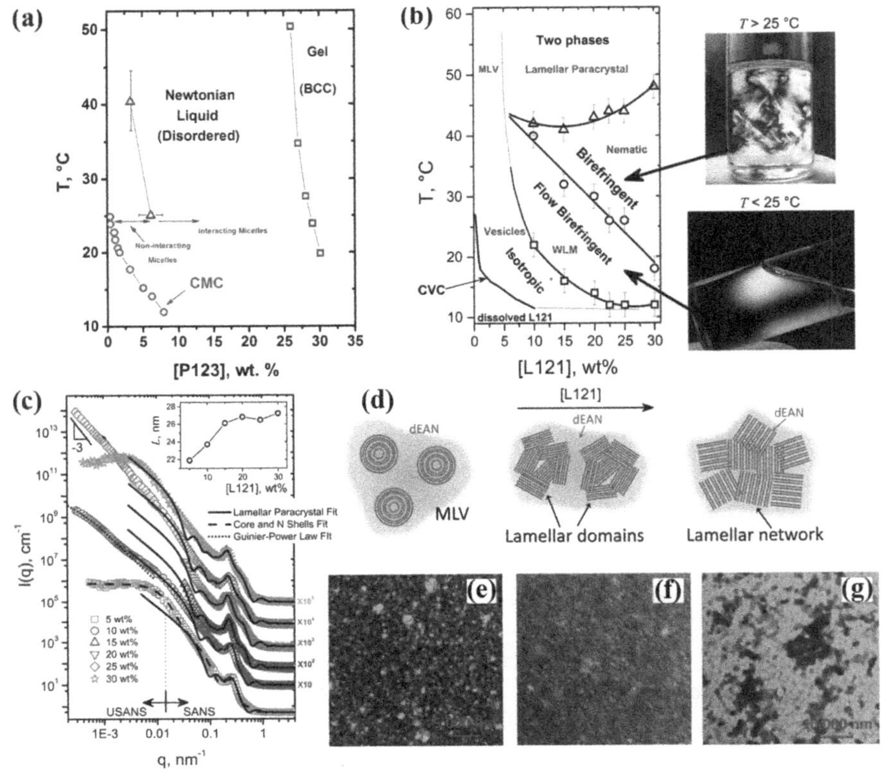

Figure 2. (a) Phase diagram of Pluronic P123 in d₃EAN. (b) Phase diagram of Pluronic L121 in d₃EAN. Right panel: photographs of vials containing a 25 wt % LdE solution at T < 25 °C (showing flow induced birefringence) and at T > 25 °C (showing permanent birefringence). (c) Combined SANS and USANS from Pluronic L121 in d₃EAN solutions with the indicated compositions, measured at 60 °C. Solid, dashed, and dotted lines are best fits to the lamellar paracrystal, the core with N shells, and the Guinier−Porod models, respectively. Data are shifted by multiplying by the indicated factors for clarity. (d) Schematics of the transitions from multilamellar vesicles to disperse lamella domains to lamella networks, upon increase of concentration. (e), (f), and (g) show confocal micrographs (taken at 60 °C) of 10, 25, and 30 wt % Pluronic L121 in d₃EAN solutions, respectively. The figure was reproduced from (47). Reprinted with permission from (47). Copyright @2014, American Chemical Society.

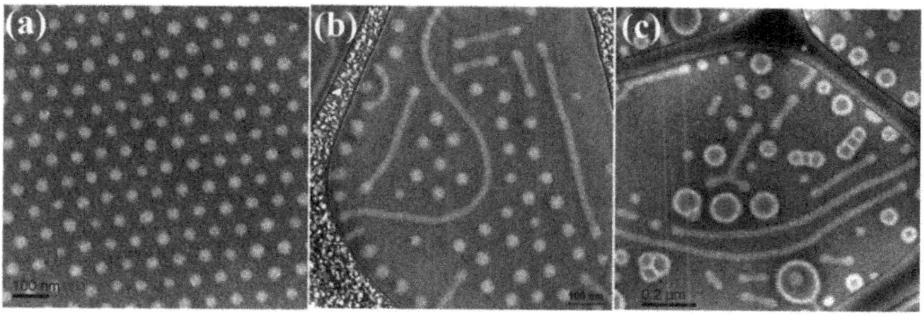

Figure 3. Cryo-TEM images for 1 wt% PB-PEO self-assembly in [BMIM][PF$_6$] solutions. (a) Bright circles, (b) bright strings, (c) bright double layered circles correspond to spherical micelles, cylindrical micelles, and bilayer vesicles, respectively. And PB lies in the micelle cores. The figure was reproduced from (60). Reprinted with permission from (60). Copyright @2006, American Chemical Society.

Following up with the dynamics and equilibrium and non-equilibrium microstructure alluded in the previous section, the nonergodic morphology, formation and relaxation kinetics, equilibrium and non-equilibrium micellar structures, as well as effect of preparation methods on self-assembly of ABC micelles in ILs was investigated by Meli *et al.* (*71*, *72*) The model system used for this study was carefully prepared PB-PEO or PS-PEO micelles in [EMIM][TFSI] or [BMIM][TFSI] using two preparation protocols: direct dissolution (ABC is directly dissolved in ILs) and cosolvent-aided dissolution (ABC and IL are dissolved first in a cosolvent followed by evaporating off the cosolvent). It was observed that steady state micelle structure are highly path dependent and consistent when changing the length of PEO corona block, the chemistry of the core block (from PB to PS), and the length of alkyl chain in imidazolium cation (from ethyl to butyl). Direct dissolution results in large and polydisperse aggregates that relax upon thermal annealing at high temperatures into smaller and monodisperse steady state spherical micelles. The rate of relaxation increases with increasing copolymer concentration, suggesting that fusion/fission is the main micelle relaxation mechanism prepared using this method. However, cosolvent-aided dissolution leads to smaller and monodisperse micelles that retain their size upon prolonged thermal annealing. These smaller spherical micelles are attribute to the lower interfacial energy between the PB or PS cores and the IL/cosolvent mixture compared to the IL alone. Kinetic stability can be derived from the steric stability of the well-solvated corona.

As mentioned in the previous section, micellization could be affected by external stimulus, such as temperature. Due to the diversity of ILs chemistry, which can be used for tuning the temperature dependent solvent quality for various polymers, many block copolymer micellar systems have been found to display interesting thermo-responsive behavior in ILs. Using homopolymer/IL system as an initial reference point, both upper critical solution temperature (UCST) and lower critical solution temperature (LCST) phase behavior have been reported: homopolymers exhibiting UCST phase behavior in IL are

PNIPAm in [EMIM][TFSA] (*73*); LCST phase behavior in ILs are PBzMA and its derivatives in [C_nMIM][TFSA] (*74*) and [EMIM][NTf_2] (*40*, *75*), PEGE in [EMIM][TFSA] (*76*), PnBMA in [C_nMIM][TFSA] (*62*) and PEO in [EMIM][BF_4] (*77*, *78*), [BMIM][BF_4] (*77*), [EMMIM][BF_4] (*78*) and [EMIM][BF_4]/[BMIM][BF_4] blends (*77*). One common characteristic of these UCST or LCST homopolymer/IL systems is that they all show asymmetric temperature-composition phase diagrams, with the critical composition shifted to low polymer concentrations excluding the case of PEO/AIL (*78*). It was postulated that hydrogen bonds between polymer/IL provide the driving force for the LCST phase behavior (*76*, *78*). Combing those homopolymers with different phase behavior could produce thermo-responsive amphiphilic block copolymers with upper critical micellization temperature (UCMT) (both blocks are consisted of UCST homopolymers), lower critical micellization temperature (LCMT) (both blocks are LCST homopolymers), and doubly thermos-sensitive block copolymer having both features (one block is UCST homopolymer and one block is LCST homopolymer). The characterization of two thermo-responsive diblock copolymers micellar systems, PEO-PNIPAm/IL and PEO-PnBMA/IL, showed that using IL blends as solvents the UCST and LCST phase behavior could be easily tuned over a wide range of temperatures by manipulating the blending ratio of two ILs, without modifying the chemical structure of the copolymers (*62*, *65*). This finding points out an additional material design strategy of simply changing the IL properties through changing the chemical structures of IL constituents or by blending different ILs (*74*, *77*), without the complexity of changing ABCs properties or customizing new ABCs. The discovery UCST or LCST phase transition of block copolymers has been utilized in creating light controlled reversible micellization (*64*), doubly thermos-sensitive micellization-unimer-inverse micellization self-assembly mechanism (*79*), micelle shuttle systems (*61*), physical ion gel (*80–82*), thermoreversible block copolymer ion gels (*83*) and the long-range-ordered regulation of ion paths by using polymer/IL composite films (*84*).

With the findings of nonergodic core-shell micellar nanostructures in ILs and LCST and UCST phase behavior in block copolymers, smart thermo-responsive micellar systems exhibiting spontaneous intact round-trip shuttling (*61*, *85*) between a hydrophobic AIL and a phase-separated aqueous phase have been explored. This research topic is significant because micelle shuttle with tunable micelle size and nanostructures may supply a simple, flexible, and scalable round-trip delivery system that has potential applications in nanoencapsulation (*86*), phase transfer (*87–89*), and biphasic catalysis (*90*, *91*) involving ILs. Increasing research efforts in this area has experimentally confirmed the phenomena with three custom-synthesized diblock and a commercially available inexpensive triblock copolymer micellar systems in two IL/water media, which are PB-PEO (*85*), PNIPAm-PEO (*92*), PNonOx-PEtOx (*93*) and P123 (*94*) micelle shuttle between [Bmim][PF_6] and H_2O, and PB-PEO between [EMIM][TFSI] and H_2O (*61*). Generally speaking, the micellar shuttle mechanisms are fully thermoreversible, repeatable and quantitative, with preservation of micelle size and structure between the two solvents. From material design perspective, the design of such a micellar shuttle system requires a polymer (PEO in this case) that is nearly equally soluble in two immiscible solvents (water and AILs in this case) (*85*). The driving force for the thermo-controlled micelle shuttle is the well-known LCST phase behavior (solubility decreases upon heating) of PEO (*95*) corona block in water. As the relative affinity of the two solvents to the corona chains is temperature dependent, the solvent quality of water for PEO is deteriorated at elevated temperature, whereas the AILs remains as a good solvent. This underlying physics was fully manifested in experimental observed micelle shuttle mechanism, where micelles transfer spontaneously from aqueous phase at room temperature to both phases at the transfer temperature, and finally reside in the hydrophobic IL phase at elevated temperature (*85*). To be noted, the thermodynamics and mechanism of micelle shuttle was studied using doubly-thermosensitive PNIPAm-PEO diblock copolymers in [Bmim][PF_6]. Aside from the LCST phase behavior of PEO-corona in water, the PNIPAm-core has a LCST phase behavior in water yet an upper critical solution temperature (UCST) phase behavior in [Bmim][PF_6] (*73–76*). This doubly-thermosensitive micelle shuttle system achieved the goal of controlling loading and unloading in both phases and creating a more advanced reversible micellization-transfer-demicellization shuttle mechanism shown in Figure 4 (*92*). Therefore, the transfer property can be effectively controlled and tuned by changing the relative solubility of PEO in water and ILs, such as adding additives to the aqueous phase to adjust the solubility of PEO in water (*61*). using hydrophobic AILs with varying anions or cations to tailor the solubility in AIL or synthesizing multi-thermosensitive block copolymers with PEO-corona. With this prior knowledge as a basis, PB-PEO vesicles in water with interiors filled with [EMIM][TFSI] (*96*) and [EMIM][TFSA] (*97*) were also successfully designed.

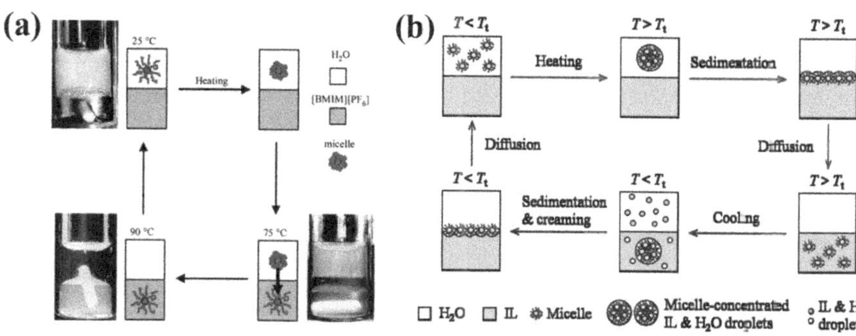

Figure 4. (a) Schematic illustration of the round trip PB-PEO micelle shuttle between [BMIM][PF_6] and water. The figure was reproduced from (85). Reprinted with permission from (85). Copyright @2006, American Chemical Society. (b) Schematic illustration of mechanism of the micelle shuttle. The figure was reproduced from (98). Reprinted with permission from (98). Copyright @ 2009, American Chemical Society.

Summary and Outlooks

Even though the micellization features of ABCs in PILs and AILs are often qualitatively consistent with comparable observations in aqueous solutions, the low volatility and tunable solvent properties of ILs allow for new examinations of significant quantitative differences in the self-assembly behavior over a wide range of temperatures. Since the micellization of ABCs in ILs is still a developing field, many interesting areas are still waiting to be conquered and can be summarized from three perspectives. From the perspective of polymer thermodynamics relating to closely to micelle structure and thermodynamics, the applicability of mean-field theories and scaling theories in morphology control has not yet been thoroughly explored. From the viewpoint of micelle dynamics, given that ABCs can form equilibrium or non-equilibrium microstructures depending on the solution preparation methods, so a coherent understanding of micelle dynamics during micelle formation and structural transitions, as well as careful optimization of the preparation conditions are critical for formulating and processing well-defined, uniform and reproducible microstructures (*28*, *29*). Lastly, achieving controllable stability in the self-assembly is also an important subject for desired applications. Indeed, given the reported phase diagrams and material design properties of ABCs in ILs, controlling the self-assembly of these systems will open up new opportunities for formulating hierarchical structured soft matter, as discussed further in the next section.

Concentrated Solutions of Block Copolymers in Ionic Liquids

Concentrated solutions of ABCs in IL have more diverse phase behavior than dilute solutions. At fixed temperature, the addition of IL to ABCs results in a rich sequence of lyotropic phase transitions due to the multiple independently varying interaction parameters (interactions between the amphiphilic blocks and the interactions between blocks and solvents) present in binary or multi-component systems. The temperature dependence of those interaction parameters results in thermotropic phase transitions. However, work in this field has been much less explored compared to the investigations of dilute solution self-assembly.

Microstructure-Property Relationship at Quiescent State

The lyotropic phase behavior of concentrated ABCs/IL mixtures resembles that of mixtures of ABCs in selective molecular solvents. Due to the selectivity of the selective solvent to solvophilic block(s), the factors affecting lyotropic phase behavior of ABC in selective solvents are the relative volume fraction of solvophilic (or solvophobic) block(s) in ABC, volume fraction of ABC in mixture and the selectivity of the solvent (*99*). It was nicely demonstrated by Lai *et al.* that order-disorder transition temperature and domain spacing of the ordered morphology decreases with increasing solvent concentration when solvent selectivity is small, and increases with increasing solvent concentration when solvent selectivity is large (*100*). Further the domain spacings have been shown to follow a scaling law parametrized by volume fraction of ABC in mixture, solvent selectivity and the morphology (*101*). The driving force for the lyotropic mesophase transitions induced by the addition of a selective solvent is the combination effect of changes in the effective volume fractions of each phase and degree of segregation between domains (*99*, *102*).

Concentrated solutions of block copolymers have similar self-assembled microstructures as the bulk state (*9*, *103*). The commonly found microstructures in diblock copolymers bulk states include cubic lattices of spheres (either BCC or FCC), hexagonally packed cylinders (HEX), the bicontinuous gyroid (G), and lamellae (LAM) (*103*). The lyotropic phases observed in four diblock copolymer/AIL model systems are PB-PEO in [EMIM][TFSI] or [BMIM][PF_6] (BCC, HEX, LAM, regions of coexistence of LAM and HEX, and a disordered network structure consisting of branched PB cylinders in a matrix of PEO/ILs) (*104*), PS-PEO/[EMIM][TFSI] (G, coexistence between LAM and G, coexistence between LAM, HEX and G microphases and coexistence of LAM phases with two different domain sizes) (*102*), and PS-P2VP/[Im][TFSI] (BCC, FCC, HEX, LAM, coexistence of LAM phases with two different domain sizes) (*105*, *106*). In these systems, the AILs selectively solvate the solvophilic PEO or the P2VP blocks of the copolymers, resulting in the lyotropic phase behavior. These interesting coexistence phase was proposed to be caused by two lamellar phases contain unequal amounts of ionic liquid, or perhaps due to cocrystallization of the polymer and ionic liquid in the case of the PS-PEO/[EMIM][TFSI]. The coexistence of G microphase in PS-PEO/[EMIM][TFSI] was attributed to the strong segregation strength of those systems. Further, the reported domain spacing scaling law also applied in PS-PEO/[EMIM][TFSI] and PS-PEO/[BMIM][PF_6] systems, revealing that [EMIM][TFSI] is a less selective solvent than [BMIM][PF_6] (*104*).

On the other hand, with the favorable interaction between AIL [BMIM][PF_6] and PEO block, concentrated triblock copolymers Puronic F127, F108 and F77 in [BMIM][PF_6] demonstrate their ability to form ordered morphologies (*107*). Further increasing IL concentration activates the order-to-order transition from cylindrical to spherical PPO microdomains, as well as a large melting point depression of the PEO blocks. However, blending [BMIM][PF_6] with these Pluronic copolymers shows no signs of liquid crystals formation, which is very different from previous work with P123/[BMIM][PF_6] (*108*) and P123/EAN (*109*) and later work of Pluronic P123 and F127/d_3EAN (*47*), where the formation of liquid crystalline phases is manifested by the birefringent textures when observed under a polarized optical microscope.

The model system for studying concentrated solutions of ABCs in PILs is Pluronic in EAN. Concentrated P123/EAN was identified to contain a series of lyotropic mesophases including normal micellar cubic (I_1) (FCC in this case, 28-30% P123), normal hexagonal (H_1), Lamellar Phase (L_α), and reverse bicontinuous cubic (V_2) using SAXS and optical microscopy (*109*). Such self-assembly behavior of P123 in EAN is similar to those observed in H_2O (*22*) or [BMIM][PF_6] (*108*) systems except for the presence of the V_2 phase in EAN and the absence of the I_1 phase in [BMIM][PF_6]. The additional V_2 phase in P123/EAN system was hypothesized to be owing to the lower solvophobicity of the PPO blocks to EAN than to water, which might reduce the effective area of the solvophilic headgroup and increase the volume of the solvophobic part. Furthermore, similar lyotropic phase behavior observed in EAN, H_2O and [BMIM][PF_6] suggests that those three solvents have similar solvent properties in their roles both as self-assembly media and PEO solvation.

Taking a detailed look at the I_1 regime, McConnell and co-workers used a series of PS-PI diblock copolymers in decane (preferential solvent for PI) to demonstrate that FCC lattices are favored in the limit of thin micelle corona, whereas BCC crystals are preferred in systems with thick corona (*110*). These results are expected with the observation that micelles with thin coronas resemble hard spheres, and micelles with large coronas should be considered to be soft spheres (*47*). With this information in mind, compared concentrated Pluronic P123 (thinner corona) and F127 (thicker corona) self-assembly in solution, the formation of FCC lattice should be anticipated for P123, as consistent with P123/EAN (*109*), and BCC lattice for F127. However, unexpected observations of FCC lattices in F127/d_3EAN (*47*) and F127/D_2O (*111*), BCC lattices in P123/d_3EAN (*47*) were obtained. However, using a purified commercial F127, Mortensen et al. observed a BCC phase in concentrated F127/D_2O (*111*). It was postulated that the dissolved F127 chains may form unimer clusters (as previously shown in PEO/D_2O and Pluronic/ D_2O solutions) that fill the intermicellar space and enhance stability of FCC lattice by interacting with PEO corona (*47*). Therefore, it was concluded that the diblock impurity within the commercially purchased triblock copolymers and/or partial solubility promotes the formation of a FCC lattice ordering by hardening the intermicellar potential (*47*). The origin of the unexpected BCC phase for P123 is still unexplained.

With these prior knowledge about concentrated systems of neat P123 and F127 in d_3EAN (*47*) and co-micellization of these two ABCs (*112*), seven concentrated mixtures of P123 and F127 in d_3EAN, ranging from "hard-sphere like" P123 with BCC microstructure to "soft" with FCC microstructure at quiescent state, were used to form a series of concentrated solutions with tunable rheological signatures (*113*). It's worthwhile to mention that 90/10 PFdE marked the supramolecular equilibrium microstructure transition from BCC micellar crystals (P123/d_3EAN) to FCC micellar crystals (F127/d_3EAN). Therefore, it was concluded that co-micellization could be used as a potentially means to trigger the BCC-to-FCC transition evidenced by SANS.

Thermotropic phases in concentrated ABCs/IL have also interesting properties. Hamley *et al.* showed that the thermotropic properties of PS-PI in diethyl phthalate is due to the temperature dependence on polymer/solvent interaction parameter or solvent selectivity (*101*). The previously mentioned temperature dependent scaling law in PS-PI/decane system was originated from the effect of temperature on the solvent selectivity, attributing to the changes in solvent portioning between the PS and PI microphases as temperature varies (*101*). Systematic investigation of Pluronics in water has led to the observation of strong temperature dependence of micelle formation in these systems (*22*). This strong temperature dependence is depicted by the thermotropic behavior of Pluronic L121 (10-30 wt%) in d_3EAN, where reversible phase transitions from vesicles, WLM, nematic to lamellar paracrystalline phase could be observed upon heating and concentrated Pluronic P123 or F127 in d_3EAN, where BCC or

FCC to disordered micellar phases could be achieved by lowering temperature, termed “inverse melting” (*47*, *114*). The thermotropic phase behavior in PS-P2VP/[IM][TSFI] system is described by the irreversible transitions from ordered LAM, HEX, or coexisting LAM/HEX phases to disordered micelles upon heating around 250 °C, while neat PS-P2VP demonstrates a reversible order-disorder transition around the same temperature (*106*).

Microstructure-Property Relationship under Flow Deformation

As noted, processing can be critical for achieving a desired nanostructured morphology in ABC/IL systems. Hence, fundamental studies of processing effects are of significant value. In particular, shear flows can be particularly effective for creating long-range ordering and orientation under specific conditions. In addition, self-assembled polymers in solutions also constitute an excellent soft colloidal model system, which can often exhibit remarkable rheological responses to imposed flow fields. This originates in part from their ability to undergo conformational changes and elastic deformations, as well as the effects of deformation on their dynamical behavior. The wide range of macroscopic rheological behavior observed is a consequence of changes in the underlying microstructure that can span mesoscopic to nanoscale organization. Most prevalent to date is research investigating the microstructure of micellar cubic crystals in molecular solvents under steady shear flow. Progress in the experimental capability of measuring the microstructure-property relationship has been aided by the development of a large number of real and reciprocal space techniques qualified in simultaneously measuring microstructural and rheological properties (see, for example, the reviews by Walker (*115*) and others on rheo-SANS (*116*)). The experimental studies for ABC/IL can be broadly categorized into two main topics, which are the exploration of the shear-induced order-order transitions, at low shear rates and understanding of the crystal shear-melting at high shear rates (*114*, *117–122*). The characteristic rheological response of FCC micellar crystals in molecular solvents under steady shear flow is shear thinning, originating from the formation of two-dimensional hexagonal close-packed (HCP) layers, which can orient to as to have glide planes commensurate with laminar shear flow, thus facilitating flow (*123*). Furthermore, the HCP layers are arranged with layers normal parallel to the velocity gradient direction and the close-packed direction parallel to the velocity direction (*121*, *124*). Those general flow and microstructure behaviors of micellar crystals were also found in dispersions of charged colloids (*125*) and single component “nearly hard” colloidal spheres (*123*). To measure the HCP layered structures, most of those studies have utilized small angle scattering experimental techniques to measure the HCP layers only in one of the three-dimensional planes, the velocity-vorticity (or 1-3) plane of flow. In this plane of flow, the in-plane HCP layered structure is visualized as a sixfold symmetric scattering pattern in reciprocal space. However, since this plane is perpendicular to the HCP layers, the information about layer staking sequence could not be obtained.

Similar studies of the microstructure under flow for self-assembled ABC/IL solutions are still rare and include two primary systems, micellar crystals and WLM IL solutions. Viscoelastic block copolymer micellar cubic crystals in ILs are formed by hierarchical self-assembly of ABCs into spherical micelles that further order into micellar crystals at sufficiently high concentrations or temperatures (Pluornic F127 FCC micellar crystals in d_3EAN) (*126*). In that work, a combination of rheo-SANS (*127*) (Figure 5(a), measuring 1-3 and 2-3 planes of flow) and spatially resolved flow-SANS (*128*) (Figure 5(b), measurements in the 1-2 plane of flow) were utilized to investigate the flow-induced crystallization, HCP layering and shear melting of the micellar crystals under steady shear flow (*126*). In addition, time-resoled oscillatory rheo-SANS (tOrSANS, measuring 1-3 plane of flow) was used to quantify the cyclic melting and recrystallization process under large-amplitude oscillatory share (LAOS) flows (*129*). As shown in Figure 5(d), the microstructure-property relationship during the flow-induced crystallization, HCP layers formation and sliding, as well as the shear melting processes in PIL are in qualitative agreement with the behavior in molecular solvents, charged colloids and hard sphere systems. The highlight of this work is the 3D microstructure measuring capability pictured in Figure 5(c), in addition to the commonly measured 1-3 plane of flow, the measurement on the 2-3 plane of flow provide information about the stacking sequence in the shear-induced HCP layers, and the gap-resolved measurements on the 1-2 plane of flow reveals the microstructural homogeneity across the measuring gap during shear flow. With the gap-resolved 1–2 flow-SANS measurements, crystal melting and HCP layer sliding was confirmed to be non-uniformly across the gap, which suggests the existence of an inhomogeneous shear rate distribution in the gradient direction. The LAOS measurements reveal the microstructural origins of the intracycle strain-stiffening and the intra-cycle shear-thinning, which was found to originate from the extremely slow relaxation of the micellar gel, similar to concentrated hard-sphere suspensions and colloidal gels.

Surfactant WLM solutions, which have been extensively reviewed (*130–133*), are "living" polymers, where continuous random breakage and reformation lead to a Poisson distribution of length scales at equilibrium for simple cases. This results in Maxwellian viscoelasticity in the limit where breakage is significantly more rapid than relaxation process. However, non-ionic block copolymer wormlike micellar aqueous solution have comparatively long lifetimes, which is manifested by the non-equilibrium size distribution of WLMs (*132*). Such solutions can also exhibit flow instabilities, such as shear banding, with the addition of salts (*134*). Recently, investigations have been performed for a non-ionic block copolymer WLM solution in ILs comprised of Pluronic L121 in d_3EAN, which was reported to exhibit flow birefringence and shear thinning (Figure 2(a)) (*47*). To interrogate the relationship between self-assembled WLM microstructures and non-linear rheology, further work is required to explore the nature of the rheological properties of these long-threadlike micelles and the qualitative and quantitative differences and similarities to the more heavily studied surfactant WLMs. The exploration of the structure-property relationships briefly summarized here serves as a reference for formulating and processing non-ionic WLMs in ionic liquids to achieve specific structures. It also provides insights useful for further applications in synthesis of iono-elastomers via crosslinking block copolymer WLMs in ILs, as discussed next.

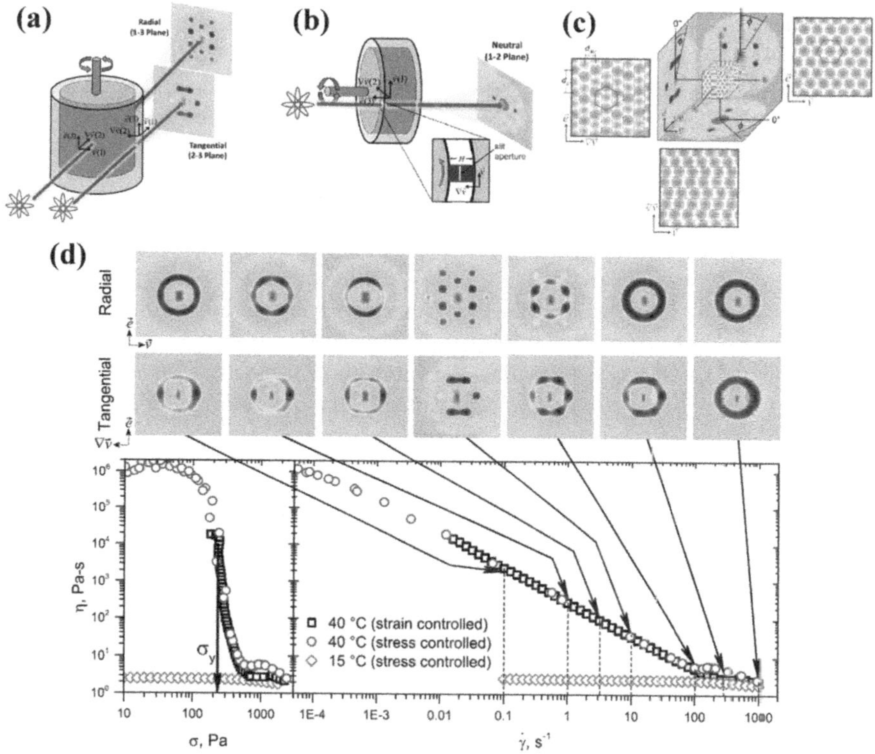

Figure 5. Schematic diagrams of (a) the rheo-SANS and (b) the flow cell. Both devices have Couette flow geometryvwith rotating inner cylinder. The beam in (a) is directed both in the radial and tangential directions, perpendicular to the 1–3 and 2–3 planes of flow, respectively. In (b), the beam is directed perpendicular to the 1–2 plane of flow and is reduced to a 0.1mm wide slit via a cadmium slit aperture for spatial resolution. (c) Schematic cartoon of the three-dimensional layered HCP structure and the corresponding 2D scattering profiles in the three perpendicular planes of flow obtained by shearing a 24 wt. % F127/dEAN solution with a shear rate of 10 s^{-1}. (d) Steady rheo-SANS data from 24 wt. % F127/dEAN solution measured at 40 °C. Lower panel: Steady shear viscosity versus stress and versus shear rate. Both strain-controlled measurements (black squares) and stress-controlled measurements (red circles) are shown. Viscosity data measured at 15 °C. (gray diamonds) are included for comparison. Upper panel: 2D SANS patterns measured at the shear rates indicated by the arrows in the radial and tangential direction, corresponding to the flow-vorticity ($\vec{v} - \vec{e}$ or 1–3) and the gradient-vorticity ($\nabla\vec{v} - \vec{e}$ or 2–3) planes of flow, respectively. The figure was reproduced from (126). Reprinted with permission from AIP Publishing LLC.

Summary and Outlooks

Concentrated ABC/IL solutions self-assemble and order into a rich variety of lyotropic and thermotropic phases in the quiescent state with a broad range of rheological properties. Under simple shear flow, these systems undergo many of the transitions observed for ABC in molecular media as well as for surfactant solutions. ABC/IL concentrated structured systems can exhibit soft solid behavior and it has been demonstrated that shear can be used to anneal, orient, melt, and even create phases nonexistent at equilibrium (*129*). However, these studies are at an early stage and there are many issues waiting to be explored. A major issue already discussed in the context of dilute ABC/IL micellar solutions, i.e., whether equilibrium is achieved and the role of non-equilibrium microstructures, path dependent microstructures, and long-term stability issues, is even more relevant when considering the role of processing on these processes. For example, a better understanding of kinetic versus thermodynamic driving forces for self-assembly in concentrated ABCs/ILs may allow for more control over the self-assembly in these complex systems (*99*), which is valuable for controlling the morphology of ion gels or membranes under static and deformations, as described in the next section.

Ion Gels

ILs possess several traits, such as electro-mechanical stability and high ion conductivity, which make them excellent candidates as electrolyte materials. However, many applications or devices require a solid electrolyte materials rather than a liquid electrolyte. Incorporating ILs into solid or quasi-solid microstructures provided by self-assembled ABCs can lead to the formation of solid ABC/IL binary composite materials, referred to as “ion gels”. Systems where ILs are solidified (or gelled) by polymers are soft solids with significant ion conductivity often comparable to that of neat ILs (*80*, *136*). Here again, formulating ion gels for specific applications will benefit from robust structure-property relationship for both the quiescent state and under deformations. This section provides a brief overview on the emerging body of research on ion gels comprised of non-ionic ABC in ILs. Space limitations prevent us from reviewing work on other polymer/IL electrolyte materials (*7*) and their electrochemical applications as well as corresponding factors affecting ion transport (*8*).

Microstructure-Electrical Property Relationship of Ion Gels at Quiescent State

Four categories of ABC/IL ion gels can be identified based primarily on their synthetic route. First, blending ILs with polymers can produce ion gels. The typical preparation methods for preparing ion gel membranes are self-assembly with solvent casting, or hot-pressing. Ion gels synthesized using this method are mostly reported for application in lithium batteries. It is proposed that the ideal polymer for ion gel electrolyte design should limit the interaction of the charged transport species with the polymer backbone and often incorporate a hydrophobic plasticizer (*137*). Typical polymers reported are PEO homopolymer (*138–140*) and PVdF-PHFP (*137*, *141–143*) diblock copolymer (Table 3). Although Park *et al.* has recently reviewed this topic for lithium battery application (*8*), it is noteworthy to pointing out that the ion conductivity for these self-assembled ion gels are usually fall within the range of 10^{-5}-10^{-2} S/cm at room temperature. Equally importantly, however, the mechanical properties vary vastly ranges from brittle to flexible behavior based on the specific choice of polymers and ILs.

Table 3. Concentrated solution of block copolymers in ionic liquids.

Line	*Block copolymers*	*Formula*	*M_n (g/mol)*
1	Pluronic F127	EO_{106}–PO_{70}–EO_{106}	12600
2	Pluronic F108		14600
3	Pluronic F77		6600
4	Pluronic P123	EO_{20}–PO_{70}–EO_{20}	7000
5		PS-P2VP	

Line	*Solvents*	*Temp(°C)*	*c(vol %)*	*c (wt%)*
1	[BMIM][PF_6]	70~80		40
2	[BMIM][PF_6]	70~80		40
3	[BMIM][PF_6]	70~80		10
4	H_2O	25		~30
		25		40-50
		25		65-85
		25		45
		25		50
		25		68
		25		75
		25		85
	[BMIM][PF_6]	25		38-52
		25		65-87
	[BMIM][PF_6]	25		45
		25		50
		25		68
		25		75
		25		85
		25		40
		25		50
		25		60
		25		70
		25		80
		25		85
	EAN	25		~30
		25		44-64
		25		66-90
		25		92-94
		25		45
		25		50

Continued on next page.

Table 3. (Continued). Concentrated solution of block copolymers in ionic liquids.

Line	*Solvents*	*Temp(°C)*	*c(vol %)*	*c (wt%)*
		25 25 25		68 75 85
5	[Im][TFSI]	145 225	3 3	

Line	*Self-assembled morphology*[a]	*d-spacing (nm)*	*Ref*
1	Well-ordered microdomainsin the melt	d-spacing = 14[b]	(*107*)
2	Well-ordered microdomainsin the melt	d-spacing = 14[b]	(*107*)
3	Well-ordered microdomainsin the melt	d-spacing = 8[b]	(*107*)
4	I_1 H_1 L_α H_1 L_α H_1 $H_1 + L_\alpha$ $H_1 + L_\alpha$ L_α L_α L_α I_1 H_1 L_α V_2	lattice-spacing = 13.57[b] lattice-spacing = 13.58[b] lattice-spacing = 12.11[b] lattice-spacing = 11.58[b] lattice-spacing = 10.94[b] lattice-spacing = 12.30[b] lattice-spacing = 11.97[b] lattice-spacing = 10.82[b] lattice-spacing = 10.50[b] lattice-spacing = 10.00[b] d-spacing ≈ 12[b] d-spacing ≈ 11.5[b] d-spacing ≈ 11.3[b] d-spacing ≈ 11[b] d-spacing ≈ 10[b] d-spacing ≈ 9.8[b] lattice-spacing = 13.10[b] lattice-spacing = 12.80[b] lattice-spacing = 11.51[b] lattice-spacing = 11.18[b] lattice-spacing = 10.59[b]	(*135*) (*108*) (*108*) (*108*) (*51*) (*109*)
5	Lα , cylindrical, disordered phases	d-spacing = 23.7 d-spacing = 23	(*105*)

[a] The self-assembled morphologies are denoted as: I1 = normal micellar cubic phase, H1 = normal hexagonal phase, Lα = lamellar phase, V2 = reverse bicontinous cubic phase.
[b] The d-spacing values were measured using SAXS.

While the conductivities of ILs increase with increasing IL concentration, the mechanical properties of the membranes tend to weaken with increasing IL concentration using the first synthetic route (*99*, *144*). One way to boost mechanical properties while largely maintaining high conductivity is to chemically cross link polymer network with ILs, which is the second synthetic route. Chemically cross-linked ion gels can be prepared via free radical polymerization of vinyl monomers in the presence of a cross-linking agent (*145–147*) or via polyaddition reaction of macromonomers with functionalized reactive groups (*148–151*). In the first report of cross-linked ion gel containing PMMA in [EMIM]TFSA] using ethylene glycol dimethacrylate as a cross-linker (*145*), the ionic conductivity of the ion gels reaching almost 6 mS/cm at the lower polymer concentration, and the storage modulus was found to be on the order of 0.1 GPa for gel with 80 mol% PMMA. Interestingly, an unexpected finding of more free charge carriers in the ion gel than in the neat IL at certain compositions was reported. It was proposed that the increased ion dissociation resulted from specific interactions between the PMMA matrix and the IL ions. Although this method was found to be fairly robust, and yielded good ion conductivity in the order of 10^{-2} S/cm at room temperature, little of the mechanical properties on binary ABC based ion gels were reported.

The third synthetic route is physical cross-linking. ABCs are especially versatile candidates in this synthetic route, because they provide more flexibility in controlling the gel structure and properties through variations of block length/composition/sequence, ABC concentration, and the choice of monomer units. Physically cross-linked ion gel could be synthesized through the gelation of ABA type triblock copolymers with IL incompatible A blocks and an IL compatible B block in ILs. Specific examples include PS-PEO-PS and PS-PMMA-PS (*80–82*) triblock copolymers in AILs with alkyl imidazolium cations and varying anions. The ABA type triblock copolymers composed of short, hard A blocks and long, soft B blocks are typically thermoplastic elastomers. The underlying principle of physically cross-linking ion gels lies in the selective solvation of ILs towards different polymer blocks. To explain this in more details, when an ABA triblock copolymer dissolves in IL, where IL is selective to B blocks but not A block, then a core-shell micelle with A in the core and B in the shell will form. At sufficient high enough concentration, ABA polymer chains will act as cross-linking points and bridge between micelles resulting in polymer network. In He *et al.*'s study of physically cross-linked PS-PEO-PS in [BMIM][PF_6], ion gels were formed with as little as 5 wt% copolymer, and the ionic conductivity was closed to that of neat IL (*80*). With a series of systematic studies on physically cross-linked ABA triblock copolymer/AIL ion gels, Zhang *et al.* generalize the material design rules (*81*). To achieve a higher modulus, the rule suggests that more polymer should be added or using a midblock with a smaller entanglement molecular weight. Once the midblock is chosen, to minimize the loss in conductivity, PS end-blocks should be relatively short but long enough to obtain a persistent gel with high enough mechanical integrity at desired temperature. Using the physically cross-linking route, the resulting ion gels have high conductivity that is comparable to the neat IL, thermal stability, sufficient mechanical strength, and a narrower distribution of the mesh size in the polymer network. As mentioned earlier, this method is promising because of its advantages in controllability and tunability in its components (polymer or ILs) and the physical reversibility of the cross-linking, which can aid in processing. More importantly, in comparison to 10-30 wt % of polymers in conventional polymer gels, much less polymer is required to form an ion gel using triblock copolymers, which can improve the ionic conductivity. Of noteworthy interest, the application of physically cross-linking method with thermos-sensitive diblock copolymer, such as previously reviewed UCST phase behavior of PNIPAm in [EMI][TFSI] and LCST phase behavior of PNIPAm/water, can produce thermos-responsive ion gels (*83*).

The fourth method to produce ion gels with enhanced mechanical properties is to combine the self-assembly provided by ABCs with subsequent chemical cross-linking steps (*136*, *152*). This synthetic route takes advantages of both chemical and physical cross-linking routes, i.e. the conductivity can be tuned for a specific application by manipulating the self-assembled microstructure, while the enhanced mechanical strength for the solid electrolyte can be achieved by subsequent cross-linking step. Gu *et al.* reported the synthesis of ion gels by self-assembly of PS-PEO-PS, of which 25% of the styrene units have a pendant azide functionality, in [EMIM][TFSA], followed by chemical cross-linking of the azide groups by thermal annealing (*152*). The latter step provides enhanced mechanical toughness to the ion gel without significant detriment on its ionic conductivity. Similarly, Miranda and co-workers synthesized PPO-PEO-PPO with crosslinkable end groups (*136*). By self-assembly in [BMIM][PF_6] and subsequent photo-cross-linking, highly conductive, solid, elastic gels was produced. López-Barrón and colleagues recently reported the synthesis of ultra-stretchable soft iono-elastomers by sequential self-assembly of inexpensive, commercially available PEO-PPO-PEO (Pluronic F127) with acrylated end groups in d_3EAN and subsequent photo-crosslinking the resulted micellar FCC lattices (*68*). These materials provide a combination of high conductivity, remarkable stretchability, tensile properties and mechano-electrical response (*153*), which will be detailed reviewed in the next section. To summarize, cross-linking of these polymers through their end groups was sufficient to impart mechanical stability to the gels, which did not have a significant effect on the conductivity or the microstructure of the gels.

Microstructure-Property Relationships of Ion Gel under Deformation

Ion gels with high conductivity and high stretchability are becoming an important area of research due to emerging technologies involving stretchable electronics (*154*). However, the work in this field is much less developed due to the complexity in sample preparation, manufacturing processes and characterization tools.

As reviewed in the previous section, the solid ion gels prepared by Gu and coworkers exhibited strain-to-break values of ~350% (*152*). López-Barrón *et al.* detailed investigated the relationship among microstructure, tensile properties and mechano-electrical responses under uniaxial deformation for synthesized Pluronic

127DA/d_3EAN ion gels (*153*). In tensile property measurement, remarkably large values of both strain to break and tensile strength were measured (Table 4) as illustrated in Figure 6(a). During uniaxial extension, shown in Figure 6(b), a pronounced Mullins effect was observed, which is a typical response of elastomer composites (*155*). The mechano-electrical response of the iono-elastomer in Figure 6(c) reveals a very surprising and counterintuitive response, namely, the resistance decreases with strain, which indicates that the intrinsic conductivity of the material increases during the stretching process. *In situ* small angle X-ray (SAXS) measurements show this to be a consequence of a strain-induced micro-structural rearrangement. Figure 6d shows the in-situ SAXS measurements, where the microstructural origin of the unique mechano-electrical response during uniaxial deformation was elucidated to be a reversible extensional strain-induced FCC-to-HCP transition. It was hypothesized that this transition is responsible for the conductivity increase during stretching, and further, that its reversibility is due to complex network structure formed during crosslinking of micellar FCC lattice (*68*).

Table 4. Examples of block copolymers containing ionic liquids: block copolymers, ion liquids, block copolymer contents and proton conductivities.

Line	*Block copolymer*	*Ionic liquid*	*Synthesis pathway*
1	PVdF-PHFP	[EMIM][Trif]	Hot pressing with PC, then freeze drying off PC[a]
			Hot pressing without cosolvent[a]
		[EMIM][BF_4]	Hot pressing with PC, then freeze drying off PC[a]
			Hot pressing without cosolvent[a]
2	Pluronic 25R4	[BMIM][PF_6]	Self-assembly[b]
3	Pluronic 64R4		
4	Pluronic 45R6		
5	Pluronic 45R6		Subsequential self-assembly and chemical cross-linking[c]
6	PS-PEO-PS	[BMIM][PF_6]	Self-assembly with cosolvent CH_2Cl_2[b]
7	PS-PEO-PS	[EMIM][TFSA]	Self-assembly with cosolvent CH_2Cl_2[b]
8	PS-PEO-PS	[EMIM][TFSA]	Self-assembly with cosolvent CH_2Cl_2[b]
9	PS-PEO-PS with azide functional end groups		Subsequential self-assembly and chemical cross-linking at elevated temperature or UV irradiation[c]
10	PS-PMMA-PS	[EMIM][TFSA]	Self-assembly with cosolvent CH_2Cl_2[b]

Continued on next page.

Table 4. (Continued). Examples of block copolymers containing ionic liquids: block copolymers, ion liquids, block copolymer contents and proton conductivities.

Line	*Block copolymer*	*Ionic liquid*	*Synthesis pathway*
11	PNIPAm-PEO-PNIPAm	[EMIM][TFSI]	Self-assembly[b]
12	Pluronic F127	d_3EAN	Self-assembly[b]
13	Pluronic F127 with acrylate functional end groups	d_3EAN	Subsequential self-assembly and chemical cross-linking by UV irradiation[c]

Line	*Cross-linking agent*	*BC/IL/CL[d] (wt%)*	*Measured temp (°C)*	*Stress to break*
1		48/52 50/50 33/67 50/50 71/29	~22	
		16/84 9/91		
		33/67		
		13/87		
2		20/80 50/50 20/80 50/50 10/90 20/80 50/50	25	
3				
4				
5	PPO-acrylate	10/90/05 20/80/10 20/80/10		
6		1/99 3/97 5/95 7/93 10/90	26.5 26.5 26.5, 80 26.5 26.5, 80	
7		10/90 40/60 50/50	25, 80 60, 160 60, 160	
8		10/90	40	≈0.75×10^5

Continued on next page.

Table 4. (Continued). Examples of block copolymers containing ionic liquids: block copolymers, ion liquids, block copolymer contents and proton conductivities.

Line	*Cross-linking agent*	*BC/IL/CL[d] (wt%)*	*Measured temp (°C)*	*Stress to break*
9		10/90	40	≈3.6x10[5]
10		10/90 30/70 40/60 50/50	25 40, 200 40, 200 40, 200	
11		10/90	20	
12		15/85 22.5/77.5	20 40 60 20 40 60	
13	1-hydroxycyclohexyl phenyl ketone	24/76/1	25	4.7±0.8 MPa

Line	*Conductivity (mS/cm)*			*Ref*
	Neat IL	*Un-crosslinked*	*Cross-linked*	
1		1.1 1.3 5.6 1.8 0.1		*(141)*
		8.0 7.3		
		5.8		
		11.0		
2	≈1.49	0.61	0.61	*(136)*
3		0.24 0.76	0.14 0.83	
4		0.37 1.24 0.79 0.25	0.31 1.22 0.76 0.22	
5		0.93 0.68 0.60	0.98 0.64 0.35	

Continued on next page.

Table 4. (Continued). Examples of block copolymers containing ionic liquids: block copolymers, ion liquids, block copolymer contents and proton conductivities.

Line	Conductivity (mS/cm)			Ref
	Neat IL	*Un-crosslinked*	*Cross-linked*	
6	1.62 1.62 1.62, 15 1.62 1.62, 15	1.6 1.55 1.47, 13 1.32 1.12, 11		*(80)*
7		6.1, 22 3.5, 18 2.4, 12		*(81)*
8	≈12	≈9.2		*(152)*
9		≈9.1	≈8.9	
10		5.5 0.85, 18 0.32, 12 0.071, 7.1	10	*(81)*
11	8	4		*(83)*
12	25.7 35.4 41.6 25.7 35.4 41.6	10.8 25.3 37.0 10.8 21.5 35.7	11.0 25.9 36.9 12.0 23.7 39.1	*(68)*
13	28		15	*(153)*

[a] The type of ion gels formed under this synthesis pathway are thin films [b] The type of ion gels formed under this synthesis pathway are physical ion gel. [c] The type of ion gels formed under this synthesis pathway are chemically cross-linked ion gels. [d] BC/IL/CL indicates Block copolymer/Ionic liquid/Cross-linker.

Summary and Outlooks

Ion gels composed of non-ionic ABCs and ILs are a growing topic of research because of their potential for such applications as membranes for batteries and stretchable electronics. ABC-based ion gels have the versatility and tunablity for designing the mechanical scaffold for supporting ion transport of the incorporated ILs. Here, we briefly summarize four synthetic routes and point out their advantages and disadvantages to aid in designing systems for specific applications. However, the challenges remain for effective formulation. On the one hand, general rules for balancing the trade-off between mechanical and electrical properties are not yet established. Although, in general, the addition of polymer and or crosslinking into ion gels increases the mechanical integrity and lowers the ionic conductivity, specific ion-polymer interactions, and as shown, self-organized microstructures, must be considered in formulation. With the nearly unbounded number of possible ion gel compositions possible, generalized guidelines for predicting the microstructural-mechanical-electrical property relationships for ion gels at rest and under deformation would be tremendously valuable for designing ion gels for specific applications. In particular, quantification of ion transport in ion gels especially is poorly understood. Park *et al.* provide a review on the diverse factors affecting the ion transport properties in polymer/IL electrolytes for lithium batteries, fuel cells and electro-active actuators (*8*), such systematic examinations are warranted for other applications of ion gels.

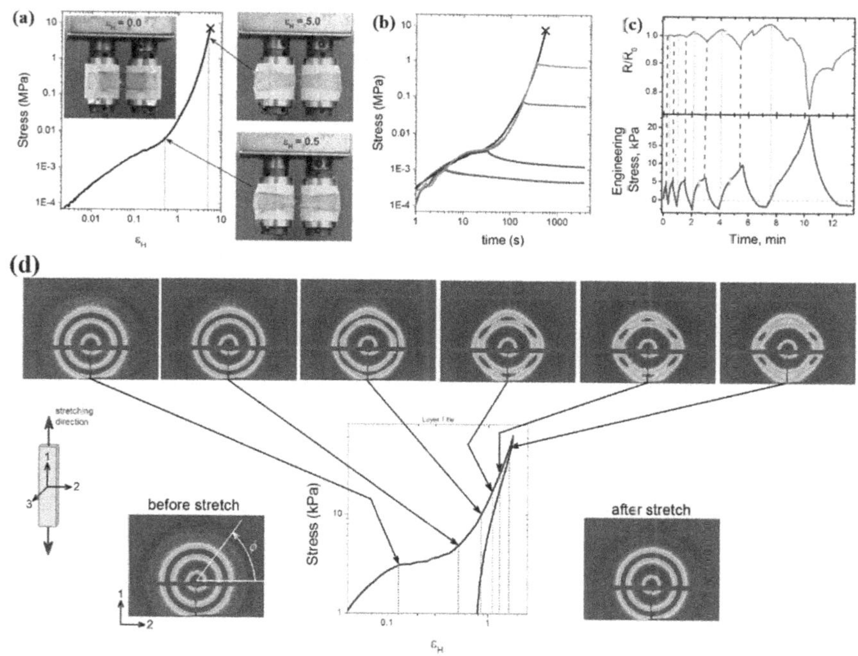

Figure 6. Tensile properties of crosslinked 24 wt% F127-DA/dEAN solution. (a) Stress-strain curve measured in the SER (with Hencky strain rate = 0.01 s-1). The photographs show the sample being stretched at the indicated Hencky strains. (b) Stress (measured in the SER, with Hencky strain rate = 0.01 s-1) as a function of time showing loading to Hencky strains of 0.032, 0.32, 2 and 4, followed by relaxation for 1 hr. (c) Electro-mechanical hysteresis tests. Normalized electrical resistance and stress as a function of time during hysteresis tests. (d) In-situ SAXS measurements during uniaxial deformation. Stress-strain curve measured in the Linkam tensile stage, and 2D SAXS profiles measured at the strain values indicated by the arrows. Also shown are the 2D SAXS profiles measured before and after loading (153).

Emerging Applications of Block Copolymer Self-Assembly in Ionic Liquids

Electrochemical Applications and Devices

Self-assembly of ABCs in ILs is of technological interest for electrochemical applications or as electrolytes in next-generation electrochemical devices (*8*), including lithium ion batteries (*156*), solar cells (*157*), electro-mechanical actuators (*158*, *159*) and electrolyte-gated transistors (*160*, *161*) and light-emitting electrochemical cells, in which microphase-separated ABCs enable the optimization of disparate properties such as mechanical stability and ion transport or light emission. As demonstrated in the previous sections and summarized in Table 5, block copolymers containing a PEO block or side-chain, a PMMA block, or a P2VP block have been frequently used as model systems in such studies due to the ability of the ether, acrylate, and pyridine chemistries to solvate a wide variety of ions (*84*). ABCs provide flexibility in tuning micelle size and nanostructure during self-assembly, yet their noncovalently-bounded structures may require further stabilization for those applications. One approach is to use highly amphiphilic block copolymers to obtain deeply metastable micelle structures by making extremely low CMC (*27*, *162*, *163*). This reduces the probability of demicellization via changing external factors such as temperature, pH, and solvent quality. Another method is to covalent cross-linking the core (*164*) or the shell (*165–167*), which will lead to the formation of ion gel in ILs with enhanced mechanical strength and greater stability as discussed earlier (*84*).

Table 5. Examples of representative block copolymers/ionic liquids system for various application.

Line	*Applications*	*Representative ABCs*
1	Lithium batteries	PEO-based polymer: PS-PEO, PEO-PMMA (*175*)
		PVdF-PHFP
2	Fuel cell	Nafion™
		PBI
		PVdF-HFP
		PMMA (and PMMA based copolymers)
		P2VP (and P2VP based copolymers)
		PSS (and PSS based copolymers)
3	Electro-active actuators	PVdF-PHFP
		Nafion™

Continued on next page.

Table 5. (Continued). Examples of representative block copolymers/ionic liquids system for various application.

Line	*Applications*	*Representative ABCs*
4	Nanoreactor	Bi or tri block ABCs that can form micelles or vesicles in ILs.
5	Wearable electronics	Triblock copolymers: PS-PEO-PS, PPO-PEO-PPO, PEO-PPO-PEO

Line	*Representative IL cations*	*Working principle*
1	AILs composed of alkyl pyrrolidinium, alkyl imidazolium, and alkyl sulfonium cations and anions can vary.	ILs can act as a plasticizer to accelerate relaxation of polymer chains, i.e., the lowering of the glass transition temperature (T_g) of the polymers.
2	AILs comprise heterocyclic diazolium or alkylimidazolium with short alkyl chains cations, and anions can vary.	Incorporation of nonvolatile and highly conductive ILs into polymer matrices is a facile means to obtain high conductivity under water-free conditions and high temperatures.
3		Incorporating IL into ionic polymer layer of the actuator can improve the performance, by reducing Young's modulus, facilitating ion transport, and enhancing electrochemical stability
4	Both AILs and PILs.	ABCs can self-assemble into micelles or vesicles in ILs, which can be used as nano delivery vehicles.
5	AILs with alkyl imidazolium cations and anions can vary. PIL reported is d_3EAN.	Iono-elastomers composed of self-assembled ABCs in ILs possess ultra-strechability and high ion conductivity can be used for wearable electronics application.

Nano Delivery Vehicles

Block copolymer micelles and vesicles have received considerable attention for their ability to encapsulate, transport, and deliver small molecules or solvents in harsh environment. With the exploration of ILs as self-assembly media, block copolymer micelles and vesicles have been extensively investigated for potential as delivery vehicles, especially in the field of catalyst nanoreactors (*168*), drug delivery (*169*) and phase transfer (*85*). However, further advancing in these technologies requires not only finding new ABC/IL systems, but also understanding the interaction between the carriers and interior capsule, and the carrier and exterior environment are pivotal. For example, the relationship between nanoreactor microstructure and catalyst properties in ILs have rarely been explored in polymeric nanoreactors. Developing these relationships requires detailed *in situ* characterization of the nanoscale structure and a deep fundamental understanding of the factors that govern the structure and dynamics of macromolecular assemblies (*29*). In the case of drug delivery, efficacy requires a systematic understanding of the vehicle structure, surface functionality, reactions between drug and vehicle, drug-loading/release procedures and solution processing conditions.

Wearable Electronics and Sensors and Smart Textiles

Self-assembly of ABCs in ILs can form aforementioned subset of ion gels (iono-elastomers), which are candidates for use in emerging technologies involving stretchable electronics, with potential applications as stretchable batteries (*170*), wearable sensors (*171*) and integrated circuits (*172*, *173*). In principle, wearable electronics and sensors can be woven into clothing, uniforms, and sporting equipment to create smart textiles (*174*) and some applications are shown in Figure 7 (*154*).

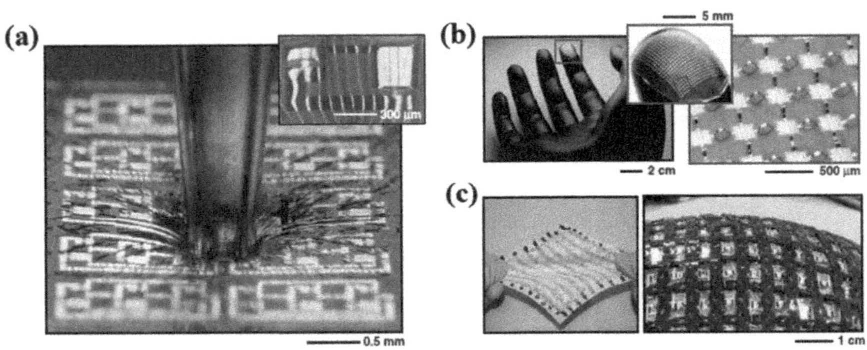

Figure 7. Examples of stretchable electronics. (a) Stretchable silicon circuit in a wavy geometry, compressed in its center by a glass capillary tube (main) and wavy logic gate built with two transistors (top right inset). (b) Stretchable silicon circuit with a mesh design, wrapped onto a model of a fingertip, shown at low (left), moderate (center) and high (right) magnification. The red (left) and blue (center) boxes indicate the regions of magnified views in the center and right, respectively. The image on the right was collected with an automated camera system that combines images at different focal depths to achieve a large depth of field. (c) Array of organic transistors interconnected by elastic conductors on a sheet of PDMS in a stretched (left) and curvilinear (right) configuration. The figure was reproduced from (154). Reprinted with permission from (154). Copyright @ 2010, American Association for the Advancement of Science.

Conclusion and Outlooks

We have taken a brief and broad view of the remarkable breadth of possible microstructure and properties that can result from the self-assembly of block copolymers in ILs. In the quiescent state, the self-assembly behavior of block copolymers in ILs is often compared to that in aqueous solutions. Under flow condition, *in situ* SANS and SAXS combining with rheology and dielectric measurements provides critical and unique spatiotemporal resolved structural information on length scales of direct relevance to the macroscopic properties for various applications. Work to date spans interests in both fundamental investigations and technical applications of block copolymer self-assembly in ionic liquids, but the wealth of possible chemistries and ability to create a plethora of hierarchically self-assembled microstructures suggests that many new discoveries await.

To further the successful application of block copolymer/IL applications, a number of challenges should be addressed. Materials genomic approaches may prove to be particularly valuable given the effectively infinite range of possible chemistries. A priori predictions of ion-polymer interactions and guidelines for selecting polymer architectures to promote specific microstructures would greatly accelerate materials design and development. Further, as ILs are often highly hygroscopic, and the self-assembly of ABCs in ILs are often time consuming and path dependent, consistent and reproducible sample preparation and processing must be established. Here, advances in micro- and nano-scale probes, and precision tools for measuring the rheological, mechanical, and electrical properties of microscopic quantities of samples to determine connections between properties and processing conditions is critical in the development of characterization techniques. Such property relationships will unequivocally require measurements of nano and mesoscale microstructure. Recent advances in instrumentation, such as the dielectric-rheo-SANS sample environment (*176*, *177*) could accelerate materials development. Meanwhile, direct visualization of self-assembled microstructure using TEM has been proved to be useful (references) but still limited by contrast and to thin samples. More reliable and applicable direct imaging techniques would be significantly beneficial to the field. Lastly, even though commercially available ABCs have been incorporated into ILs, the mass fabrication of devices will require considerations of product lifecycle and follow principles of green engineering. Specifically, low-cost ILs with human and environmental compatibility would be valuable. The rapid growth in developments in the self-assembly of ABCs in ILs promises to create hierarchically functional materials with technologically useful emergent properties. The field is inherently multi-disciplinary and these materials promise solutions across a broad range of technological needs.

Acknowledgments

The authors acknowledge the support of the National Institute of Standards and Technology (NIST), U.S. Department of Commerce, in providing the neutron research facilities used in this work. N.J.W. and R.C. acknowledge the support of cooperative agreements 70NANB12H239 and 70NANB15H260 from NIST, U.S. Department of Commerce. The statements, findings, conclusions, and recommendations are those of the authors and do not necessarily reflect the views of NIST or the US Department of Commerce. R.C. acknowledges support from the National Science Foundation Graduate Research Fellowship under Grant No.1247394.

Appendix

Chemical Names and Their Acronyms

$[BMIM][BF_4]$	1-butyl-3-methylimidazolium tetrafluoroborate
[BMIM][FAP]	1-butyl-3-methylimidazolium tris(pentafluoroethyl)trifluorophosphate
$[BMIM][PF_6]$	1-butyl-3-methylimidazolium hexafluorophosphate
CH_2Cl_2	methylene chloride
EAN	ethylamonium nitrate
d_3EAN	Partially deuterated ethylammonium nitrate
$[EMIM][BF_4]$	1-ethyl-3-methylimidazolium tetrafluoroborate
$[EMIM][NFf_2]$	1-ethyl-3-methylimidazolium bis(trifluoromethane sulfonyl)imide
[EMIM][TFSA]	1-ethyl-3-methylimidazoliumbis-(trifluoromethylsulfonyl)amide
[EMIM][TFSI]	1-ethyl-3-methylimidazolium bis(trifluoromethylsulfonyl)imide
[EMIM][Trif]	1-ethyl-3-methylimidazolium triflate
[Im][TFSI]	Imidazolium bis(trifluoromethane)sulfonamide
P2VP	Poly(2-vinylpyridine)
PAN	penthylammonium nitrate
PB-PEO	Poly(butadiene-b-ethylene oxide)
PBI	Polybenzimidazole
PBnMA-PMMA	Poly(benzyl methacrylate-b-methyl methacrylate)
PEGE-PEO	Poly(ethyl glycidyl ether-b-ethylene oxide)
PEO-PNIPAm	Poly(ethylene oxide-b-N-isopropylacrylamide)
PEO-b-P(AzoMA-r-NIPAm)	Poly(ethylene oxide-b(4-phenylazophenyl methacrylate-r-N-isopropylacrylamide))
PEO-PPO-PEO	Poly(ethylene oxide-b-propylene oxide-b-ethylene oxide)
PGPrE-PEO	Poly(glycidyl propyl ether-b-ethylene oxide)
PnBMA-PEO	poly(n-butyl methacrylate-b-ethylene oxide)
PnBMA-PMMA	poly(n-butyl methacrylate-b-methyl methacrylate)
PS-P2VP	Poly(styrene-b-2-vinylpyridine)
PS-PEO	poly(styrene-b-ethylene oxide)
PS-PMMA	poly(styrene-b-methyl methacrylate)
PS-PEO-PS	poly(styrene-b-ethylene oxide-b-styrene)
PS-PMMA-PS	poly(styrene-b-methyl methacrylate-b-styrene)
PSS	Poly(styrenesulfonate)
PNIPA, PNIPAAm, NIPA, PNIPAA or PNIPAm	Poly(N-isopropylacrylamide)
PPO-PEO-PPO	Poly(propylene oxide-b-ethylene oxide-b-propylene oxide)
PVdF-PHFP	poly(vinylidene fluoride-b-hexafluoropropylene)

References

1. Riess, G. Micellization of Block Copolymers. *Prog. Polym. Sci.* **2003**, *28* (7), 1107–1170.
2. Hayes, R.; Warr, G. G.; Atkin, R. Structure and Nanostructure in Ionic Liquids. *Chem. Rev.* **2015**, *115* (13), 6357–6426.
3. Greaves, T. L.; Drummond, C. J. Protic Ionic Liquids: Evolving Structure-Property Relationships and Expanding Applications. *Chem. Rev.* **2015**, *115* (20), 11379–11448.
4. Greaves, T. L.; Drummond, C. J. Ionic Liquids as Amphiphile Self-Assembly Media. *Chem. Soc. Rev.* **2008**, *37* (8), 1709–1726.
5. Greaves, T. L.; Drummond, C. J. Solvent Nanostructure, the Solvophobic Effect and Amphiphile Self-Assembly in Ionic Liquids. *Chem. Soc. Rev.* **2013**, *42* (3), 1096.
6. Ueki, T.; Watanabe, M. Polymers in Ionic Liquids: Dawn of Neoteric Solvents and Innovative Materials. *Bull. Chem. Soc. Jpn.* **2012**, *85*, 33–50.
7. Ye, Y.-S.; Rick, J.; Hwang, B.-J. Ionic Liquid Polymer Electrolytes. *J. Mater. Chem. A* **2013**, *1* (8), 2719–2743.
8. Park, M. J.; Choi, I.; Hong, J.; Kim, O. Polymer Electrolytes Integrated with Ionic Liquids for Future Electrochemical Devices. *J. Appl. Polym. Sci* **2013**, *129*, 2363–2376.
9. Hamley, I. W. *The Physics of Block Copolymers*; Oxford University Press: 1998.
10. Lindman, B.; Alexandridis, P. *Amphiphilic Block Copolymers : Self-Assembly and Applications*; Elsevier: 2000.
11. Hamley, I. W. *Block Copolymers in Solution : Fundamentals and Applications*; Wiley: 2005.
12. Weingärtner, H. Understanding Ionic Liquids at the Molecular Level: Facts, Problems, and Controversies. *Angew. Chem., Int. Ed.* **2008**, *47*, 654–670.
13. Atkin, R.; Warr, G. G. Structure in Confined Room-Temperature Ionic Liquids. *J. Phys. Chem. C* **2007**, *111* (13), 5162–5168.
14. Atkin, R.; Warr, G. G. The Smallest Amphiphiles: Nanostructure in Protic Room-Temperature Ionic Liquids with Short Alkyl Groups. *J. Phys. Chem. B* **2008**, *112* (14), 4164–4166.
15. Hayes, R.; Warr, G. G.; Atkin, R. At the Interface: Solvation and Designing Ionic Liquids. *Phys. Chem. Chem. Phys.* **2010**, *12* (8), 1709.
16. Hayes, R.; Imberti, S.; Warr, G. G.; Atkin, R. Amphiphilicity Determines Nanostructure in Protic Ionic Liquids. *Phys. Chem. Chem. Phys.* **2011**, *13* (8), 3237–3247.
17. Hayes, R.; Imberti, S.; Warr, G. G.; Atkin, R. Effect of Cation Alkyl Chain Length and Anion Type on Protic Ionic Liquid Nanostructure. *J. Phys. Chem. C* **2014**, *118* (25), 13998–14008.
18. Gao, J.; Wagner, N. J. Water Nanocluster Formation in the Ionic Liquid 1-Butyl-3-Methylimidazolium Tetrafluoroborate ([C4mim][BF4])-D2O Mixtures. *Langmuir* **2016**, *32* (20), 5078–5084.

19. López-Barrón, C. R.; Li, D.; Derita, L.; Basavaraj, M. G.; Wagner, N. J. Spontaneous Thermoreversible Formation of Cationic Vesicles in a Protic Ionic Liquid. *J. Am. Chem. Soc.* **2012**, *134* (51), 20728–20732.
20. Israelachvili, J. N. *Intermolecular and Surface Forces*; Academic Press: 2011.
21. Considine, R. F.; Drummond, C. J. Long-Range Force of Attraction between Solvophobic Surfaces in Water and Organic Liquids Containing Dissolved Air. *Langmuir* **2000**, *16* (2), 631–635.
22. Alexandridis, P.; Alan Hatton, T. Poly(ethylene Oxide)-Poly(propylene Oxide)-Poly(ethylene Oxide) Block Copolymer Surfactants in Aqueous Solutions and at Interfaces: Thermodynamics, Structure, Dynamics, and Modeling. *Colloids Surf., A* **1995**, *96* (1–2), 1–46.
23. Halperin, A.; Tirrell, M.; Lodge, T. P. Tethered Chains in Polymer Microstructures. *Adv. Polym. Sci.* **1992**, *100* (1), 31–71.
24. Hayes, R.; Imberti, S.; Warr, G. G.; Atkin, R. The Nature of Hydrogen Bonding in Protic Ionic Liquids. *Angew. Chem., Int. Ed.* **2013**, *52* (17), 4623–4627.
25. Jain, S. On the Origins of Morphological Complexity in Block Copolymer Surfactants. *Science* **2003**, *300* (5618), 460–464.
26. Lund, R. Small Angle Neutron Scattering as a Tool to Study Kinetics of Block Copolymer Micelles. In *Studying Kinetics with Neutrons, Springer Series in Solid-State Sciences*; Springer-Verlag: Berlin, 2009; Vol. 161, pp 213–240.
27. Won, Y.-Y.; Davis, H. T.; Bates, F. S. Molecular Exchange in PEO–PB Micelles in Water. *Macromolecules* **2003**, *36* (3), 953–955.
28. Hayward, R. C.; Pochan, D. J. Tailored Assemblies of Block Copolymers in Solution: It Is All about the Process. *Macromolecules* **2010**, *43* (8), 3577–3584.
29. Kelley, E. G. Synthesis, Solution Assembly, and Characterization of Amphiphilic Block Polymers. University of Delaware, 2014.
30. Choi, S. H.; Lodge, T. P.; Bates, F. S. Mechanism of Molecular Exchange in Diblock Copolymer Micelles: Hypersensitivity to Core Chain Length. *Phys. Rev. Lett.* **2010**, *104* (4), 47802.
31. Mok, M. M.; Thiagarajan, R.; Flores, M.; Morse, D. C.; Lodge, T. P. Apparent Critical Micelle Concentrations in Block Copolymer/ionic Liquid Solutions: Remarkably Weak Dependence on Solvophobic Block Molecular Weight. *Macromolecules* **2012**, *45* (11), 4818–4829.
32. Kelley, E. G.; Murphy, R. P.; Seppala, J. E.; Smart, T. P.; Hann, S. D.; Sullivan, M. O.; Epps, T. H. Size Evolution of Highly Amphiphilic Macromolecular Solution Assemblies via a Distinct Bimodal Pathway. *Nat. Commun.* **2014**, *5*, 3599.
33. Halperin, A.; Alexander, S. Polymeric Micelles: Their Relaxation Kinetics. *Macromolecules* **1989**, *22* (2), 2403–2412.
34. Esselink, F. J.; Dormidontova, E.; Hadziioannou, G. Evolution of Block Copolymer Micellar Size and Structure Evidenced with Cryo Electron Microscopy. *Macromolecules* **1998**, *31* (9), 2925–2932.

35. Esselink, F.; Dormidontova, E.; Hadziioannou, G. Redistribution of Block Copolymer Chains between Mixed Micelles in Solution. *Macromolecules* **1998**, *31* (15), 4873–4878.
36. Li, Z.; Dormidontova, E. E. Kinetics of Diblock Copolymer Micellization by Dissipative Particle Dynamics. *Macromolecules* **2010**, *43* (7), 3521–3531.
37. Li, Z.; Dormidontova, E. E. Equilibrium Chain Exchange Kinetics in Block Copolymer Micelle Solutions by Dissipative Particle Dynamics Simulations. *Soft Matter* **2011**, *7* (9), 4179.
38. Evans, D. F.; Yamauchi, A.; Roman, R.; Casassa, E. Z. Micelle Formation in Ethylammonium Nitrate, a Low-Melting Fused Salt. *J. Colloid Interface Sci.* **1982**, *88* (1), 89–96.
39. Evans, D. F.; Yamauchi, A.; Wei, G. J.; Bloomfield, A. Micelle Size in Ethylammonium Nitrate As Determined by Classical and Quasi-Elastic Light Scattering. *J. Phys. Chem.* **1983**, *87* (3), 3537–3541.
40. Ueki, T.; Arai, A. A.; Kodama, K.; Kaino, S.; Takada, N.; Morita, T.; Nishikawa, K.; Watanabe, M. Thermodynamic Study on Phase Transitions of Poly(benzyl Methacrylate) in Ionic Liquid Solvents. *Pure Appl. Chem.* **2009**, *81* (10), 1829–1841.
41. Evans, D. F.; Kaler, E. W.; Benton, W. J. Liquid Crystals in a Fused Salt: B, γ-Distearoylphosphatidylcholine in N-Ethylammonium Nitrate. *J. Phys. Chem.* **1983**, *87* (4), 533–535.
42. Araos, M. U.; Warr, G. G. Self-Assembly of Nonionic Surfactants into Lyotropic Liquid Crystals in Ethylammonium Nitrate, a Room-Temperature Ionic Liquid. *J. Phys. Chem. B* **2005**, *109* (30), 14275–14277.
43. Bordel Velasco, S.; Turmine, M.; Di Caprio, D.; Letellier, P. Micelle Formation in Ethyl-Ammonium Nitrate (an Ionic Liquid). *Colloids Surfac., A* **2006**, *275* (1–3), 50–54.
44. Batrakova, E.; Lee, S.; Li, S.; Venne, A.; Alakhov, V.; Kabanov, A. Fundamental Relationships between the Composition of Pluronic Block Copolymers and Their Hypersensitization Effect in MDR Cancer Cells. *Pharm. Res.* **1999**, *16* (9), 1373–1379.
45. Alexandridis, P.; Holzwarth, J. F.; Hatton, T. A. Micellization of Poly(ethylene Oxide)-Poly(propylene Oxide)-Poly(ethylene Oxide) Triblock Copolymers in Aqueous Solutions: Thermodynamics of Copolymer Association. *Macromolecules* **1994**, *27* (9), 2414–2425.
46. Wanka, G.; Hoffmann, H.; Ulbricht, W. Phase Diagrams and Aggregation Behavior of Poly(oxyethylene)-Poly(oxypropylene)-Poly(oxyethylene) Triblock Copolymers in Aqueous Solutions. *Macromolecules* **1994**, *27* (15), 4145–4159.
47. López-Barrón, C. R.; Li, D.; Wagner, N. J.; Caplan, J. L. Triblock Copolymer Self-Assembly in Ionic Liquids: Effect of PEO Block Length on the Self-Assembly of PEO-PPO-PEO in Ethylammonium Nitrate. *Macromolecules* **2014**, *47* (21), 7484–7495.
48. Kabanov, a V; Nazarova, I. R.; Astafieva, I. V; Batrakova, E. V; Alakhov, V. Y.; Yaroslavov, a a; Kabanov, V. a. Micelle Formation and Solubilization of Fluorescent-Probes in Poly(Oxyethylene-B-Oxypropylene-B-Oxyethylene) Solutions. *Macromolecules* **1995**, *28*, 2303–2314.

49. Zhang, S.; Li, N.; Zheng, L.; Li, X.; Gao, Y.; Yu, L. Aggregation Behavior of Pluronic Triblock Copolymer in 1-Butyl-3-Methylimidazolium Type Ionic Liquids. *J. Phys. Chem. B* **2008**, *112*, 10228–10233.
50. Chen, Z.; Greaves, T. L.; Caruso, R. A.; Drummond, C. J. Amphiphile Micelle Structures in the Protic Ionic Liquid Ethylammonium Nitrate and Water. *J. Phys. Chem. B* **2015**, *119* (1), 179–191.
51. Sharma, S. C.; Atkin, R.; Warr, G. G. The Effect of Ionic Liquid Hydrophobicity and Solvent Miscibility on Pluronic Amphiphile Self-Assembly. *J. Phys. Chem. B* **2013**, *117* (46), 14568–14575.
52. Atkin, R.; De Fina, L.-M.; Kiederling, U.; Warr, G. G. Structure and Self Assembly of Pluronic Amphiphiles in Ethylammonium Nitrate and at the Silica Surface. *J. Phys. Chem. B* **2009**, *113* (36), 12201–12213.
53. Mortensen, K. Structural Study on the Micelle Formation of PEO-PPO-PEO Triblock Copolymer in Aqueous Solutions. *Macromolecules* **1993**, *26*, 805–812.
54. Brown, W.; Schilliin, K.; Hvidt, S. Tribiock Copolymers in Aqueous Solution Studied by Static and Dynamic Light Scatterhg and Oscillatory Shear Measurements. Influence of Relative Block Sizes. *J. Phys. Chem.* **1992**, *9662* (75), 6038–6044.
55. Chen, Z.; Fitzgerald, P. A.; Kobayashi, Y.; Ueno, K.; Watanabe, M.; Warr, G. G.; Atkin, R. Micelle Structure of Novel Diblock Polyethers in Water and Two Protic Ionic Liquids (EAN and PAN). *Macromolecules* **2015**, *48* (6), 1843–1851.
56. Wilhelm, M.; Zhao, C. L.; Wang, Y. C.; Xu, R. L.; Winnik, M. A.; Mura, J. L.; Riess, G.; Croucher, M. D. Polymer Micelle Formation .3. Poly(Styrene-Ethylene Oxide) Block Copolymer Micelle Formation in Water - a Fluorescence Probe Study. *Macromolecules* **1991**, *24* (5), 1033–1040.
57. Villari, V.; Triolo, A.; Micali, N. Evidence of Repulsive Yukawa Tail for Copolymer Micelles in Room Temperature Ionic Liquid. *Soft Matter* **2010**, *6* (8), 1793.
58. Zheng, Y.; Won, Y.-Y.; Bates, F. S.; Davis, H. T.; Scriven, L. E.; Talmon, Y. Directly Resolved Core-Corona Structure of Block Copolymer Micelles by Cryo-Transmission Electron Microscopy. *J. Phys. Chem. B* **1999**, *103* (47), 10331–10334.
59. Won, Y. Y.; Brannan, A. K.; Davis, H. T.; Bates, F. S. Cryogenic Transmission Electron Microscopy (Cryo-TEM) of Micelles and Vesicles Formed in Water by Poly(ethylene Oxide)-Based Block Copolymers. *J. Phys. Chem. B* **2002**, *106* (13), 3354–3364.
60. He, Y.; Li, Z.; Simone, P.; Lodge, T. Self-Assembly of Block Copolymer Micelles in an Ionic Liquid. *J. Am. Chem. Soc.* **2006**, *128* (8), 2745–2750.
61. Bai, Z.; He, Y.; Lodge, T. P. Block Copolymer Micelle Shuttles with Tunable Transfer Temperatures between Ionic Liquids and Aqueous Solutions. *Langmuir* **2008**, *24* (10), 5284–5290.
62. Lee, H. N.; Lodge, T. P. Poly(n -Butyl Methacrylate) in Ionic Liquids with Tunable Lower Critical Solution Temperatures (LCST). *J. Phys. Chem. B* **2011**, *115* (9), 1971–1977.

63. Tamura, S.; Ueki, T.; Ueno, K.; Kodama, K.; Watanabe, M. Thermosensitive Self-Assembly of Diblock Copolymers with Lower Critical Micellization Temperatures in an Ionic Liquid. *Macromolecules* **2009**, *42* (16), 6239–6244.
64. Ueki, T.; Nakamura, Y.; Lodge, T. P.; Watanabe, M. Light-Controlled Reversible Micellization of a Diblock Copolymer in an Ionic Liquid. *Macromolecules* **2012**, *45* (18), 7566–7573.
65. Lee, H. N.; Bai, Z.; Newell, N.; Lodge, T. P. Micelle/inverse Micelle Self-Assembly of a PEO-PNIPAm Block Copolymer in Ionic Liquids with Double Thermoresponsivity. *Macromolecules* **2010**, *43* (22), 9522–9528.
66. Zheng, L.; Chai, Y.; Liu, Y.; Zhang, P. Controlled RAFT Synthesis of Polystyrene-B-Poly(acrylic Acid)-B-Polystyrene Block Copolymers and Their Self-Assembly in an Ionic Liquid [BMIM][PF6]. *E-Polymers* **2011**, *11* (1), 436–443.
67. DeRita, L. Self-Assembly of Pluornic Polymer Systems in Ionic Liquids. Univeristy of Delaware, 2013.
68. López-Barrón, C. R.; Chen, R.; Wagner, N. J.; Beltramo, P. J. Self-Assembly of Pluronic F127 Diacrylate in Ethylammonium Nitrate: Structure, Rheology, and Ionic Conductivity before and after Photo-Cross-Linking. *Macromolecules* **2016**, *49* (14), 5179–5189.
69. Laflèche, F.; Nicolai, T.; Durand, D.; Gnanou, Y.; Taton, D. Association of Adhesive Spheres Formed by Hydrophobically End-Capped PEO. 2. Influence of the Alkyl End-Group Length and the Chain Backbone Architecture. *Macromolecules* **2003**, *36*, 1341–1348.
70. Anderson, J. A; Lorenz, C. D.; Travesset, A. Micellar Crystals in Solution from Molecular Dynamics Simulations. *J. Chem. Phys.* **2008**, *128* (18), 184906.
71. Meli, L.; Lodge, T. P. Equilibrium vs Metastability: High-Temperature Annealing of Spherical Block Copolymer Micelles in an Ionic Liquid. *Macromolecules* **2009**, *42* (3), 580–583.
72. Meli, L.; Santiago, J. M.; Lodge, T. P. Path-Dependent Morphology and Relaxation Kinetics of Highly Amphiphilic Diblock Copolymer Micelles in Ionic Liquids. *Macromolecules* **2010**, *43* (4), 2018–2027.
73. Ueki, T.; Watanabe, M. Upper Critical Solution Temperature Behavior of Poly(N-Isopropylacrylamide) in an Ionic Liquid and Preparation of Thermo-Sensitive Nonvolatile Gels. *Chem. Lett.* **2006**, *35* (8), 964–965.
74. Kodama, K.; Nanashima, H.; Ueki, T.; Kokubo, H.; Watanabe, M. Lower Critical Solution Temperature Phase Behavior of Linear Polymers in Imidazolium-Based Ionic Liquids: Effects of Structural Modifications. *Langmuir* **2009**, *25* (6), 3820–3824.
75. Ueki, T.; Watanabe, M. Lower Critical Solution Temperature Behavior of Linear Polymers in Ionic Liquids and the Corresponding Volume Phase Transition of Polymer Gels. *Langmuir* **2007**, *23* (3), 988–990.
76. Tsuda, R.; Kodama, K.; Ueki, T.; Kokubo, H.; Imabayashi, S.; Watanabe, M. LCST-Type Liquid-Liquid Phase Separation Behaviour of Poly(ethylene Oxide). *Chem. Commun.* **2008** (40), 4939–4941.

77. Lee, H.-N.; Lodge, T. P. Lower Critical Solution Temperature (LCST) Phase Behavior of Poly(ethylene Oxide) in Ionic Liquids. *J. Phys. Chem. Lett.* **2010**, *1* (13), 1962–1966.
78. Lee, H.-N.; Newell, N.; Bai, Z.; Lodge, T. P. Unusual Lower Critical Solution Temperature Phase Behavior of Poly(ethylene Oxide) in Ionic Liquids. *Macromolecules* **2012**, *45* (8), 3627–3633.
79. Ueki, T.; Watanabe, M.; Lodge, T. P. Doubly Thermosensitive Self-Assembly of Diblock Copolymers in Ionic Liquids. *Macromolecules* **2009**, *42* (4), 1315–1320.
80. He, Y.; Boswell, P. G.; Bühlmann, P.; Lodge, T. P. Ion Gels by Self-Assembly of a Triblock Copolymer in an Ionic Liquid. *J. Phys. Chem. B* **2007**, *111* (18), 4645–4652.
81. Zhang, S.; Lee, K. H.; Sun, J.; Frisbie, C. D.; Lodge, T. P. Viscoelastic Properties, Ionic Conductivity, and Materials Design Considerations for Poly(styrene-B-Ethylene Oxide-B-Styrene)-Based Ion Gel Electrolytes. *Macromolecules* **2011**, *44* (22), 8981–8989.
82. Zhang, S.; Lee, K. H.; Frisbie, C. D.; Lodge, T. P. Ionic Conductivity, Capacitance, and Viscoelastic Properties of Block Copolymer-Based Ion Gels. *Macromolecules* **2011**, *44* (4), 940–949.
83. He, Y.; Lodge, T. P. A Thermoreversible Ion Gel by Triblock Copolymer Self-Assembly in an Ionic Liquid. *Chem. Commun. (Camb).* **2007**, *16* (26), 2732–2734.
84. Virgili, J. M. Studies of Block Copolymer Thin Films and Mixtures with an Ionic Liquid. Univeristy of California, Berkeley, 2009.
85. He, Y.; Lodge, T. P. The Micellar Shuttle: Thermoreversible, Intact Transfer of Block Copolymer Micelles between an Ionic Liquid and Water. *J. Am. Chem. Soc.* **2006**, *128* (39), 12666–12667.
86. Meier, W. Polymer Nanocapsules. *Chem. Soc. Rev.* **2000**, *29* (5), 295–303.
87. Cooper, A.; Londono, J.; Wignall, G.; McClain, J.; Samulski, E.; Lin, J.; Dobrynin, A.; Rubinstein, M.; Burke, A.; Frechet, J.; DeSimone, J. Extraction of a Hydrophilic Compound from Water into Liquid CO2 Using Dendritic Surfactants. *Nature* **1997**, *389* (6649), 368–371.
88. Sunder, A.; Krämer, M.; Hanselmann, R.; Mülhaupt, R.; Frey, H. Molecular Nanocapsules Based on Amphiphilic Hyperbranched Polyglycerols. *Angew. Chem., Int. Ed.* **1999**, *38* (23), 3552–3555.
89. Chechik, V.; Zhao, M.; Crooks, R. M. Self-Assembled Inverted Micelles Prepared from a Dendrimer Template: Phase Transfer of Encapsulated Guests. *J. Am. Chem. Soc.* **1999**, *121*, 4910–4911.
90. Vriezema, D. M.; Aragonès, M. C.; Elemans, J. A. A. W.; Cornelissen, J. J. L. M.; Rowan, A. E.; Nolte, R. J. M. Self-Assembled Nanoreactors. *Chem. Rev.* **2005**, *105*, 1445–1489.
91. Dwars, T.; Paetzold, E.; Oehme, G. Reactions in Micellar Systems. *Angew. Chem., Int. Ed.* **2005**, *44* (44), 7174–7199.
92. Bai, Z.; He, Y.; Young, N. P.; Lodge, T. P. A Thermoreversible Micellization-Transfer-Demicellization Shuttle between Water and an Ionic Liquid. *Macromolecules* **2008**, *41* (18), 6615–6617.

93. Guerrero-Sanchez, C.; Wouters, D.; Hoeppener, S.; Hoogenboom, R.; Schubert, U. S. Micellar Dye Shuttle between Water and an Ionic Liquid. *Soft Matter* **2011**, *7* (8), 3827–3831.
94. Bai, Z.; Lodge, T. P. Pluronic Micelle Shuttle between Water and an Ionic Liquid. *Langmuir* **2010**, *26* (11), 8887–8892.
95. Saeki, S.; Kuwahara, N.; Nakata, M.; Kaneko, M. Upper and Lower Critical Solution Temperatures in Poly (Ethylene Glycol) Solutions. *Polymer (Guildf).* **1976**, *17* (8), 685–689.
96. Bai, Z.; Lodge, T. P. Polymersomes with Ionic Liquid Interiors Dispersed in Water. *J. Am. Chem. Soc.* **2010**, *132* (45), 16265–16270.
97. Bai, Z.; Zhao, B.; Lodge, T. P. Bilayer Membrane Permeability of Ionic Liquid-Filled Block Copolymer Vesicles in Aqueous Solution. *J. Phys. Chem. B* **2012**, *116* (28), 8282–8289.
98. Bai, Z.; Lodge, T. P. Thermodynamics and Mechanism of the Block Copolymer Micelle Shuttle between Water and an Ionic Liquid. *J. Phys. Chem. B* **2009**, *113* (43), 14151–14157.
99. Hoarfrost, M. L. Ion Transport in Nanostructured Block Copolymer/Ionic Liquid Membranes. University of California, Berkeley, 2012.
100. Lai, C.; Russel, W. B.; Register, R. A. Scaling of Domain Spacing in Concentrated Solutions of Block Copolymers in Selective Solvents. *Macromolecules* **2002**, *35* (10), 4044–4049.
101. Hanley, K. J.; Lodge, T. P.; Huang, C. I. Phase Behavior of a Block Copolymer in Solvents of Varying Selectivity. *Macromolecules* **2000**, *33* (16), 5918–5931.
102. Simone, P. M.; Lodge, T. P. Phase Behavior and Ionic Conductivity of Concentrated Solutions of Polystyrene-Poly(ethylene Oxide) Diblock Copolymers in an Ionic Liquid. *ACS Appl. Mater. Interfaces* **2009**, *1* (12), 2812–2820.
103. Park, J.-W.; Kim, H.; Han, M. Polymeric Self-Assembled Monolayers Derived from Surface-Active Copolymers: A Modular Approach to Functionalized Surfaces. *Chem. Soc. Rev.* **2010**, *39* (8), 2935–2947.
104. Simone, P. M.; Lodge, T. P. Lyotropic Phase Behavior of Polybutadiene-Poly (Ethylene Oxide) Diblock Copolymers in Ionic Liquids. *Macromolecules* **2008**, *41* (5), 1753–1759.
105. Virgili, J. M.; Hexemer, A.; Pople, J. A.; Balsara, N. P.; Segalman, R. A. Phase Behavior of Polystyrene-Block-poly(2-Vinylpyridine) Copolymers in a Selective Ionic Liquid Solvent. *Macromolecules* **2009**, *42* (13), 4604–4613.
106. Virgili, J. M.; Hoarfrost, M. L.; Segalman, R. A. Effect of an Ionic Liquid Solvent on the Phase Behavior of Block Copolymers. *Macromolecules* **2010**, *43* (12), 5417–5423.
107. Miranda, D. F.; Russell, T. P.; Watkins, J. J. Ordering in Mixtures of a Triblock Copolymer with a Room Temperature Ionic Liquid. *Macromolecules* **2010**, *43* (24), 10528–10535.
108. Wang, L.; Chen, X.; Chai, Y.; Hao, J.; Sui, Z.; Zhuang, W.; Sun, Z. Lyotropic Liquid Crystalline Phases Formed in an Ionic Liquid. *Chem. Commun.* **2004**, *6* (24), 2840–2841.

109. Zhang, G.; Chen, X.; Zhao, Y.; Ma, F.; Jing, B.; Qiu, H. Lyotropic Liquid-Crystalline Phases Formed by Pluronic P123 in Ethylammonium Nitrate. *J. Phys. Chem. B* **2008**, *112* (21), 6578–6584.
110. McConnell, G. A.; Gast, A. P.; Huang, J. S.; Smith, S. D. Disorder-Order Transitions in Soft Sphere Polymer Micelles. *Phys. Rev. Lett.* **1993**, *71* (13), 2102–2105.
111. Mortensen, K.; Batsberg, W.; Hvidt, S. Effects of PEO - PPO Diblock Impurities on the Cubic Structure of Aqueous PEO - PPO - PEO Pluronics Micelles : Fcc and Bcc Ordered Structures in F127. *Macromolecules* **2008**, *41*, 1720–1727.
112. Chen, R.; López-Barrón, C. R.; Wagner, N. J. Co-Micellization of Binary Mixture of PEO-PPO-PEO Tri-Block Copolymer P123 and F127 in Ethylammonium Nitrate. **2016**, manuscript in preparation.
113. Chen, R.; López-Barrón, C. R.; Wagner, N. J. Tunable Rheology, Microstructure and Shear Alignment of Thermo-Reversible Soft Micellar Crystals in Ionic Liquid. *Prep.* **2016**, manuscript in preparation.
114. Mortensen, K.; Brown, W.; Nordén, B. Inverse Melting Transition and Evidence of Three-Dimensional Cubatic Structure in a Block-Copolymer Micellar System. *Phys. Rev. Lett.* **1992**, *68* (15), 2340–2343.
115. Walker, L. M. Rheology and Structure of Worm-like Micelles. *Curr. Opin. Colloid Interface Sci.* **2001**, *6* (5–6), 451–456.
116. Eberle, A. P. R.; Porcar, L. Flow-SANS and Rheo-SANS Applied to Soft Matter. *Curr. Opin. Colloid Interface Sci.* **2012**, *17* (1), 33–43.
117. Phoon, C. L.; Higgins, J. S.; Allegra, G.; Leeuwen, P. V.; Staples, E. Shear Induced Crystallization and Melting of a Micellar Solution. *Proc. R. Soc. A* **1993**, *442* (1914), 221–230.
118. McConnell, G. A.; Lin, M. Y.; Gast, A. P. Long-Range Order in Polymeric Micelles Under Steady Shear. *Macromolecules* **1995**, *28* (20), 6754–6764.
119. Berret, J.-F.; Molino, F.; Porte, G.; Diat, O.; Lindner, P. The Shear-Induced Transition between Oriented Textures and Layer-Sliding-Mediated Flows in a Micellar Cubic Crystal. *J. Phys.: Condens. Matter* **1996**, *8* (47), 9513–9517.
120. Slawecki, T.; Glinka, C.; Hammouda, B. Shear-Induced Micellar Crystal Structures in an Aqueous Triblock Copolymer Solution. *Phys. Rev. E* **1998**, *58* (4), R4084–R4087.
121. Jiang, J.; Burger, C.; Li, C.; Li, J.; Lin, M. Y.; Colby, R. H.; Rafailovich, M. H.; Sokolov, J. C. Shear-Induced Layered Structure of Polymeric Micelles by SANS. *Macromolecules* **2007**, *40* (11), 4016–4022.
122. Jiang, J.; Malal, R.; Li, C.; Lin, M. Y.; Colby, R. H.; Gersappe, D.; Rafailovich, M. H.; Sokolov, J. C.; Cohn, D. Rheology of Thermoreversible Hydrogels from Multiblock Associating Copolymers. *Macromolecules* **2008**, *41* (10), 3646–3652.
123. Ackerson, B. J.; Pusey, P. N. Shear-Induced Order in Suspensions of Hard Spheres. *Phys. Rev. Lett.* **1988**, *61* (8), 1033–1036.
124. Chen, L. B.; Zukoski, C. F.; Ackerson, B. J.; Hanley, H. J. M.; Straty, G. C.; Barker, J.; Glinka, C. J. Structural-Changes and Orientational Order in a Sheared Colloidal Suspension. *Phys. Rev. Lett.* **1992**, *69* (4), 688–691.

125. Wu, Y. L.; Derks, D.; van Blaaderen, A.; Imhof, A. Melting and Crystallization of Colloidal Hard-Sphere Suspensions under Shear. *Proc. Natl. Acad. Sci. U. S. A.* **2009**, *106* (26), 10564–10569.
126. López-Barrón, C. R.; Wagner, N. J.; Porcar, L. Layering, Melting, and Recrystallization of a Close-Packed Micellar Crystal under Steady and Large-Amplitude Oscillatory Shear Flows. *J. Rheol. (N. Y. N. Y).* **2015**, *59* (3), 793–820.
127. Porcar, L.; Pozzo, D.; Langenbucher, G.; Moyer, J.; Butler, P. D. Rheo-Small-Angle Neutron Scattering at the National Institute of Standards and Technology Center for Neutron Research. *Rev. Sci. Instrum.* **2011**, *82* (8), 83902.
128. Gurnon, A. K.; Godfrin, P. D.; Wagner, N. J.; Eberle, A. P. R.; Butler, P.; Porcar, L. Measuring Material Microstructure under Flow Using 1-2 Plane Flow-Small Angle Neutron Scattering. *J. Visualized Exp.* **2014**, *84*, e51068.
129. López-Barrón, C. R.; Porcar, L.; Eberle, A. P. R.; Wagner, N. J. Dynamics of Melting and Recrystallization in a Polymeric Micellar Crystal Subjected to Large Amplitude Oscillatory Shear Flow. *Phys. Rev. Lett.* **2012**, *108* (25), 258301.
130. Dreiss, C. A. Wormlike Micelles: Where Do We Stand? Recent Developments, Linear Rheology and Scattering Techniques. *Soft Matter* **2007**, *3* (8), 956.
131. Cates, M. E.; Fielding, S. M. Rheology of Giant Micelles. *Adv. Phys.* **2006**, *55*, 799–879.
132. Zana, R.; Kaler, E. W. *Giant Micelles : Properties and Applications*; CRC Press: 2007.
133. Berret, J. F. Rheology of Wormlike Micelles: Equilibrium Properties and Shear Banding Transitions. In *Molecular Gels: Materials with Self-Assembled Fibrillar Networks*; Springer-Verlag: Berlin/Heidelberg, 2006; pp 667–720.
134. Castelletto, V.; Parras, P.; Hamley, I. W.; Bäverbäck, P.; Pedersen, J. S.; Panine, P. Wormlike Micelle Formation and Flow Alignment of a Pluronic Block Copolymer in Aqueous Solution. *Langmuir* **2007**, *23* (13), 6896–6902.
135. Holmqvist, P.; Alexandridis, P.; Lindman, B. Modification of the Microstructure in Block Copolymer−Water−"Oil" Systems by Varying the Copolymer Composition and the "Oil" Type: Small-Angle X-Ray Scattering and Deuterium-NMR Investigation. *J. Phys. Chem. B* **1998**, *102* (7), 1149–1158.
136. Miranda, D. F.; Versek, C.; Tuominen, M. T.; Russell, T. P.; Watkins, J. J. Cross-Linked Block Copolymer/ionic Liquid Self-Assembled Blends for Polymer Gel Electrolytes with High Ionic Conductivity and Mechanical Strength. *Macromolecules* **2013**, *46* (23), 9313–9323.
137. Sutto, T. E. Hydrophobic and Hydrophilic Interactions of Ionic Liquids and Polymers in Solid Polymer Gel Electrolytes. *J. Electrochem. Soc.* **2007**, *154* (10), P101.
138. Lewandowski, A.; Świderska, A. New Composite Solid Electrolytes Based on a Polymer and Ionic Liquids. *Solid State Ionics* **2004**, *169* (1–4 SPEC. ISS.), 21–24.

139. Zhu, C.; Cheng, H.; Yang, Y. Electrochemical Characterization of Two Types of PEO-Based Polymer Electrolytes with Room-Temperature Ionic Liquids. *J. Electrochem. Soc.* **2008**, *155* (8), A569.
140. Kim, Y. H.; Cheruvally, G.; Choi, J. W.; Ahn, J. H.; Kim, K. W.; Ahn, H. J.; Choi, D. S.; Song, C. E. Electrochemical Properties of PEO-Based Polymer Electrolytes Blended with Different Room Temperature Ionic Liquids. In *Macromolecular Symposia*; WILEY-VCH Verlag: 2007; Vol. 249–250, pp 183–189.
141. Fuller, J.; Breda, A. C.; Carlin, R. T. Ionic Liquid-Polymer Gel Electrolytes. *J. Electrochem. Soc.* **1997**, *144* (4), L67–L70.
142. Singh, B.; Sekhon, S. S. Ion Conducting Behaviour of Polymer Electrolytes Containing Ionic Liquids. *Chem. Phys. Lett.* **2005**, *414* (1–3), 34–39.
143. Yeon, S. H.; Kim, K. S.; Choi, S.; Cha, J. H.; Lee, H. Characterization of PVdF(HFP) Gel Electrolytes Based on 1-(2-Hydroxyethyl)-3-Methyl Imidazolium Ionic Liquids. *J. Phys. Chem. B* **2005**, *109* (38), 17928–17935.
144. Scott, M. P.; Brazel, C. S.; Benton, M. G.; Mays, J. W.; Holbrey, J. D.; Rogers, R. D. Application of Ionic Liquids as Plasticizers for Poly(methyl Methacrylate). *Chem. Commun.* **2002**, *3* (13), 1370–1371.
145. Susan, M. A. B. H.; Kaneko, T.; Noda, A.; Watanabe, M. Ion Gels Prepared by in Situ Radical Polymerization of Vinyl Monomers in an Ionic Liquid and Their Characterization as Polymer Electrolytes. *J. Am. Chem. Soc.* **2005**, *127* (13), 4976–4983.
146. Seki, S.; Susan, M. A. B. H.; Kaneko, T.; Tokuda, H.; Noda, A.; Watanabe, M. Distinct Difference in Ionic Transport Behavior in Polymer Electrolytes Depending on the Matrix Polymers and Incorporated Salts. *J. Phys. Chem. B* **2005**, *109* (9), 3886–3892.
147. Jana, S.; Parthiban, A.; Chai, C. L. L. Transparent, Flexible and Highly Conductive Ion Gels from Ionic Liquid Compatible Cyclic Carbonate Network. *Chem. Commun.* **2010**, *46* (9), 1488–1490.
148. Klingshirn, M. A.; Spear, S. K.; Subramanian, R.; Holbrey, J. D.; Huddleston, J. G.; Rogers, R. D. Gelation of Ionic Liquids Using a Cross-Linked Poly(ethylene Glycol) Gel Matrix. *Chem. Mater.* **2004**, *16* (16), 3091–3097.
149. Néouze, M. A.; Le Bideau, J.; Gaveau, P.; Bellayer, S.; Vioux, A. Ionogels, New Materials Arising from the Confinement of Ionic Liquids within Silica-Derived Networks. *Chem. Mater.* **2006**, *18* (17), 3931–3936.
150. Matsumoto, K.; Endo, T. Confinement of Ionic Liquid by Networked Polymers Based on Multifunctional Epoxy Resins. *Macromolecules* **2008**, *41* (19), 6981–6986.
151. Matsumoto, K.; Endo, T. Synthesis of Ion Conductive Networked Polymers Based on an Ionic Liquid Epoxide Having a Quaternary Ammonium Salt Structure. *Macromolecules* **2009**, *42* (13), 4580–4584.
152. Gu, Y.; Zhang, S.; Martinetti, L.; Lee, K. H.; McIntosh, L. D.; Frisbie, C. D.; Lodge, T. P. High Toughness, High Conductivity Ion Gels by Sequential Triblock Copolymer Self-Assembly and Chemical Cross-Linking. *J. Am. Chem. Soc.* **2013**, *135* (26), 9652–9655.

153. López-Barrón, C. R.; Chen, R.; Wagner, N. J. Ultra-Stretchable Iono-Elastomers with Mechano-Electrical Responses. *ACS Macro Lett.* **2016**, *5*, 1332–1338.
154. Rogers, J. A.; Someya, T.; Huang, Y. Materials and Mechanics for Stretchable Electronics. *Science* **2010**, *327* (5973), 1603–1607.
155. Mullins, L. Softening of Rubber by Deformation. *Rubber Chem. Technol.* **1969**, *42* (1), 339–362.
156. Seki, S.; Kobayashi, Y.; Miyashiro, H.; Ohno, Y.; Usami, A.; Mita, Y.; Kihira, N.; Watanabe, M.; Terada, N. Lithium Secondary Batteries Using Modified-Imidazolium Room-Temperature Ionic Liquid. *J. Phys. Chem. B* **2006**, *110* (21), 10228–10230.
157. Zakeeruddin, S. M.; Grätzel, M. Solvent-Free Ionic Liquid Electrolytes for Mesoscopic Dye-Sensitized Solar Cells. *Adv. Funct. Mater.* **2009**, *19* (14), 2187–2202.
158. Lu, W.; Fadeev, A. G.; Qi, B.; Smela, E.; Mattes, B. R.; Ding, J.; Spinks, G. M.; Mazurkiewicz, J.; Zhou, D.; Wallace, G. G.; MacFarlane, D. R.; Forsyth, S. A.; Forsyth, M. Use of Ionic Liquids for Pi-Conjugated Polymer Electrochemical Devices. *Science* **2002**, *297* (5583), 983–987.
159. Ding, J.; Zhou, D.; Spinks, G.; Wallace, G.; Forsyth, S.; Forsyth, M.; MacFarlane, D. Use of Ionic Liquids as Electrolytes in Electromechanical Actuator Systems Based on Inherently Conducting Polymers. *Chem. Mater.* **2003**, *15* (12), 2392–2398.
160. Cho, J. H.; Lee, J.; Xia, Y.; Kim, B.; He, Y.; Renn, M. J.; Lodge, T. P.; Frisbie, C. D. Printable Ion-Gel Gate Dielectrics for Low-Voltage Polymer Thin-Film Transistors on Plastic. *Nat. Mater.* **2008**, *7* (11), 900–906.
161. Cho, J. H.; Lee, J.; He, Y.; Kim, B.; Lodge, T. P.; Frisbie, C. D. High-Capacitance Ion Gel Gate Dielectrics with Faster Polarization Response Times for Organic Thin Film Transistors. *Adv. Mater.* **2008**, *20* (4), 686–690.
162. Zhang, L.; Eisenberg, A. Multiple Morphologies of "Crew-Cut" Aggregates of Polystyrene-B-Poly(acrylic Acid) Block Copolymers. *Science* **1995**, *268* (5218), 1728–1731.
163. Creutz, S.; van Stam, J.; De Schryver, F. C.; Jérôme, R. Dynamics of Poly((dimethylamino)alkyl Methacrylate-Block-Sodium Methacrylate) Micelles. Influence of Hydrophobicity and Molecular Architecture on the Exchange Rate of Copolymer Molecules. *Macromolecules* **1998**, *31* (3), 681–689.
164. Loppinet, B.; Sigel, R.; Larsen, A.; Fytas, G.; Vlassopoulos, D.; Liu, G. Structure and Dynamics in Dense Suspensions of Micellar Nanocolloids. *Langmuir* **2000**, *16* (16), 6480–6484.
165. Discher, B. M.; Won, Y. Y.; Ege, D. S.; Lee, J. C. M.; Bates, F. S.; Discher, D. E.; Hammer, D. A. Polymersomes: Tough Vesicles Made from Diblock Copolymers. *Science* **1999**, *284*, 1143–1146.
166. O'Reilly, R. K.; Joralemon, M. J.; Wooley, K. L.; Hawker, C. J. Functionalization of Micelles and Shell Cross-Linked Nanoparticles Using Click Chemistry. *Chem. Mater.* **2005**, *17* (24), 5976–5988.

167. Joralemon, M. J.; O'Reilly, R. K.; Hawker, C. J.; Wooley, K. L. Shell Click-Crosslinked (SCC) Nanoparticles: A New Methodology for Synthesis and Orthogonal Functionalization. *J. Am. Chem. Soc.* **2005**, *127* (48), 16892–16899.
168. Blach, D.; Pessêgo, M.; Silber, J. J.; Correa, N. M.; García-Río, L.; Falcone, R. D. Ionic Liquids Entrapped in Reverse Micelles as Nanoreactors for Bimolecular Nucleophilic Substitution Reaction. Effect of the Confinement on the Chloride Ion Availability. *Langmuir* **2014**, *30* (41), 12130–12137.
169. Marrucho, I. M.; Branco, L. C.; Rebelo, L. P. N. Ionic Liquids in Pharmaceutical Applications. *Annu. Rev. Chem. Biomol. Eng.* **2014**, *5*, 527–546.
170. Xu, S.; Zhang, Y.; Cho, J.; Lee, J.; Huang, X.; Jia, L.; Fan, J. A.; Su, Y.; Su, J.; Zhang, H.; Cheng, H.; Lu, B.; Yu, C.; Chuang, C.; Kim, T.-I.; Song, T.; Shigeta, K.; Kang, S.; Dagdeviren, C.; Petrov, I.; Braun, P. V; Huang, Y.; Paik, U.; Rogers, J. A. Stretchable Batteries with Self-Similar Serpentine Interconnects and Integrated Wireless Recharging Systems. *Nat. Commun.* **2013**, *4*, 1543.
171. Kaltenbrunner, M.; Sekitani, T.; Reeder, J.; Yokota, T.; Kuribara, K.; Tokuhara, T.; Drack, M.; Schwödiauer, R.; Graz, I.; Bauer-Gogonea, S.; Bauer, S.; Someya, T. An Ultra-Lightweight Design for Imperceptible Plastic Electronics. *Nature* **2013**, *499* (7459), 458–463.
172. Sekitani, T.; Noguchi, Y.; Hata, K.; Fukushima, T.; Aida, T.; Someya, T. A Rubberlike Stretchable Active Matrix Using Elastic Conductors. *Science* **2008**, *321* (5895), 1468–1472.
173. Kim, D.-H.; Ahn, J.-H.; Choi, W. M.; Kim, H.-S.; Kim, T.-H.; Song, J.; Huang, Y. Y.; Liu, Z.; Lu, C.; Rogers, J. A. Stretchable and Foldable Silicon Integrated Circuits. *Science* **2008**, *320* (5875), 507–511.
174. Stoppa, M.; Chiolerio, A. Wearable Electronics and Smart Textiles: A Critical Review. *Sensors* **2014**, *14* (7), 11957–11992.
175. Appetecchi, G. B.; Zane, D.; Scrosati, B. PEO-Based Electrolyte Membranes Based on LiBC4O8 Salt. *J. Electrochem. Soc.* **2004**, *151* (9), A1369.
176. Richards, J. J.; Gagnon, C. V. L.; Krzywon, J. R.; Wagner, N. J.; Butler, P. D. Dielectric RheoSANS –Simultaneous Interrogation of Impedance, Rheology and Small Angle Neutron Scattering of Complex Fluids. *J. Vis. Exp.* **2017**, *122*, e55318.
177. Richards, J. J.; Hipp, J.; Wagner, N. J.; Butler, P. D. Dielectric RheoSANS: The Study of Electrical and Mechanical Properties of Complex Fluids. **2016**, in preparation.

Chapter 6

Multi-Purpose Cellulosic Ionogels

Chip J. Smith II,[1] Durgesh V. Wagle,[1] Hugh M. O'Neill,[2] Barbara R. Evans,[3] Sheila N. Baker,[1] and Gary A. Baker*,[1]

[1]Department of Chemistry, University of Missouri-Columbia, Columbia, Missouri 65201, United States
[2]Biology and Soft Matter Division, Oak Ridge National Laboratory, Oak Ridge, Tennessee 37831, United States
[3]Chemical Sciences Division, Oak Ridge National Laboratory, Oak Ridge, Tennessee 37831, United States
***E-mail: bakergar@missouri.edu**

The immobilization of ionic liquid into a support matrix for practical applications is frequently inefficient (i.e., low loading capacity) or otherwise compromises the attractive properties of the sequestered ionic liquid phase. One promising strategy for ionic liquid immobilization entails the formation of an ionogel, although reported ionogels sometimes suffer from solvent/matrix incompatibility, limited liquid loading capacity, and the development of optical opacity or physical embrittlement. In this chapter, we introduce a straightforward procedure for preparing bacterial cellulose ionogels (BCIGs) using an ethanol co-solvent exchange process to achieve ionic liquid loadings of up to 99 weight percent. The resulting ionogels are transparent, stable, flexible, size- and shape-tunable, and can host a range of chemistries toward multi-purpose applications.

Introduction

While ionic liquids (ILs) possess many advantageous properties, these designer solvents suffer in their use as analytical, electrochemical, and separation platforms due to their viscous nature and/or poor containment within the devices in which they are utilizied (i.e., IL leakage) (*1–3*). However, immobilizing the ILs within or onto a solid support matrix, which results in the formation of a gel, can remedy these daunting problems that ILs face, whilst still retaining their unique properties. To date, several different methods have been explored for immoblizing ILs such as IL chemical tethering to a framework, polymerization of the IL to form a gel, and IL encapsulation within a solid support through diffusion or polymerization (*2*, *4*, *5*). Of particular interest though, is the confinement of ILs within a porous matrix to produce what is known as an ionogel. Since the liquid state of the IL is retained within these gels, a quasi-solid platform is created that has advantages in areas such as wound healing, electrochemical operation, and analytics due to the added ionic conductivity as well as the ability to dissolve analytes.

Historically, ionogels have been prepared in many different solid supports with silica being a predominant component (*1*, *5*, *6*). Within the last decade, the use of biopolymers as the hosting porous matrix has brought forth greener alternatives to conventional ionogels (*7*, *8*) and since the advent of this class of ionogel, additional developments of biopolymer-based ionogels have emerged (*9–12*). Despite their greener nature, these materials typically suffer from a lack of thermal, mechanical, and/or chemical stability as well as miscibility issues with ILs. More specifically, the biopolymers are plagued with poor IL loading (i.e., low weight percentages, wt%), are limited to water soluble ILs, or a chemical modification step of the biopolymer is required.

One such example is ligno-cellulose ionogels, which require purification from the recalcitrant biopolymers lignin and hemicellulose that typically encase the cellulose micro-fibrils of plant cell wall. After purification, the ionogels are formed via dissolution of the cellulose in 15 wt% 1-butyl-3-methylimidazolium chloride and then allowed to gel over the course of 7 days (*7*). Other ionogels have been prepared using methyl cellulose as the support matrix and were able to hold up to 97 wt% IL unless decorated with silica which resulted in 98 wt% IL loading (*12*, *13*). Despite these high loadings, as mentioned above, these biopolymer gels require extensive purification and chemically modifation of the cellulose prior to use. To avoid the need for cellulose dissolution, which destroys its natural fibrous macro-molecular structure, a more unadultered form of cellulose that can retain large amounts of solvent is required. Bacterial cellulose (BC), as the name implies, is cellulose synthesized within pores on the cell surface of specific bacteria genera (i.e., *Acetobacter*, *Sarcina*, and *Agrobacterium*) as an extracellular, three-dimensional network that forms a protective envelope around the cells to serve biologically-relavent functions such as maintaining an aerobic environment, aiding in flocculation, and allowing attachment to plants (*14*). BC is advantageous because the bacteria synthesize solely cellulose (i.e., no side product formation) in a very pure, crystalline form (crystal index ~ 0.89) that is mechanically resilient (Young's modulus 16.9 GPa) and can hold ~ 99 wt% water (*15–17*), which makes it a promsing and rising platform in the formation of composite materials that span a large array of applications (*18*, *19*).

Herein, the preparation of IL-loaded BC ionogel composite materials possessing a plethora of properties that were conducive for their use in chemosensory applications, is discussed. In addition, these materials have possibile applications in separation membranes and electrochemistry.

Bacterial Cellulose Ionogel Production

Bacterial Cellulose Production and Preparation

BC was produced in a modified Hestrin Schramm (HS) medium using the bacterial strain *Gluconacetobacter xylinus* (ATCC 700178). Modification of the HS medium was chosen because of the benefits that could be attained by changing the formulation (e.g. yield and culture speed). The original HS medium consisted of 2% m/v glucose (carbon source), 0.5% m/v peptone, 0.5% yeast extract (nitrogen source), 0.27% disodium phosphate (buffer), and 0.115% citric acid (boosting antioxidant additive) which would be adjusted to pH 6.0 and autoclaved (*20*). In the process of developing methods for making cellulose the addition of a 1% v/v ethanol, using a 2 µm syringe filter after autoclaving, was added to this procedure. Ethanol has been shown to boost production of cellulose in the literature due to its ability to function as an additional energy source in the hexose monophosphate pathway, reducing the byproduct glycerol (*21*, *22*). Besides the addition of ethanol, mannitol was used as the replacement carbon source, using the same wt% as glucose in the HS medium. Mannitol was shown to boost BC growth above that of glucose and other carbon sources earlier in the fermentation process, which influenced its use in this work (*23*).

Other factors to consider when growing cellulose are the pH and temperature. The *G. xylinus* strain used in this work has been conditioned to work optimally at pH 6.0 in a standard HS media; this strain is vigorous enough, however, to survive at pH ~ 6.0 ± 1.0. The pH 6.0 was chosen ultimately due to the propensity of HS medium spiked with the mannitol carbon source to drop in pH as fermentation progressed and the higher yields of cellulose obtained at pH ranges <6.0 compared to pH ranges >6.0 (*23*). The culture temperature of 30 °C was used as suggested by ATCC, even though cellulose cultures were able to be grown at room temperature, only marginally slower than cultures grown at 30 °C.

The culturing method was also important within the scope of the project based on the needed size and speed for cellulose growth. It is well documented that by varying the surface to volume ratio of the culture you can directly affect the BC yield in culture (*24*). In this study when yield was the primary concern a culture of 450–500 mL would be inoculated inside a 14.5 diameter crystallization dish giving a surface to volume ratio of approximately 0.73–0.66 cm^{-1}. When speed and BC yield were not as important (i.e. thin cellulose films), higher surface to volume ratios could be used, which Hestrin and Schramm had originally used for rapid appearance of the BC pellicle (*20*). With the appropriate culture conditions set, a continuous culture could be prepared for BC growth. Continuous culture works best with mid-range surface to volume ratios of media because with larger volumes there is no major pH changes that occur, but continuous culturing can be accomplished with the lower and higher ratios as well (Figure 1), requiring more oversight.

Figure 1. Static culturing of G. xylinus, to generate bacterial cellulose pellicles can be performed in any arbitrary container. For reference, examples are shown for (from left to right) 50 mL falcon tubes, a one-liter pyrex bottle, and 200 mL tissue culture flasks.

Pellicles were able to be plucked from the continuous culture every 2–7 days using sterilized forceps, and directly after the harvesting of cellulose, fresh medium would be added to the culture in order to "refresh" the culture of used nutrients and buffer. However, there are caveats to this method: (1) overtime, the pH does slowly lower in the culture, so cultures must be restarted after approximately 30–60 days depending on the time between harvesting and refreshing the medium, (2) continuous culture always runs the risk of contamination, which is why multiple continuous cultures are grown alternating at least 2–3 days between harvesting, and (3) the first pellicle of the culture is generally very uneven in its growth and must be removed and discarded.

Once the cellulose was grown and plucked from the medium, the harvested pellicle (Figure 2) was then washed in a 1 wt% solution of NaOH at 95 °C for 1 h to remove cellular debris and culture medium. The 0.1 wt% NaOH treatment was chosen because the low wt% of NaOH and short heating time have been shown to remove cellular debris with no adverse effects on the cellulose structure (*25*). After alkali treatment was finished, the pellicles were transferred from the alkali treatment vessel and washed by soaking in deionized water, changing the water every 2 h until the pellicle's pH was neutral and all residual color was removed from the gel. The pellicle was placed in a large, excess volume of pure ethanol to produce an alcohol containing gel (alcogel) from the water containing gel (hydrogel). Cellulose pellicles were stored in ethanol for the convenience of having BC with co-solvent ready for ionogel preparation, as well as the increased shelf life of alcogels compared to hydrogels due to the reduced ability to grow bacteria and mold in alcohol.

Figure 2. Representative geometries for pellicles harvested from the container types shown in Figure 1 prior to cellulose purification. The brown color originates from the culture medium and from bacteria entrapped in the pellicle.

Preparation of Bacterial Cellulose Ionogels (BCIGs)

To prepare an ionogel from a bacterial cellulose alcogel, the alcogel (Figure 3) was cut into the desired size and shape using a scalpel and cardiac scissors, followed by weighing the alcogel and measuring its thickness. A sampling of pieces from the same alcogel pellicle was dried and weighed to obtain the average mass fraction of cellulose within the alcogel. The pre-cut alcogels were converted to ionogels by first incubating with 0.5 mL of an ethanolic stock of the desired IL (i.e., [bmpy][Tf_2N] or [emim][Tf_2N] at 10 to 200 wt% relative to the ethanol). After a 12 h soaking period, the vial cap was removed to allow for ambient alcohol evaporation.

The negligible vapor pressure of the IL allowed for the evaporation of the co-solvent alcohol without the complete collapse of the cellulose structure. At this point, it was very important to control the ethanol evaporation rate. If too much surface area was exposed or the vial was kept in a cross-breeze, the ethanol would evaporate at a faster rate than the rate of diffusion of the IL into the gel. If the rate of evaporation was too fast, then the gel would begin to collapse and not incorporate the IL completely.

Humidity also played a factor in the gel development process. When humidity was high, gels would start to become opaque due to gel collapse from the IL not sufficiently incorporating into the gel. At the onset of opacity, 0.2 mL of ethanol was added to the vial and the vial was capped in order to arrest and reverse cellulose collapse. When the opacity had vanished from the gel, the vial cap would again be removed and the evaporation process resumed. Once the ethanol had evaporated (~ 2–7 days), the resulting gels were weighed to determine the mass of incorporated IL.

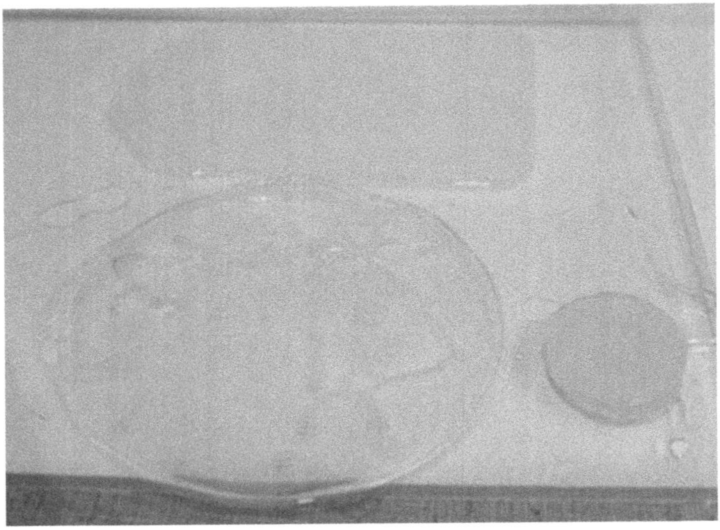

Figure 3. Washed pellicles resulting from purification of those shown in Figure 2 using an NaOH solution at 95 °C for 1 h to remove culture remnants and bacteria. The cleanup method results in pure cellulosic pellicles which are clear and transparent.

Ionogels prepared with BC (BCIGs) were shown to have tunable thickness, ranging in size from sub-millimeter to millimeter scales (Figure 4). The BCIG thickness was tuned in two different ways. The first entailed the bacterial culture itself. The bacteria in the culture grow the pellicle in a downward fashion, allowing for tunable thickness of the obtained gel (*20*). The gels were also able to be tuned in thickness post-culture given the propensity for the cellulose to collapse on itself given various degrees of dehydration (*26*). This affect was able to be used, along with the negligible vapor pressure of the IL, to allow collapsing of the cellulose structure to varying degrees by incorporating incremental amounts of IL. The cross-sectional gel shape can also be tuned by changing the size and shape of the cultures vessels or using molds, as has been previously demonstrated by making tubes and blood vessel replacements from BCs (*27*, *28*).

Natural cellulose iongels are known to not be very amenable to making high-loading ionogels without the need for dissolution and/or chemical modification, but BCIGs are able to produce gels that can hold up to 99 wt% IL (*7*, *12*). Despite having either large or small amounts of IL incorporated, optically transparent or at least mostly transparent ionogels were obtained.

The co-solvent evaporation method used to produce BCIGs made it possible to incorporate other molecules (e.g., reporter dyes) into the BC matrix. Reporter dyes incorporated into the matrix were able to be excited and monitored, due to the transparency and conductivity of the gel, in the presence of various analytes This lends credence to the future use of BCIGs in chemosensory and optical display applications.

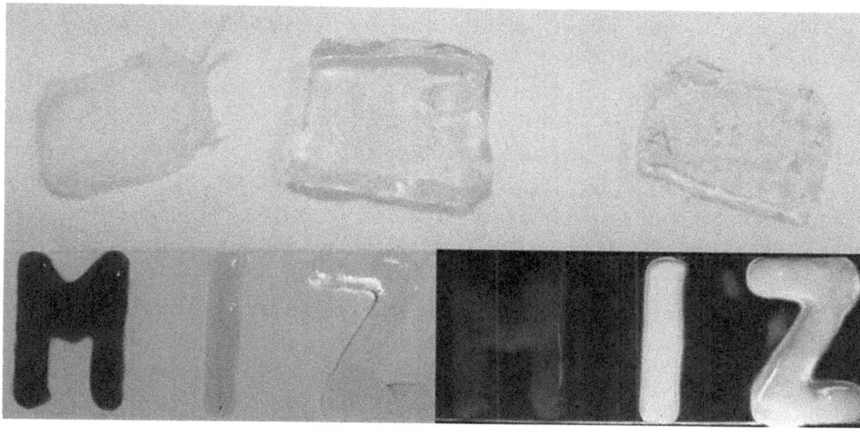

Figure 4. Ionogels prepared using the IL [bmpy][Tf_2N]. Gels can be tuned in thickness by controlling the wt% of IL used and/or by using different initial pellicle thicknesses (top). Ionogels of arbitrary shape are accessible by growing cellulose in various molds (bottom left) and can be doped with various fluorescent dyes (bottom right) to prepare sensory materials.

BCIG Characterization

IL Chemical Structure

No-deuterium (no-D) ^{1}H NMR was used to make direct comparisons between the neat IL and the IL confined within the BCIG. ^{1}H NMR experiments were completed on an Oxford AS600 NMR magnet with a Bruker AVIII HD 600 MHz console using a 5 mm CPTCI cryo-probe. Liquid NMR was able to be used in this case due to the large amount of IL that was incorporated in the BCIG coupled with the flexibility of the BCIG. To prepare the samples, BCIG slices were pushed to the bottom of a 5 mm NMR tube, removing as many air bubbles as possible without smashing the gel. Samples were measured without spinning and in the absence of deuterated solvent for the purposes of measuring the neat liquid and the confinement on the liquid due to the cellulose. When the neat IL was measured the characteristic peaks of the pyrrolidinium and imidazolium ILs were verified with the correct splitting. However, ILs incorporated in BCIGs exhibited peak broadening with the loss of splitting in the peaks, a singularity similar to ILs in silica-based ionogels, which has been shown to be due to confinement of the IL (*29*, *30*).

IL and BC Thermal Properties

To study the thermal properties of BCIGs, differential scanning calorimetry (DSC) and thermogravimetric analysis (TGA) were employed. Thermal studies have been used in previous reports as auxiliary methods for studying confinement within ionogels (*5*). Here we employ these methods in order to study the confinement, thermal stability, and thermodynamic properties of [bmpy][Tf_2N], BC, and the ensuing BCIGs.

Differential scanning calorimetry (DSC) was measured under a nitrogen atmosphere using a TA Instruments model DSC Q100 fitted with a liquid nitrogen cooling system. Samples weighing 8–12 mg were hermetically sealed in aluminum pans and heated to 80 °C followed by cooling to –150 °C and finally heated back to 80 °C, all at a rate of 10 °C min^{-1}. In order to eliminate effects from thermal history, DSC thermograms from the second heating cycle are presented and used for analysis (Figure 5(a)). The crystallization temperature, T_{cr}, and the melting temperature, T_m, were taken at the onset of the respective transitions (*31*). BCIGs with 88–99 wt% [bmpy][Tf_2N] displayed a decreased T_m as large as 6 °C where the loading was >98 wt% while the lower loadings remained close to that of the bulk IL. T_{cr} was shown to increase as the amount of IL decreased in the BCIG by approximately 4 °C. These changes to the thermodynamic properties of the [bmpy][Tf_2N] would suggest that the IL is confined when inside the cellulose structure. This is supported further by the disappearance of the second transition at the T_m in ionogels containing larger amounts of IL, which is present in the bulk IL.

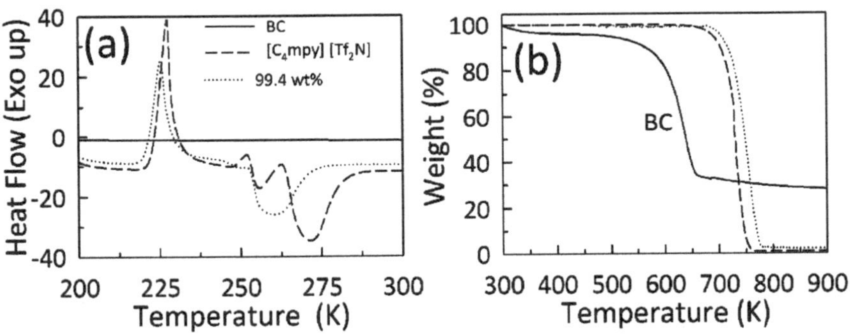

Figure 5. Thermograms collected using (a) DSC as a means to characterize T_{cr} and T_m transitions and (b) TGA to determine thermal stability (T_{dcp}). The thermal transitions were determined as described previously (31).

TGA measurements were performed on BCIGs (containing 88–99 wt% IL) with a Q50 analyzer (TA Instruments, Inc.) ramping from room temperature up to 600 °C at a constant heating rate of 10 °C min^{-1} under nitrogen flow. The decomposition temperature, T_{dcp}, was taken at the onset of decomposition which is defined as the temperature at which 10% mass loss had occurred (*31*). It was seen in the T_{dcp} of the BCIGs that as the IL wt% was increased the T_{dcp} increased as high as 15 °C higher than the bulk IL at 98.8 wt% IL (Figure 5(b)). At the lower loadings T_{dcp} was lower in the BCIG than that of the bulk IL by as much as 13 °C. This large range of temperature difference shows that at the lower loadings the onset of decomposition starts to move towards that of the BC matrix while confinement of the IL lends higher T_{dcp} at higher loadings of IL.

BC Structure

Previous reports characterize BC structurally using powder X-ray diffraction, usually finding at least 3 primary crystallographic peaks ([110], $[1\bar{1}0]$, and [200]) (*32*). By using the peak fitting method, 5 crystallographic peaks and the amorphous peak of BC were able to be elucidated (Figure 6) and the crystallinity index was able to be determined for all samples without the overestimation that occurs using the peak height method (*33*). XRD measurements were performed on a Bruker Prospector instrument with an Apex II CCD detector and an IMuS micro-focus Cu tube, from $2\theta = 5–45°$ with a 0.1° step size using a polyimide capillary sample holder. XRD measurements carried out on a BC aerogel, neat [bmpy][Tf_2N], and BCIGs containing 14–90 wt% [bmpy][Tf_2N], were fit to Voigt functions using PeakFit® version 4.12.

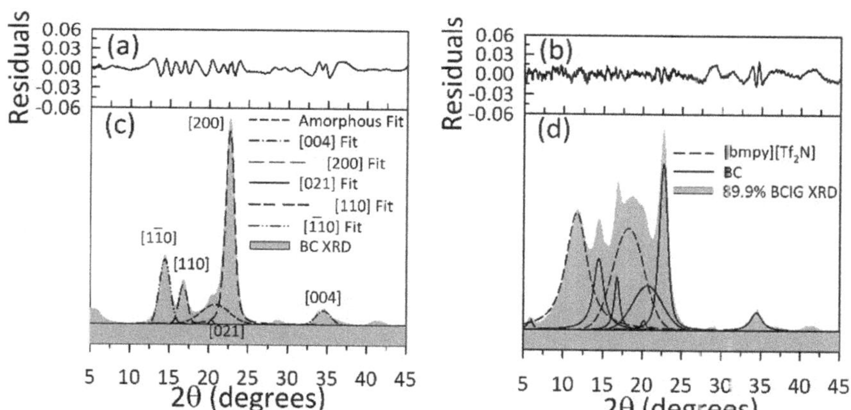

Figure 6. XRD patterns measured for (c) BC and (d) 89.9 wt% BCIG. In peak fitting analysis, a Voigt function was used for all peaks. Cellulose displayed five crystalline peaks ($[1\bar{1}0]$, [110], [200], [021], [004]) and one amorphous peak. Panels (a) and (b) show the corresponding residuals.

The crystallinity index (CrI), crystallite size (τ), and d-spacing (d_{hkl}) were calculated from the fitted data in order to understand what effect the IL has on the cellulose structure (*32–35*). It was observed that the crystallinity of the cellulose is reduced with increasing IL wt%. Similarly, d-spacing in the $[1\bar{1}0]$ plane was shown to decrease with increasing IL wt% while τ in the [110] plane increased. These changes suggest that there is a change in the cellulose fiber that is localized to the $[1\bar{1}0]$ and [110] planes that is caused by the increase or decrease in the IL loading; most likely these changes in the crystallographic planes are due to the increase in the amorphous character (decrease in CrI) which decreases the d-spacing of the $[1\bar{1}0]$ plane pushing together fibers in the [110] plane, increasing τ.

Chemosensory Applications

After characterization was complete, BCIGs were shown, as proof of concept, to be amenable to chemosensory applications (Figure 7). The pH sensitive IL trihexyl(tetradecyl)phosphonium 8-hydroxypyrene-1,3,6-trisulfonate ($[P_{14,6,6,6}]_3$[HPTS]) was incorporated into the BCIG platform by first mixing it in $[P_{14,6,6,6}][Tf_2N]$ and then incorporating the mixture into the gel using the co-solvent evaporation method discussed earlier. When putting together these chemosensory platforms, choice of the appropriate solvating IL had to be considered in order to have reversibility and dye solubility. For example, for this particular application [bmpy]$[Tf_2N]$ was unable to be used to supply the needed reversibility for the chemosensory application, so $[P_{14,6,6,6}][Tf_2N]$ was used instead. The dye infused ionogels performed similarly for NH_3 sensing to the neat IL studied previously (*36*). Upon complete saturation, it was found that the BCIG platform could be subjected to a vacuum for several hours to completely remove the NH_3 from the structure and "reset" the sensor. With reversible NH_3 (g) sensing, it is clearly shown that the BCIG is very amenable to chemosensory function, even in toxic and corrosive chemical environments.

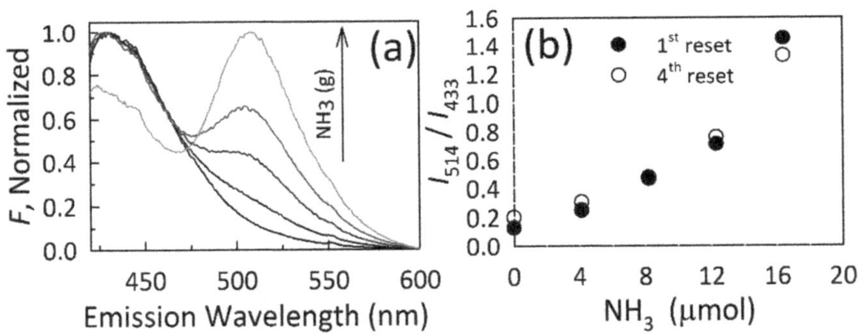

Figure 7. NH_3(g) sensing using ratiometric fluorescence. (a) Fluorescence spectral response to the incremental addition of NH_3 gas and (b) the ratiometric intensity response for the initial and 4th "reset". Following ammonia sensing, the ionogel sensor was "reset" under vacuum to yield similar response over multiple iterations.

Conclusions

For practical applications, it is important to find new ways of incorporating ILs into quasi-solid-state formats whilst maintaining their many attractive fluid properties. Ionogels represent one way of integrating ILs within a solid porous matrix, while maintaining or enhancing the unique IL properties (*5*, *6*). This work takes this effort further by demonstrating the confinement of ~99 wt% of an IL within a biopolymer film, lending high flexibility, transparency, and molecular tunability to its application. The produced bacterial cellulose ionogel (BCIG) maintains the fluidity of the IL inside the cellulose without IL leakage from the BCIG. Despite tremendous amounts of IL within the BCIG, very little crystallinity is lost from the cellulose structure and the characteristic BC web-like structure is fully maintained. With the newly developed ionogel platform, chemosensory function was demonstrated by the marked optical response of BCIGs modified with an analyte-sensitive dye to an interrogated gas stream. Given the molecular, shape, and thickness tunability of BCIGs, these ionogel platforms will undoubtedly be applied as green alternatives in applications such as gas separation/capture, electrochemical assays, (bio)chemical sensing, biomedical films, catalysis, and electrolyte membranes.

Acknowledgments

Financial support from Research Cooperation for Science Advancement to G.A.B. is gratefully acknowledged. C.J.S. was supported by an IGERT trainee fellowship at the Univerisity of Missouri (NSF Grant No. DGE-1069091).

References

1. Néouze, M.-A.; Bideau, J. L.; Gaveau, P.; Bellayer, S.; Vioux, A. Ionogels, New Materials Arising from the Confinement of Ionic Liquids within Silica-Derived Networks. *Chem. Mater.* **2006**, *18*, 3931–3936.
2. Valkenberg, M. H.; Decastro, C.; Holderich, W. F. Immobilisation of Ionic Liquids on Solid Supports. *Green Chem.* **2002**, *4*, 88–93.
3. Mehnert, C. P. Supported Ionic Liquid Catalysis. *Chem. Eur. J.* **2005**, *11*, 50–56.
4. Marr, P. C.; Marr, A. C. Ionic Liquid Gel Materials: Applications in Green and Sustainable Chemistry. *Green Chem.* **2016**, *18*, 105–128.
5. Singh, M. P.; Singh, R. K.; Chandra, S. Ionic Liquids Confined in Porous Matrices: Physicochemical Properties and Applications. *Prog. Mater. Sci.* **2014**, *64*, 73–120.
6. Le Bideau, J.; Viau, L.; Vioux, A. Ionogels, Ionic Liquid Based Hybrid Materials. *Chem. Soc. Rev.* **2011**, *40*, 907–925.
7. Kadokawa, J. I.; Murakami, M. A.; Kaneko, Y. A Facile Preparation of Gel Materials from a Solution of Cellulose in Ionic Liquid. *Carbohydr. Res.* **2008**, *343*, 769–772.
8. Vidinha, P.; Lourenco, N. M.; Pinheiro, C.; Bras, A. R.; Carvalho, T.; Santos-Silva, T.; Mukhopadhyay, A.; Romao, M. J.; Parola, J.; Dionisio, M.; Cabral, J. M.; Afonso, C. A.; Barreiros, S. Ion Jelly: A Tailor-made Conducting Material for Smart Electrochemical Devices. *Chem. Commun.* **2008**, 5842–5844.
9. Trivedi, T. J.; Rao, K. S.; Kumar, A. Facile Preparation of Agarose-chitosan Hybrid Materials and Nanocomposite Ionogels Using an Ionic Liquid Via Dissolution, Regeneration and Sol-gel Transition. *Green Chem.* **2014**, *16*, 320–330.
10. Trivedi, T. J.; Srivastava, D. N.; Rogers, R. D.; Kumar, A. Agarose Processing in Protic and Mixed Protic-aprotic Ionic Liquids: Dissolution,

Regeneration and High Conductivity, High Strength Ionogels. *Green Chem.* **2012**, *14*, 2831–2839.

11. Sharma, A.; Rawat, K.; Solanki, P. R.; Aswal, V. K.; Kohlbrecher, J.; Bohidar, H. B. Internal Structure and Thermo-viscoelastic Properties of Agar Ionogels. *Carbohydr. Polym.* **2015**, *134*, 617–626.
12. Mantravadi, R.; Chinnam, P. R.; Dikin, D. A.; Wunder, S. L. High Conductivity, High Strength Solid Electrolytes Formed by in Situ Encapsulation of Ionic Liquids in Nanofibrillar Methyl Cellulose Networks. *ACS Appl. Mater. Interfaces* **2016**, *8*, 13426–13436.
13. Buchtova, N.; Guyomard-Lack, A.; Le Bideau, J. Biopolymer Based Nanocomposite Ionogels: High Performance, Sustainable and Solid Electrolytes. *Green Chem.* **2014**, *16*, 1149–1152.
14. Jonas, R.; Farah, L. F. Production and Application of Microbial Cellulose. *Polym. Degrad. Stab.* **1998**, *59*, 101–106.
15. Iguchi, M.; Yamanaka, S.; Budhiono, A. Bacterial Cellulose—A Masterpiece of Nature's Arts. *J. Mater. Sci.* **2000**, *35*, 261–270.
16. Huang, Y.; Zhu, C.; Yang, J.; Nie, Y.; Chen, C.; Sun, D. Recent Advances in Bacterial Cellulose. *Cellulose* **2014**, *21*, 1–30.
17. Lee, K.-Y.; Buldum, G.; Mantalaris, A.; Bismarck, A. More Than Meets the Eye in Bacterial Cellulose: Biosynthesis, Bioprocessing, and Applications in Advanced Fiber Composites. *Macromol. Biosci.* **2014**, *14*, 10–32.
18. Qiu, K.; Netravali, A. N. A Review of Fabrication and Applications of Bacterial Cellulose Based Nanocomposites. *Polym. Rev.* **2014**, *54*, 598–626.
19. Shah, N.; Ul-Islam, M.; Khattak, W. A.; Park, J. K. Overview of Bacterial Cellulose Composites: A Multipurpose Advanced Material. *Carbohydr. Polym.* **2013**, *98*, 1585–1598.
20. Hestrin, S.; Schramm, M. Synthesis of Cellulose by *Acetobacter xylinum*. 2. Preparation of Freeze-dried Cells Capable of Polymerizing Glucose to Cellulose. *Biochem. J.* **1954**, *58*, 345–352.
21. Naritomi, T.; Kouda, T.; Yano, H.; Yoshinaga, F. Effect of Ethanol on Bacterial Cellulose Production from Fructose in Continuous Culture. *J. Ferment. Bioeng.* **1998**, *85*, 598–603.
22. Li, Y.; Tian, C.; Tian, H.; Zhang, J.; He, X.; Ping, W.; Lei, H. Improvement of Bacterial Cellulose Production by Manipulating the Metabolic Pathways in which Ethanol and Sodium Citrate Involved. *Appl. Microbiol. Biotechnol.* **2012**, *96*, 1479–1487.
23. Mikkelsen, D.; Flanagan, B. M.; Dykes, G. A.; Gidley, M. J. Influence of Different Carbon Sources on Bacterial Cellulose Production by *Gluconacetobacter xylinus* Strain ATCC 53524. *J. Appl. Microbiol.* **2009**, *107*, 576–583.
24. Krystynowicz, A.; Czaja, W.; Wiktorowska-Jezierska, A.; Gonçalves-Miśkiewicz, M.; Turkiewicz, M.; Bielecki, S. Factors Affecting the Yield and Properties of Bacterial Cellulose. *J. Ind. Microbiol. Biotechnol.* **2002**, *29*, 189–195.
25. George, J.; Ramana, K. V.; Sabapathy, S. N.; Jagannath, J. H.; Bawa, A. S. Characterization of Chemically Treated Bacterial (*Acetobacter xylinum*)

Biopolymer: Some Thermo-mechanical Properties. *Int. J. Biol. Macromol.* **2005**, *37*, 189–194.
26. Clasen, C.; Sultanova, B.; Wilhelms, T.; Heisig, P.; Kulicke, W. M. Effects of Different Drying Processes on the Material Properties of Bacterial Cellulose Membranes. *Macromol. Symp.* **2006**, *244*, 48–58.
27. Putra, A.; Kakugo, A.; Furukawa, H.; Gong, J. P.; Osada, Y. Tubular Bacterial Cellulose Gel with Oriented Fibrils on the Curved Surface. *Polymer* **2008**, *49*, 1885–1891.
28. Klemm, D.; Schumann, D.; Udhardt, U.; Marsch, S. Bacterial Synthesized Cellulose — Artificial Blood Vessels for Microsurgery. *Prog. Polym. Sci.* **2001**, *26*, 1561–1603.
29. Hoye, T. R.; Eklov, B. M.; Ryba, T. D.; Voloshin, M.; Yao, L. J. No-D NMR (No-Deuterium Proton NMR) Spectroscopy: A Simple Yet Powerful Method for Analyzing Reaction and Reagent Solutions. *Org. Lett.* **2004**, *6*, 953–956.
30. Le Bideau, J.; Gaveau, P.; Bellayer, S.; Neouze, M. A.; Vioux, A. Effect of Confinement on Ionic Liquids Dynamics in Monolithic Silica Ionogels: 1H NMR Study. *Phys. Chem. Chem. Phys.* **2007**, *9*, 5419–5422.
31. Jin, H.; O'Hare, B.; Dong, J.; Arzhantsev, S.; Baker, G. A.; Wishart, J. F.; Benesi, A. J.; Maroncelli, M. Physical Properties of Ionic Liquids Consisting of the 1-Butyl-3-Methylimidazolium Cation with Various Anions and the Bis(trifluoromethylsulfonyl)imide Anion with Various Cations. *J. Phys. Chem. B* **2008**, *112*, 81–92.
32. Czaja, W.; Romanovicz, D.; Brown, R. M. Structural Investigations of Microbial Cellulose Produced in Stationary and Agitated Culture. *Cellulose* **2004**, *11*, 403–411.
33. Park, S.; Baker, J. O.; Himmel, M. E.; Parilla, P. A.; Johnson, D. K. Cellulose Crystallinity Index: Measurement Techniques and Their Impact on Interpreting Cellulase Performance. *Biotechnol. Biofuels* **2010**, *3*, 10.
34. French, A.; Santiago Cintrón, M. Cellulose Polymorphy, Crystallite Size, and the Segal Crystallinity Index. *Cellulose* **2013**, *20*, 583–588
35. Poletto, M.; Pistor, V.; Zattera, A. J. Structural Characteristics and Thermal Properties of Native Cellulose. In *Cellulose - Fundamental Aspects*; InTech: Rijeka, 2013; Ch. 2.
36. Yung, K. Y.; Schadock-Hewitt, A. J.; Hunter, N. P.; Bright, F. V.; Baker, G. A. 'Liquid Litmus': Chemosensory pH-responsive Photonic Ionic Liquids. *Chem. Commun.* **2011**, *47*, 4775–4777.

Chapter 7

Liquid–Liquid Extraction of f-Block Elements Using Ionic Liquids

Jérémy Dehaudt,[*,1] Chi-Linh Do-Thanh,[1] Huimin Luo,[2] and Sheng Dai[*,1,3]

[1]Joint Institute for Advanced Materials, Department of Chemistry, The University of Tennessee, Knoxville, Tennessee 37996, United States
[2]Energy and Transportation Science Division, Oak Ridge National Laboratory, Oak Ridge, Tennessee 37831, United States
[3]Chemical Sciences Division, Oak Ridge National Laboratory, Oak Ridge, Tennessee 37831, United States
[*]E-mail: jdehaudt@utk.edu; dais@ornl.gov

F-elements liquid–liquid extraction has been studied for decades in molecular diluents, but ionic liquids have been more recently considered promising as new solvents for liquid–liquid extraction. In this chapter, the properties of ionic liquids for liquid–liquid extraction are investigated. The different extraction mechanisms are also discussed. Then, focus is made on the design and properties of "task-specific ionic liquids." Finally, advanced selective separations are examined in the last paragraph.

Introduction

The f-block of the periodic table comprises the lanthanides and the actinides. Lanthanides have become ubiquitous in our society for a variety of applications. In the meantime, actinides are responsible for the long term radiotoxicity in spent nuclear fuel. Therefore, many efforts have been devoted to the separation of these elements.

Rare earth elements (REEs) consist of the 15 lanthanides plus scandium and yttrium. REEs play a prominent role for numerous applications such as lighting, magnets, wind turbines, or batteries in hybrid cars (*1*). However, REEs have recently been considered to be critical materials by the US Department of Energy (DOE) (*2*) and the European Commission (*3*).

China currently produces over 90% of the global REEs market (*4*). A reduction of its export quota is therefore of great concern and could result in a potential shortage and prompt the rest of the world to find alternative supplies. Rare earths can be divided between light (LREE) and heavy (HREE) rare earths, the latter being the less abundant. Consequently, the REEs industry also has to face the "balance problem" (*5*) since LREE production from mining largely exceeds the demand. Therefore, recycling is now considered an alternative to the supply of REEs from mined ores (*6*). Even though many efforts have been devoted to the recycling of REEs from end-of-life products such as permanent magnets, nickel–metal hydride batteries, and lamp phosphors, less than 1% of REEs are currently recycled (*7*).

Nuclear waste reprocessing is another challenge that implies separation of f-elements (*8*). Long-term radiotoxicity of spent nuclear fuel is mainly due to plutonium and minor actinides (Am, Cm, and Np). Plutonium can be recovered along with uranium using the Purex process and converted into a new fuel. Further separation of minor actinides from the other fission products such as lanthanides could dramatically reduce the amount of long term radiotoxic remaining waste and lead to a better management of disposal in deep underground repository. Transmutation of minor actinides is another ambitious option to manage spent nuclear fuel (*9*). In this process, minor actinides are subjected to a neutron flux leading to the formation of short-living or stable isotopes. However, transmutation requires the separation of minor actinides from lanthanides since the latter are neutron scavengers.

Liquid–liquid extraction (LLE) of f-block elements has been extensively studied for decades using a variety of extractants in different solvents (*10*, *11*). Ionic liquids (ILs) have more recently been regarded as a green alternative to molecular solvents due to their low volatility, low flammability, and thermal stability. Besides these attractive properties, ILs have been proven to be very effective solvents for LLE (*12*).

Properties of Ionic Liquids for f Element Extraction

Generalities

Although ammonium (*13*), phosphonium (*14*), pyrrolidinium (*15*), pyridinium, or piperidinium (*16*) cations have been investigated, alkyl methylimidazolium is the most extensively studied family of ionic liquid for metal extraction. Denoted as $C_n mim^+$, with n indicating the number of carbons in the alkyl chain, the longer this alkyl chain, the more viscous and the more hydrophobic the IL is (*17*). A variety of anions are available to be combined with the cations, offering a large library of ILs. The anion nature has as well consequences on the properties of the IL. Trifluoromethanesulfonimide (also denoted as NTf_2^- or $TFSI^-$) is by far the most employed anion. Other widely used anions in the literature include tetrafluoroborate (BF_4^-) and hexafluorophosphate (PF_6^-). However, the latter has been lately neglected because this anion is hydrolytically unstable and prone to decompose and release HPO_2F_2, H_2PO_3F, H_3PO_4, and HF (*18*). BF_4^- also has a propensity to hydrolysis in aqueous medium (*19*). Bis[(perfluoroethyl)sulfonyl]imide ($BETI^-$) (*20*) (Figure 1) is another fluorinated anion capable of good efficiency due to its high hydrophobicity. Not only can the combination of cations and anions lead to an improvement of the extraction efficiency, but they can also provide a better understanding of the mechanistic aspect of LLE.

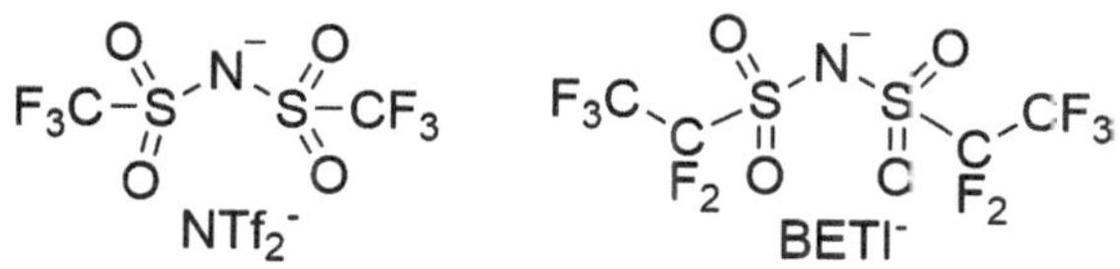

Figure 1. Structures of NTf_2^- and $BETI^-$

Water and IL Mutual Solubility

Although ionic liquids are considered hydrophobic liquids, they exhibit not negligible water solubility. Water–IL mutual solubility is therefore an important issue in LLE, leading to a significant modification of the properties, such as density or viscosity. Furthermore, the aqueous phase can be polluted by the presence of IL. Consequently, it is important to determine and quantify this mutual solubility.

The solubility of water in ILs is mainly governed by hydrogen bonding. Therefore, anion structure has the more important effect on water solubility. Hydrophobicity of the principal anions increase in the following order: $[BF_4]^-$ < $[C(CN)_3]^-$ < $[PF_6]^-$ < $[NTf_2]^-$ (*21*), while halogens can be considered to be hydrophilic anions, and their corresponding ILs are therefore generally water soluble.

Cation structures have, however, an impact on solubility too. First, increasing the alkyl chain length will lead without surprise to a more hydrophobic cation. Also, the nature of the cation has an influence, and the solubility of water in IL can be ranked in the following order: $[C_nmim]^+$ > pyridinium $[Py^+]$ > pyrrolidinium $[Pyr^+]$ > piperidinium $[Pip^+]$, while the IL solubility in water decreases in the following order: $[C_nmim]^+ > [Pyr]^+ > [Py]^+ > [Pip]^+$ (*22*). Aromatic ILs exhibit a higher solvation capability for water, and the solubility of ILs in water seems to be primarily controlled by the cation size, and, to a lower extent, by their aromaticity (*23*).

From a thermodynamic point of view, the solubility of ILs in water is related to the Gibbs free energy of the transfer of the ions constituting the ILs into water (*24*). Therefore, the hydrophobicity of ILs can be determined using ion transfer electrochemistry (*25*). Using this technique, Stockmann *et al.* evaluated the relative hydrophobicity of various phosphonium ionic liquids (*26*).

The presence of inorganic salts such as NaCl, KCl (*27*), or K_3PO_4 (*28, 29*) in the aqueous phase has a strong influence on the solubility of ILs in water. At low concentrations, the IL solubility in water tends to increase (salting-in), while at high concentrations it will decrease (salting-out) and lead to a demixing phenomenon due to a water-structuring effect of the inorganic salt.

Viscosity

Viscosity is another important parameter in liquid–liquid extraction since it has consequences on mass transfer and extraction kinetics. The high viscosity of ILs is one of their major inconveniences compared to molecular solvents for LLE. For example, a viscosity of 52 cP was determined for $[C_4mim][NTf_2]$ (*30*) while dodecane has a viscosity of 1.36 cP (*31*, *32*) at room temperature and atmospheric pressure. Consequently, extraction equilibrium is typically obtained in several hours (*33*) in ILs. The viscosity of ILs decreases with increasing temperature (*34*), therefore, LLE in ILs can be performed at higher temperatures to facilitate mass transfer (*35*). Viscosity in ILs is essentially governed by van der Waals interactions and hydrogen bonding (*30*). The presence of water (*36*, *37*) or molecular solvents (*38*, *39*) significantly reduces the viscosity of ILs. Indeed, due to its high dielectric constant, water is capable of dissociating the ion pairs of ILs. Furthermore, water is prone to form strong hydrogen bonds with anions (*40*). The presence of organic solvents with high dielectric constants leads as well to a significant reduction of electrostatic interactions between ions, thus contributing to decrease viscosity (*41*). Increasing the alkyl chain length of the cation can raise its viscosity. However, alkoxy chains contribute to reduce viscosity because they are less prone to aggregation than aliphatic chains (*42*, *43*). It is also believed that the alkoxy chains increase the free volume due to a greater flexibility (*44*) compared to alkyl chains (*45*). Both cation and anion structures influence viscosity. The viscosity for anions increases in the following order: $[NTf_2]^- < [OTf]^- < [BF_4]^- < [PF_6]^- < [OAc]^-$ (*46*). For the cations, the viscosity increases in the following order: $[Im]^+ < [Py]^+ < [Pyr]^+ < [Pip]^+ < [N1888]^+$ (*47*) holding $[NTf_2]^-$as the counter anion.

Radiolytic Stability

Since actinides are ionizing radiation emitters, radiolytic stability is an important feature to consider for both solvents and extracting molecules. Radiolysis studies have been performed with molecular solvents such as dodecane (*48*). Stability and radiolysis products of extracting molecules have been studied as well (*49*). With the increasing importance of ILs in LLE, radiochemists have become eager to answer the question of stability of these solvents under ionizing radiation. Allen *et al.* (*50*) examined the radiolytic stability of alkylimidazolium chloride and nitrate ILs. They investigated the effect of alpha, beta, and gamma radiations on these ILs (0.4 MGy). They concluded that these ILs were very stable with less than 1% undergoing radiolysis. The radiations led to a darkening of the solution. The radiolytic yield *G*(H2) = 0.72 molecules per 100 eV was determined for $[C_6mim]Cl$, which would be equivalent to *G*(H2) = 1.7 molecules per 100 eV, assuming only the alkyl side chain was subjected to radiolysis. This value is lower than the values of *G*(H2) reported for alkanes (5–6 molecules per 100 eV). The radiolysis of the $[C_nmim]X$ liquids reflects a combination of the properties of a salt, an alkane, and an aromatic. In general, aromatic compounds are more stable than aliphatic chains because the aromatic ring can absorb energy and relax it non-dissociatively.

Le Rouzo *et al.* (*51*) have investigated the stability of [C_4mim]X ILs (X = Tf_2N^-, TfO^-, PF_6^- and BF_4^-) under gamma irradiation for high irradiation doses (up to 2.0 MGy). Using gas chromatography and ESI-MS, they identified the radiolytic products. Gaseous products CO_2, H_2O, COS, SO_2, CO_2, and N_2 were the major gases observed. Fluorinated gases were also observed from the radiolysis of NTf_2^- or TfO^-. Tf_2N^- and TfO^- exhibited approximately the same stability. PF_6^- appeared to have approximately the same stability as compared to BF_4^- which underwent lower radiolytic damage. Cation direct radiolysis occurs by C–H bond dissociations on the imidazolium ring and on the butyl chain and C–N bond dissociations of the methyl and butyl groups. Regarding anions, C–F, N–S, or S–C are the main dissociations. The combination of these radicals or indirect radiolysis leads to the formation of a variety of new species (Scheme 1).

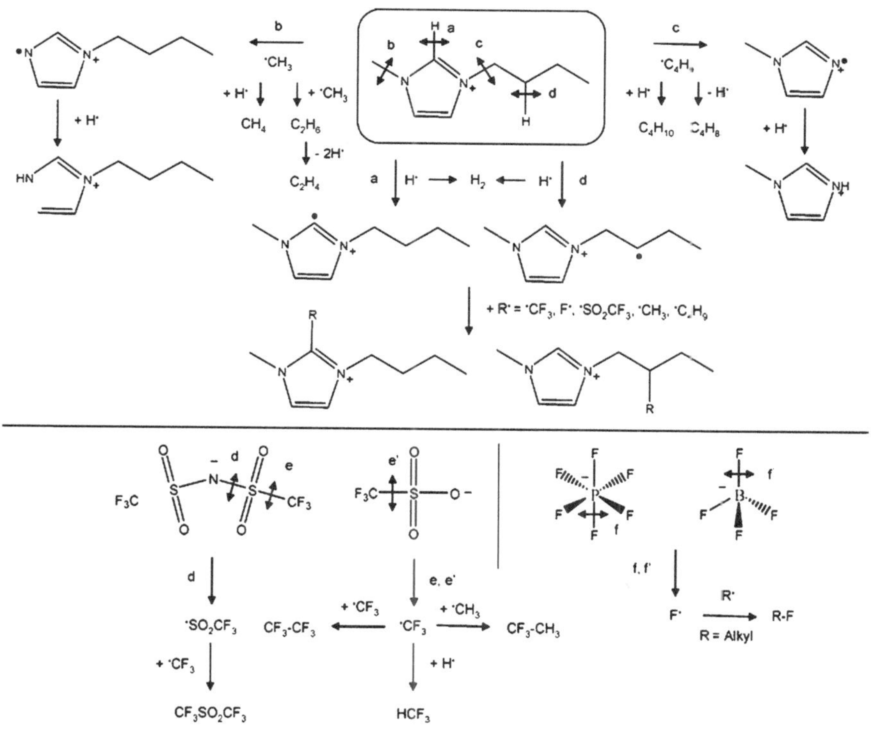

Scheme 1. Cations and anions degradation pathways under gamma radiolysis. (Reproduced with permission from reference (51). Copyright 2009 The Royal Society of Chemistry).

Interestingly, the water content in IL does not seem to have an effect on radiolysis. The stability of an ammonium IL ([$MeBu_3N$][NTf_2]) was also investigated by Bossé *et al.* (*52*). This IL was found to be also very resistant to radiation. A number of different degradation products were observed, essentially by recombination of fluorinated radicals with cations, but in low quantities.

Berthon *et al.* (*53*) investigated the physicochemical changes of imidazolium ILs under radiation (1200 kGy). Densities, surface tensions, and refraction indices remained unchanged but the viscosities increased (which led to a decrease of conductivity). Rao *et al.* (*54*) made the same observation in another study (700 kGy).

The effect of α and γ radiations on [C_4mim][NTf_2] were compared by Ao *et al.* (*55*). The He^+ beam had a smaller effect than γ radiations, which is consistent with what is generally observed in molecular solvents due to recombinations in track by high LET (Linear Energy Transfer) radiations (*56, 57*). The extraction of Dy^{3+} by camphor-bistriazinyl pyridine (CA-BTP) was affected by the dose: over 50 kGy, the extractability decreased due to protonation of the ligand and inhibition of the cation exchange mechanism.

Solvation and Speciation of f-Ions in IL

In water, Ln^{3+} ions are known to form nine-coordinate $[Ln(OH_2)_9]^{3+}$ species for the light lanthanides and eight-coordinate $[Ln(OH_2)_8]^{3+}$ for the heavy lanthanides (*58*). The solvation of Ln(III) with coordinating solvents was also examined by X-ray diffraction (XRD): eight or nine-coordinate complexes were mostly observed with DMF (*59*) or acetonitrile (*60*). Chaumont *et al.* have studied the solvation of lanthanoids using different ILs by computational calculations (*61, 62*). Ln(III) were found to be surrounded by 6 (PF_6^-) to 8 anions ($AlCl_4^-$). This first shell was surrounded by 11 to 13 imidazolium cations, leading to an onion type solvation. The smaller cations are better solvated, but since the solvation involves the same number of anions, the difference from a cation to one another is weaker than in water in which 8 or 9 coordination numbers are observed.

Since ILs are capable of accommodating lanthanoid cations, it is not surprising that significant extraction has been observed using an IL phase without any extracting ligand. Thus, the cations are distributed between the aqueous phase and the IL, depending on their solubility. A striking example is the extraction of Ce^{4+} using pure [C_8mim][PF_6] (*63*). In this report, the authors demonstrated the selective extraction of Ce(IV) over Ln(III) and Th(IV). They explained this selectivity by the formation of an anionic complex $Ce(NO_3)_6^{2-}$. These results highlight the speciation as a major factor governing extraction. In another study, U(VI) was found to be extracted from nitric acid medium by [C_4mim][PF_6]. Even though the extraction was moderate for [HNO_3] = 10^{-2} M (D_U = 0.004 = 0.4% extraction), the extractability was significant at [HNO_3] = 8.0 M (E% = 30%) (*64*). However, extraction of f-ions using a pure IL phase remains scarce, except if the latter is functionalized by an extracting moiety (*i.e.* task-specific ionic liquid) (*65, 66*). As such, specifically designed extracting ligands are required, such as TODGA (*N,N,N',N'*-Tetraoctyl diglycolamide) (*67*), malonamide (*68*) or CMPO (carbamoylmethylphosphine oxide) (*69*) (Figure 2).

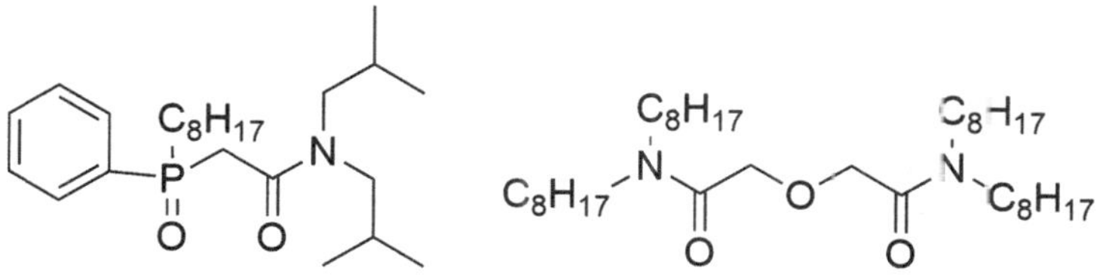

Figure 2. Structures of CMPO and TODGA ligands

These hard donor ligands have been successfully employed in molecular solvents. Different techniques are employed to determine the speciation in liquid–liquid extraction. The slope analysis provides valuable information (*70*) but is limited to the number of extracting molecules involved in the extraction process. However, in order to achieve a comprehensive study of the extraction mechanisms, it is important to determine the nature of the extracted species. X-Ray diffraction solid state structures give invaluable insights on the formed complexes, but the structure in solution can differ. *In situ* techniques have been developed to analyze the structure of the coordination spheres of complexes in liquid–liquid extraction. Extended X-ray absorption fine structure (EXAFS) spectroscopy has been intensively used for this purpose (*71*, *72*). The coordination of uranyl by CMPO was investigated in IL and dodecane using this technique (*73*). In a nitric acid/dodecane system, the formation of an hexagonal bipyramidal complex $UO_2(NO_3)_2(CMPO)_2$ was observed, with two coordinated bidentate nitrate anions and two monodentate CMPO (*via* phosphoryl groups) molecules coordinated equatorially. In $[C_4mim][PF_6]$ and $[C_4mim][NTf_2]$ only a single equatorial U–O was observed with an average coordination number of about 4.5 oxygen atoms. Thanks to the use of EXAFS and slope analysis, the authors determined the formation of $UO_2(NO_3)(CMPO)^+$ in ILs. The coordination of uranyl by tributylphosphate (TBP) was also studied using EXAFS (*74*). The formation of $[UO_2(TBP)_2]^{2+}$ at low HNO_3 concentration and $[UO_2(NO_3)(HNO_3)(TBP)_2]^+$ at high acidity were determined. Consequently, the authors were able to propose an extraction mechanism on the whole acidic range.

Mechanisms of Extraction

ILs are capable of solubilizing neutral or charged complexes, unlike molecular solvents in which the extracted species are neutral or have to form ion pairs in order to be neutral. This versatility can explain the high efficiency of these systems in metal extraction. Indeed, a dramatic increase of extraction efficiency is often observed in IL in comparison with molecular solvents using the same ligand.

However, if charged species can be extracted in the IL phase, they have to be counterbalanced in the aqueous phase to respect electroneutrality. Thus, diverse mechanisms can be encountered in ILs when only neutral extraction is observed in molecular solvents.

Cation Exchange

Cation exchange is a common phenomenon in IL-based LLE. In this mechanism, a cationic complex is extracted from the aqueous phase into the IL. Consequently, a cation from the IL phase has to be transferred into the aqueous phase. Cation exchange is well-studied and documented in ionic liquid. This mechanism can lead to a great enhancement of extraction. This cation exchange is typical in ILs and has been observed using a large array of extractants such as CMPO (*73*) or TODGA (*75, 76*).

Shimojo *et al.* (*75*) compared the extraction of lanthanoids in ILs to an isooctane system. As the nitric acid concentration in the aqueous phase increased, the extraction of lanthanide was improved. This behavior is typical in a neutral solvent: anions are required to compensate for the charge of the target cation and thus extract a neutral complex or an ion pair. Therefore, the concentration of nitrate in the aqueous phase becomes a driving force for the extraction. According to slope analysis measurements, TODGA forms 1:3 or 1:4 complexes depending on the extracted cation. The neutral extraction equilibrium can be written as follows:

Ln^{3+}_{aq} + (3 or 4) $TODGA_{org}$ + 3 $NO_3^-{}_{aq}$ ↔ $Ln(TODGA)_{(3\ OR\ 4)}(NO_3)_{3\ org}$ (Equation 1)

In $[C_2mim][NTf_2]$ an opposite trend was observed, and the distribution values were reduced as $[HNO_3]$ increased. Therefore, nitrate ions do not assist in the extraction of lanthanides. However, the extraction is not independent of the nitric acid concentration. The extraction decrease is attributed by the authors to the nitric acid extraction by the IL phase. Indeed, it has been demonstrated that nitric acid uptake in imidazolium IL was far from negligible (*77*). Billard *et al.* (*78*) determined that starting from an initial concentration of 7.4 mol·L^{-1}, the final nitric acid concentration of the aqueous phase was only at 5.4 mol·L^{-1} after equilibration with $[C_4mim][NTf_2]$. Consequently, the authors suggested a cation exchange mechanism, as observed in many reports dealing with extraction in IL. In this particular case, the slope analysis revealed the formation of 1:3 complexes for the three studied cations, ranging from light to heavy lanthanides (La, Eu and Lu). The equilibrium equation can be written as follows:

$Ln^{3+}_{aq} + 3\ TODGA_{IL} + 3\ C_2mim^+_{IL} \leftrightarrow Ln(TODGA)_3{}^{3+}_{IL} + 3\ C_2mim^+_{aq}$ (Equation 2)

This mechanism was further confirmed by adding imidazolium cations to the aqueous phase using water soluble $[C_2mim]Br$ salt. When the concentration of the imidazolium cations increased in the aqueous phase, the extractability of Eu^{3+} decreased in a linear manner which is in agreement with Equation 2 according to Le Châtelier's principle. Moreover, the C_2mim^+ cation was substituted by C_4mim^+, C_6mim^+, and C_8mim^+ in the IL phase (*76*). As the alkyl chain length increased, the extraction was reduced due to the hydrophobicity of the imidazolium cation, which impeded its transfer into the aqueous phase (Figure 3). This methodology has been utilized in many reports to validate a cation exchange mechanism.

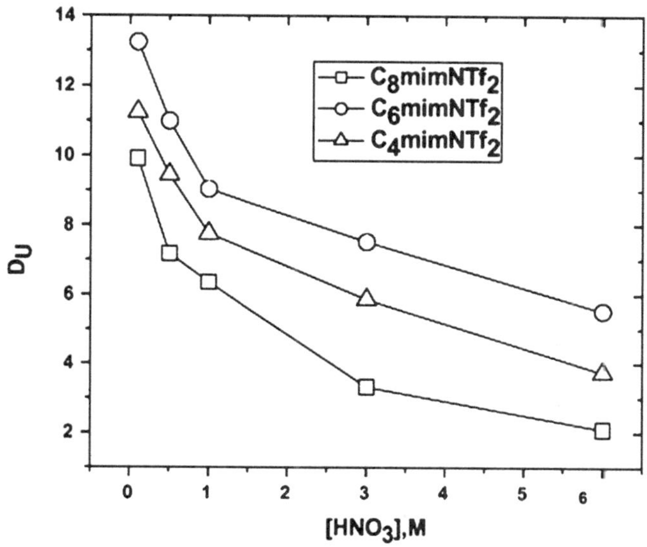

Figure 3. Extraction of U(VI) by TODGA in [$C_n$mim][NTf_2] (n = 4, 6, 8) at different feed nitric acid concentrations. (Reproduced with permission from reference (76). Copyright 2012 Elsevier).

Cation exchange is however often considered an issue since it changes the nature of the IL phase but also leads to the contamination of the aqueous layer. Hydrophobic cations can mitigate this exchange while hydrophilic anions increase this phenomenon (*79*, *80*). However, in most cases, long chain imidazolium cations will not totally suppress ion exchange, and they are associated with toxicity (*81*, *82*) and high viscosity (*17*). Using a sacrificial hydrophilic cation in the IL phase is another strategy to alleviate this ion exchange problem. This strategy was successfully employed for the extraction of Cs by calixarene crown ethers using $NaBPh_4$ as a sacrificial ion exchanger (*83*). The addition of $NaBPh_4$ decreased the loss of [C_4mim][NTf_2] by about 24%, and UV spectra of aqueous phase confirm the substitution of C_4mim^+ by Na^+.

Anion Exchange

While cation exchange is the most observed ion exchange observed in IL-based LLE, some examples of anion exchange mechanisms have been elucidated. The most studied example is the extraction of cations using 2-thenoyltrifluoroacetone (Htta) (*84*). Such ligands have been examined for the extraction of lanthanoids. The neutral complexes $Ln(tta)_3(H_2O)_x$ and $Ln(tta)_3$(Htta) are usually formed in nonpolar molecular organic solvents. Jensen *et al.* have examined the mechanism of extraction of Nd^{3+} and Eu^{3+} by using Htta in [C_4mim][NTf_2] using equilibrium thermodynamics, optical spectroscopies, EXAFS, and molecular dynamics calculations. Slope analysis revealed a 1:4 coordination with lanthanoids and the release of 4 protons in the aqueous phase. A linear increase of NTf_2^- concentration was observed in the aqueous phase as a function of europium extraction. Taking these results into account, the authors proposed the following mechanism:

Ln^{3+} + 4 $Htta_{IL}$ + $[C_4mim^+][NTf_2^-]_{IL}$ ↔ $[C_4mim^+][Ln(tta)_4^-]_{IL}$ + 4 H^+_{aq} + $NTf_2^-{}_{aq}$ (Equation 3)

Though C_4mim^+ can be regarded in this equation as a spectator cation, it is believed by the author that the extracted $Ln(tta)_4^-$ forms a weak ionic pair, becoming part of the ionic liquid without altering the IL phase.

Few other systems involving anion exchange in ILs have been studied to date. For example, the extraction of Pu(IV) and U(VI) were examined using a phosphonate task-specific ionic liquid (*85*) or, more recently, tertiary amines (*86*).

Neutral Extraction

Neutral extraction is the mechanism observed in molecular solvents. This mechanism can also be observed in ILs even though an identical behavior compared to molecular solvents is very scarce. The extraction of UO_2^{2+} by tributylphosphate (TBP) was investigated in ILs by mass spectrometry (*87*). The results show that the formation of $[UO_2(TBP)_2NO_3]^+$ cation is more likely to occur in $[C_4mim][PF_6]$ but using the more hydrophobic cation C_8mim^+, both $[UO_2(TBP)_2NO_3]^+$ and $UO_2(TBP)_2(NO_3)_2$ complexes can be observed. Cocalia *et al.* have studied the extraction of uranyl and trivalent f-ions by dialkylphosphoric or dialkylphosphinic acids in $[C_{10}mim][NTf_2]$ and dodecane systems (*88*). UV–visible and EXAFS measurements were used to demonstrate that the same species were formed in IL and dodecane. In addition, the distribution ratios were almost the same in both media: this confirms that UO_2^{2+} is extracted *via* the same extraction mechanism in both solvents. On the other hand, one could question the benefit of using an expensive and viscous IL if it does not improve the extraction efficiency. Of course, ion exchanges result in the pollution of the aqueous phase and modification of the IL phase, but they are often also responsible of the dramatic enhancement of the extractability.

Mixed Mechanisms

The nature of the IL phase, in particular its hydrophobicity, has consequences on the extraction mechanism. The modification of the aqueous phase is also significant and can lead to different mechanisms. Dietz and Stepinski (*89*) examined the extraction of UO_2^{2+} by TBP in ILs. As the nitric acid concentration increased, two distinct domains could be observed. At low acidity, a decrease in the distribution ratios was observed, indicating a cation exchange, followed by a steep increase at higher acidity, indicating a switchover of the mechanism to neutral extraction according to the authors. The limit between the two domains is accentuated by the utilization of less hydrophobic ILs which favor the cation exchange (Figure 4).

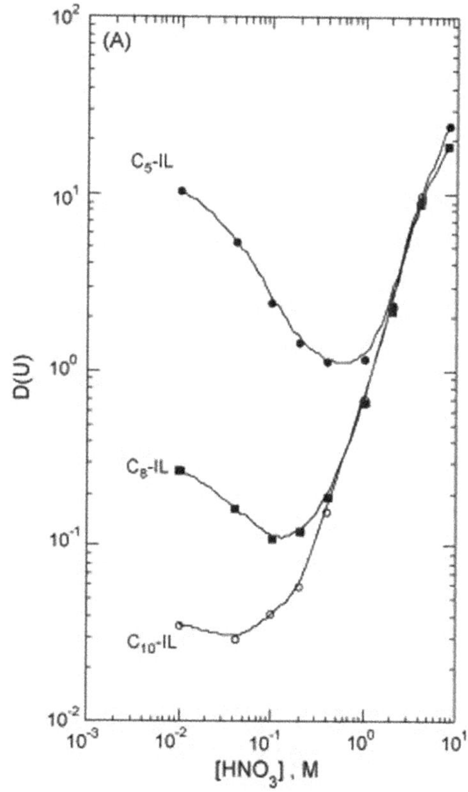

Figure 4. Extraction of UO_2^{2+} as a function of HNO_3 concentration at constant (1.2 M) TBP concentration in several $[C_nmim][NTf_2]$ ILs. (Reproduced with permission from reference (89). Copyright 2008 Elsevier).

Billard *et al.* (*78*) made the same experimental observations but arrived at another mechanism involving anion exchange at higher acidity. According to UV-vis experiments, the authors determined the formation of $UO_2(NO_3)_3(TBP)_m^-$ species, excluding neutral extraction. Such mechanism switchovers have been demonstrated by the same group using malonamide ligands (*90*). They proposed that at low nitric acid concentration, UO_2^{2+} was extracted by a cation exchange between UO_2^{2+} and 2 protons, while it would occur at high HNO_3 concentration by an anion exchange between $UO_2(NO_3)_3^-$ and a NTf_2^- anion. However, in a more recent study supported by EXAFS and UV-vis measurements, they observed the formation of a neutral complex at high acidity (*91*). These discrepancies highlight the difficulty to elucidate the extraction mechanisms, especially when varying parameters that can affect both phases.

Task-Specific Ionic Liquids

Task-specific ionic liquids (TSILs), also called functionalized ionic liquids (FILs), are ILs covalently tethered to specifically tailored functional groups. TSILs have been designed for many applications (*92*) such as catalysis (*93*), organic synthesis (*94*), CO_2 adsorption (*95*), or luminescent materials (*96*). TSILs have also been designed for metal extraction (*16, 97*) and especially for trivalent f-ions extraction. This class of ligands was expected to promote better extraction efficiency but also increase their solubility in ILs.

Cationic TSILs

Cationic ligands are the most prevalent type of TSILs. They are generally synthesized by appending an imidazolium, ammonium (*98*), or pyridinium (*99*) cation to the structure of a reference ligand. This class of ligands has been also investigated for f-ions complexation (*98, 100, 101*). Ouadi *et al.* (*66*) synthesized a 2-hydroxybenzylamine TSIL extractant by reacting salicylaldehyde and *N*-(3-aminopropyl)imidazole. The latter reagent has been extensively used as a synthon for the design of TSILs. The imidazolium moiety of the resulting imine was quaternized using butyl bromide. The ionic liquid imine was reduced to an amine using sodium borohydride. Finally, the trifluoromethanesulfonylimide and the hexafluorophosphate salts were obtained by metathesis from $LiNTf_2$ and KPF_6, respectively (Scheme 2). The NTf_2^- salt had a sufficiently low viscosity (2070 cP) to be used as a pure IL phase without dilution. However, The PF_6^- salt was too viscous (257 000 cP) and had to be diluted but exhibited good solubility in imidazolium IL. The two TSILs have been examined along with their neutral analog for the extraction of Am^{3+}. The TSILs extractants were more efficient than a neutral complexant dissolved in IL.

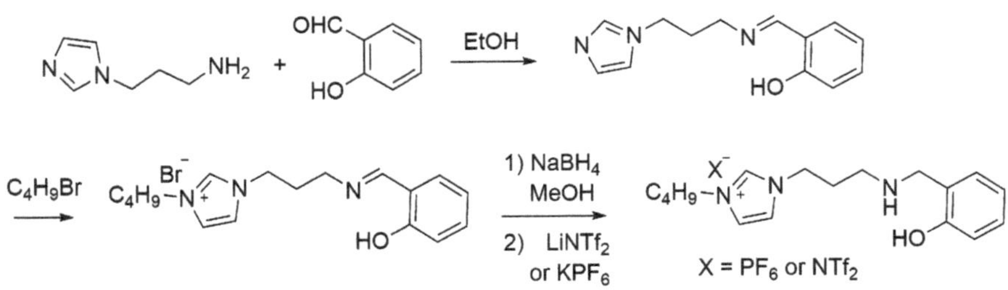

Scheme 2. Synthesis of 2-hydroxybenzylamine TSIL

However, TSILs do not always lead to an improvement of extraction. A similar study was carried out with CMPO TSILs ligands (*102*). In this case, the PF_6^- salt gave a notably poor efficiency, and only the NTf_2^- salt was assessed (Figure 5). Surprisingly, even if the extractability of various f-ions was better compared to CMPO in dodecane, the distribution ratios were systematically inferior compared to the neutral CMPO in IL. This behavior could be rationally explained by the coulombic repulsion between the target cation and the imidazolium moiety (*103*). However, this hypothesis is in contradiction with the study on 2-hydroxybenzylamine derivatives. Furthermore, in another study, malonamide derivatives were examined for the extraction of UO_2^{2+} (*90*) (Figure 5). A dramatic increase of D_{UO2^+} was observed with the TSIL ligand compared to its neutral analog. Out of the extraction efficiency, the authors observed a similar behavior for both ligands and concluded that TSILs do not need to be classified apart from classical molecular extractants when diluted in an IL phase.

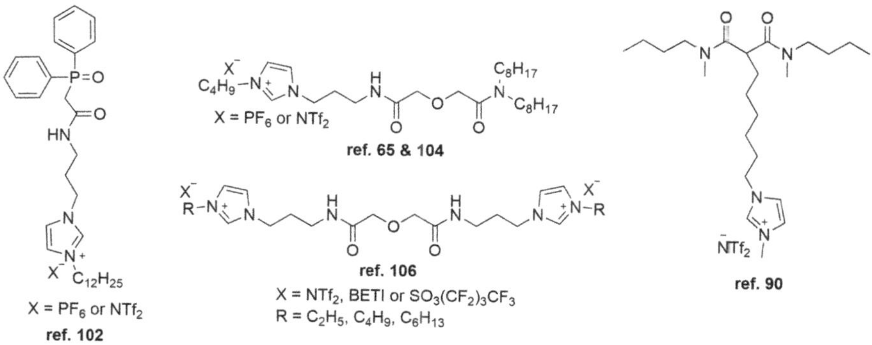

Figure 5. Cationic TSILs

However, using a TSIL as a pure IL phase leads to different observations. Two diglycolamide TSILs (TODGA analogs) have been examined by Mohapatra *et al.* (*65, 104*) and compared to its neutral analog in dodecane and IL phases (Figure 5). Used as a pure IL phase, the *D* values of various f-ions were increased by two orders of magnitude compared to the TODGA/IL system. The kinetics of extraction were, however, slow due to the high viscosity of the TSIL. The extraction profiles were similar to those of the TODGA/[$C_n$mim]X system with a decrease in *D* values with increasing acidity, suggesting a nitric acid uptake in both systems. In terms of speciation, in TODGA/[$C_n$mim]X, 1:3 or 1:4 complexes were determined (*75*) while in dodecane, the extraction mechanism of TODGA involves the formation of reverse micelles and the formation of 1:4 complexes (*105*). For the pure TSIL phase, the slope analysis revealed the formation of 1:2 complexes, including an additional nitrate ligand. The TSIL pure phase has therefore a specific behavior. Efficient stripping of the extracted cations and superior radiolytic stability compared to the dodecane and TODGA/[$C_n$mim]X systems are additional interesting features of the pure TSIL phase.

Dicationic diglycolamide TSIL ligands have also been synthesized and investigated for the extraction of lanthanides in imidazolium-based ILs (*106*) (Figure 5) . Unfortunately, no extraction was possible with these kinds of ligands since the resulting complexes were soluble in the aqueous phase.

Anionic TSILs

Even though fewer anionic TSILs have been designed for cation extraction, it seems more intuitive to use negatively charged ligands to interact with cationic species. Some anionic TSILs have been studied. They show strong pH dependence, since their design is generally based on the deprotonation of a neutral extractant.

A 1,3-diketonate Nd complex has been prepared by Mehdi *et al.* (*107*) by mixing 1-butyl-3-methylimidazolium hexafluoroacetylacetonate [C_4mim][hfac] (Figure 6), with a $Nd(NTf_2)_3$ aqueous solution. A crystal structure of [C_4mim][$Nd(hfac)_4$] was obtained, and the following mechanism was proposed:

$$Nd(Tf_2N)_{3,aq} + 4\ [C_4mim][hfac]_{IL} \leftrightarrow [C_4mim][Nd(hfac)_4]_{IL} + 3\ [C_4mim][NTf_2]_{IL} \quad \text{(Equation 4)}$$

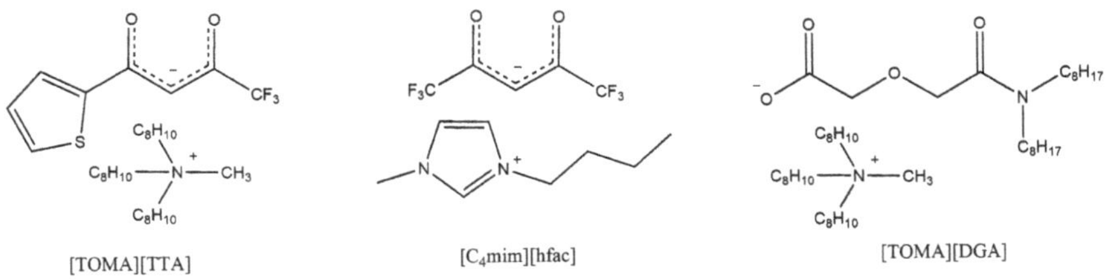

Figure 6. Anionic TSILs

An interesting feature of this equilibrium is that every constituent of the reaction is transferred to the IL phase.

Another diketonate derivative, tri-*n*-octylmethylammonium thenoyltrifluoroacetonate ([TOMA][TTA]) (Fig. 6) has been examined for the extraction of Pu(IV) using xylene as the organic phase (*108*). While the neutral extractant 2-thenoyltrifluoroacetone (HTTA) is more efficient at slightly alkaline pH, Pu(IV) extraction was enhanced at higher nitric acid concentration. This was attributed to the formation of neutral extractant 2-thenoyltrifluoroacetone (and tri-*n*-octylmethylammonium nitrate = [TOMA][NO_3]) in the organic phase driven by the concentration of nitrate in the aqueous phase:

$$[TOMA][TTA]_{org} + HNO_3 \leftrightarrow [TOMA][NO_3]_{org} + HTTA_{org} \quad \text{(Equation 5)}$$

This TSIL extractant is therefore particularly interesting since LLE from nuclear waste is performed from highly acidic feeds.

Trioctylmethylammonium dioctyl diglycolamate [TOMA][DGA] (Figure 6), was examined and compared to the neutral analog DGAH (*109*). The TSIL exhibited better extractability of Nd^{3+} in trioctylmethylammonium nitrate compared to DGAH at pH > 2. This behavior has been attributed to a synergistic effect between the ion pair of the IL and the formation of more stable and hydrophobic complexes (*110*). At pH < 2, the extraction was about the same for both neutral and TSIL extractants. The following equilibrium can explain this phenomenon:

$$[TOMA][DGA]_{IL} + HNO_3 \leftrightarrow TOMA_{IL} + HDGA_{IL} \quad \text{(Equation 6)}$$

At lower pH, the DGA^- anion is protonated to give HDGA, and both systems behave in the same manner. In addition, TSIL dialkylphosphonate extractants have been studied, and strong pH dependence was also observed (*111*).

Advanced Separations

TALSPEAK Process

Developed at Oak Ridge National Laboratory in the 1960s, The TALSPEAK process (Trivalent Actinide Lanthanide Separation by Phosphorus Reagent Extraction from Aqueous Komplex) (*112–114*) was initially used for the separation of lanthanides from actinides. This system is constituted by DTPA (diethylenetriamine-*N,N,N',N'',N''*-pentaacetic acid), a holdback reagent which selectively retains actinides in a buffered aqueous phase and HDEHP (di(2-ethylhexyl)phosphoric acid) as the extracting molecule typically solubilized in diisopropylbenzene (DIPB) or dodecane (Figure 7).

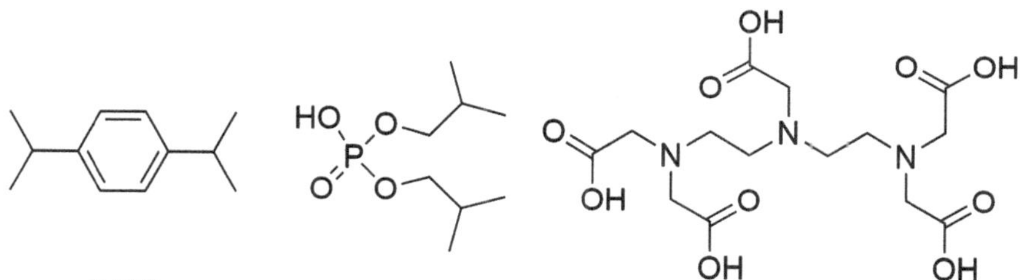

Figure 7. Structures of the reagents used in the TALSPEAK process

Besides lanthanide/actinide separation, this process has been found useful for the separation of the lanthanides series. Indeed, parabolic dependence of the distribution ratios is often observed as a function of the ionic radii of the trivalent lanthanide cations (*115*).

The selectivity along the lanthanide series has as well been studied in ILs. In a first report, the DIPB was replaced by different imidazolium and pyrrolidinium ILs with more or less long alkyl chains (*116*). Compared to DIPB, the extractability was much higher in ILs. The nature of the buffer had a strong influence on the selectivity: while citric acid promoted extraction of heavy lanthanides, the selectivity was enhanced for light lanthanides using glycolic acid. The extraction was more effective using less hydrophobic ILs, demonstrating a typical cation exchange mechanism (Figure 8).

Though promising, HDEHP was, however, found to exhibit low solubility in ILs. Therefore, different anionic TSILs were prepared by deprotonation of HDEHP, forming an ion pair (*117*). The idea was to increase the solubility according to the "like dissolves like" principle. Three different ammonium and phosphonium TSILs were prepared and compared to HDEHP as extractants. Both extractability and selectivity were increased with the TSILs ligands. The mechanism of extraction was therefore investigated (*118*). Surprisingly, while the extraction of REEs by HDEHP followed a cation exchange mechanism, in the case of TSILs, the effect of the alkyl chain or the anion structure of the IL diluents did not have much of an effect on the extraction or selectivity. The ion exchange was therefore suppressed by the use of these TSILs ligands, which can explain the differences in terms of selectivity (Figure 9). If the cation structure of the TSIL did not have a strong influence, then the anion structure is the determining factor since it is in this case the extracting part of the molecule. The alkyl chain length of ammonium DEHP-based TSILs derivatives had for example only a slight effect on the extraction of REEs (*119*). However, replacing the DEHP anion by bis(2,4,4-trimethylpentyl)dithiophosphinite (BTMPP) resulted in a drop in extraction.

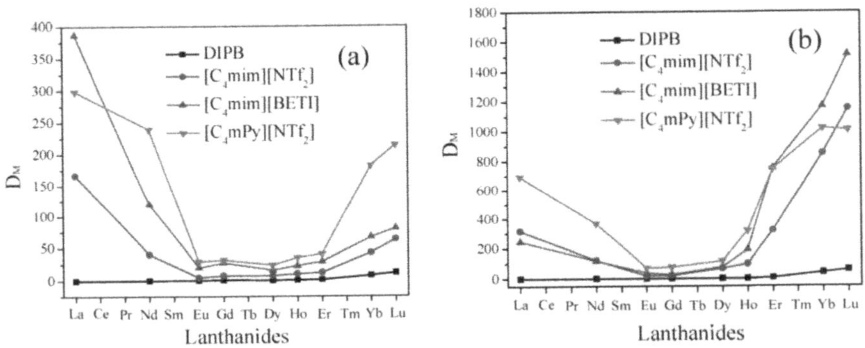

Figure 8. D_M values for lanthanides in different buffered solutions. (a) 50 mM glycolic acid and (b) 50 mM citric acid. (Reproduced with permission from reference (116). Copyright 2011 The Royal Society of Chemistry).

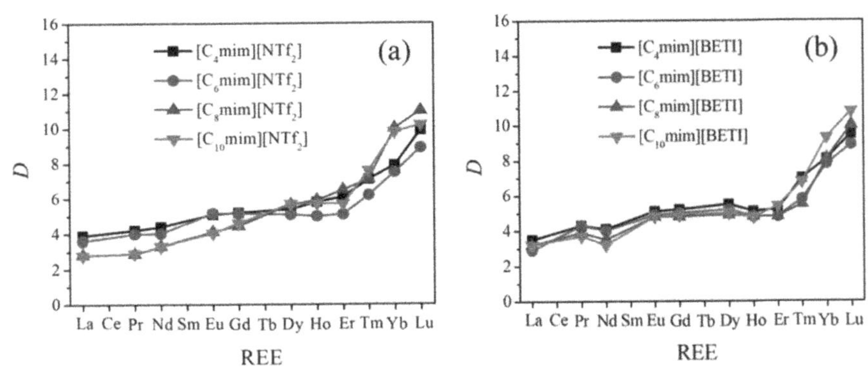

Figure 9. The extraction behaviors of [TOMA][DEHP] in [$C_n$mim][NTf_2] (a) and [$C_n$mim][BETI] (b) (n = 4, 6, 8, 10) for REEs. [TOMA][DEHP] = 0.1 M, REE^{3+} =0.84 mM for each rare earth ion. Adapted with permission from reference (118). Copyright 2013 The Royal Society of Chemistry).

The TALSPEAK process has also been studied in ILs for its initial purpose: the separation of lanthanides from actinides. Rout *et al.* (*120*) replaced the organic solvent with [C_8mim][NTf_2]. In some experimental conditions, the IL promoted higher selectivity for Eu^{3+} over Am^{3+}, especially when the DTPA concentration was kept very low (> 10^{-4}M). As a matter of fact, a $SF_{Eu/Am}$ of 150 can be reached in such an extraction system, but the concentration of the complexants and the pH have to be finely tuned. Shkrob *et al.* (*121*) revisited the TALSPEAK process by integrating *in situ* a DTPA moiety to functionalized ILs that is immiscible with an organic phase containing HDEHP. In this system, a $SF_{Eu/Am}$ up to 270 is reported.

Actinides/Lanthanides Separation

The TALSPEAK process presents several inconveniences. First, there is a strong dependence on the holdback reagent concentration, and pH sensitivity is a major issue. Furthermore, denitrification of the strong acidic spent nuclear fuel is required. Consequently, the opposite approach of a selective extraction of actinides over lanthanides has to be considered.

Soft donor ligands such as polynitrogen ligands have been studied extensively for the extraction of actinides over lanthanides (*122*, *123*). Recently, bis-triazinyl pyridine (R-BTP) (*124*) derivatives have been studied in ILs. D_{Am} > 2000 and $SF_{Am/Eu}$ >3000 were obtained in [C_8mim][NTf_2] ([HNO_3] = 0.1 M). For comparison, in dodecane D_{Am} did not exceed 213 (*125*), and a separation factor $SF_{Am/Eu}$ = 41 (*126*) was determined in another study. It is interesting to notice that less hydrophobic ILs lead to a decrease in extraction and selectivity, suggesting that the extraction does not follow a cation exchange mechanism but an ion-pair extraction. Remarkable speciation differences were also determined. While in molecular solvents 1:1 coordination complexes were reported, 1:3 complexes were extracted here. Consequently, the less hindered Me-BTP led to a better extraction compared to ethyl and isopropyl derivatives. In the case of Me-BTP, the authors determined the formation of an unusual 1:4 stoichiometry with Am^{3+} which could be the reason of this high selectivity.

A separation factor $SF_{Am/Eu}$ > 7800 was obtained using Me-BTphen (2,9-Bis(5,6-dimethyl-1,2,4-triazin-3-yl)-1,10-phenanthroline) in [C_4mim][NTf_2] (*72*). A cationic exchange mechanism was determined with the formation of a nine coordinate $[Eu(Me\text{-}BTPhen)_2(H_2O)]^{3+}$ according to EXAFS measurements associated with DFT calculations. Since BTPhen ligands usually lead to a 1:2 coordination with an additional bidentate nitrate in molecular solvent, the selectivity in IL can be rationally explained by the formation of this nine coordinate complex leading to shorter ion–nitrogen distances and therefore to a stronger binding.

Another approach is to combine soft *N* donors which induce An(III)/Ln(III) selectivity and hard *O* donors which improve the extraction of minor actinides. This strategy was successfully applied to a variety of polynitrogen ligands functionalized by amide moieties (*127–129*). Furthermore, the withdrawing effect of the amide moieties decreases the basicity of the ligand and improves the extraction from highly acidic feeds. 1,10-Phenanthroline-2,9-dicarboxamide complexants decorated with alkyl chains (neutral ligand) or imidazolium cations

(TSILs) have been studied for the selective extraction of actinides over lanthanides in imidazolium ILs ([$C_n$mim][NTf_2], n = 4,6,8) (*130*). Interestingly, the extraction of Am^{3+} was very low with the neutral ligand while $D_{Am^{3+}}$ increased up to two orders of magnitude using TSILs ligands with a selectivity $SF_{Am/Eu} > 50$ at [HNO_3] = 1 M. According to computational calculations, strong H-bonding occurs between the secondary amide hydrogen and the sulfuryl group of NTf_2^- (Figure 10). The insertion of this anion between the imidazolium cation and the phenanthroline ring is believed to stabilize the positive charge of the extracted f-ion and be responsible of such a high extractability.

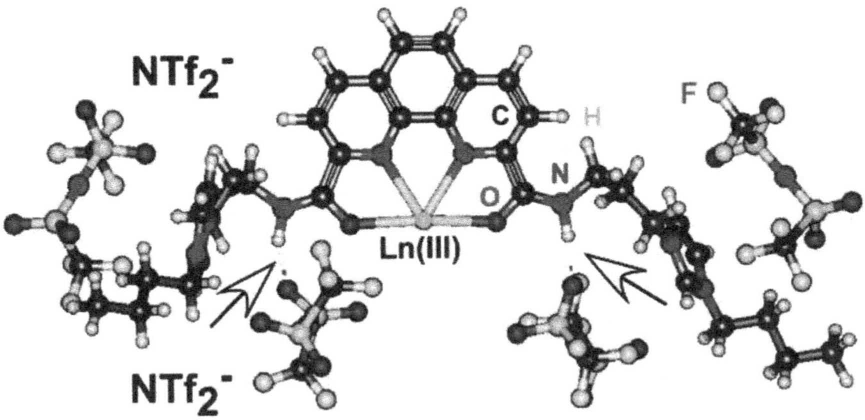

Figure 10. Molecular structure of the TSIL-phenanthroline diamide interacting with four NTf_2^- anions (taken from the optimized geometry gas phase Gd complex). The arrows indicate hydrogen bonds between the amide groups of the ligand and the sulfuryl groups of the IL anions. (Reproduced from reference (130). Copyright 2016 The Royal Society of Chemistry).

Although only three examples of actinide-selective extractions have been described to date in ILs, these results are promising, and other soft donor ligands will likely be studied in the future. Another interesting feature of these ILs systems is that no phase modifier or additive was added.

Conclusion

LLE of f-ions for the recycling of lanthanides and the reprocessing of spent nuclear fuel has been reviewed in this chapter. ILs exhibit different physicochemical properties compared to classic molecular solvents. One of the most interesting features of these solvents is their ability to dissolve ions. Consequently, unlike molecular solvents, which are only able to dissolve neutral complexes or ion pairs, different extraction mechanisms involving ion exchanges can be observed.

A large library of anions and cations is available to tailor ILs with desired properties, such as hydrophobicity, viscosity, or radiation hardness. Indeed, ILs are often considered as designer solvents. TSILs are another class of ILs that can be used as ligands for LLE. A variety of these functionalized ILs have been synthesized and examined as extractants. In some cases, these molecules led to a dramatic improvement in extraction by several orders of magnitude. TSILs can be used as a pure ILs phase and they can also be dissolved in ILs.

ILs have been used in advanced separations such as the TALSPEAK process for the selective separation of minor actinides over lanthanides. In both cases, great improvements have been made compared to molecular solvents. Not only were the distribution ratios dramatically increased, but also, the selectivity was synergistically improved. As such, it is even possible to obtain large selectivity along the lanthanide series. ILs have been proven to be quite effective in LLE on a small scale. However, one might wonder if they can be realistically used on a large, industrial scale. ILs are considered as green solvents, with low flammability and volatility. Additionally, they are very resistant to ionizing radiations. Stripping tests were also carried out in several ILs systems, and the resulting back extraction of cations was very effective (*66, 76, 111*). However, some issues can be pinpointed, and several questions have to be answered. First, are their efficiencies worth their prohibitive cost? The slow kinetics of extraction, and the volume and density variations due to the water solubility can be additional issues. Finally, the ion exchange is also often regarded as a problem of pollution of the aqueous phase that has to be suppressed while it is paradoxically often responsible for the good efficiencies of ILs.

Acknowledgments

This work is supported by the U.S. Department of Energy, Office of Science, Basic Energy Sciences, Chemical Sciences, Geosciences, and Biosciences Division.

References

1. Charalampides, G.; Vatalis, K. I.; Apostoplos, B.; Ploutarch-Nikolas, B. Rare Earth Elements: Industrial Applications and Economic Dependency of Europe. *Procedia Econ. Finance* **2015**, *24*, 126–135.
2. U.S. Department of Energy. *2011 Critical Materials Strategy*; 2011.
3. European Commission. *Report on Critical Raw Materials for the EU - Report of the Ad hoc Working Group on Defining Critical Raw Materials*; 2014.
4. Borra, C. R.; Blanpain, B.; Pontikes, Y.; Binnemans, K.; Van Gerven, T. Recovery of Rare Earths and Other Valuable Metals from Bauxite Residue (Red Mud): A Review. *J. Sustain. Metallurgy* **2016**, *2* (4), 365–386.
5. Binnemans, K.; Jones, P. T. Rare Earths and the Balance Problem. *J. Sustain. Metallurgy* **2015**, *1* (1), 29–38.
6. Binnemans, K.; Jones, P. T.; Blanpain, B.; Van Gerven, T.; Pontikes, Y. Towards Zero-Waste Valorisation of Rare-Earth-Containing Industrial Process Residues: A Critical Review. *J. Clean. Prod.* **2015**, *99*, 17–38.

7. Binnemans, K.; Jones, P. T.; Blanpain, B.; Van Gerven, T.; Yang, Y.; Walton, A.; Buchert, M. Recycling of Rare Earths: A Critical Review. *J. Clean. Prod.* **2013**, *51*, 1–22.
8. Mathur, J. N.; Murali, M. S.; Nash, K. L. Actinide Partitioning - A Review. *Solvent Extr. Ion Exch.* **2001**, *19* (3), 357–390.
9. Poinssot, C.; Rostaing, C.; Baron, P.; Warin, D.; Boullis, B. Main Results of the French Program on Partitioning of Minor Actinides, A Significant Improvement Towards Nuclear Waste Reduction. *Procedia Chem.* **2012**, *7*, 358–366.
10. Reddy, M. L. P.; Prasada Rao, T.; Damodaran, A. D. Liquid-Liquid Extraction Processes for the Separation and Purification of Rare Earths. *Miner. Process. Extr. Metall. Rev.* **1993**, *12* (2–4), 91–113.
11. Paiva, A. P.; Malik, P. Recent Advances on the Chemistry of Solvent Extraction Applied to the Reprocessing of Spent Nuclear Fuels and Radioactive Wastes. *J. Radioanal. Nucl. Chem.* **2004**, *261* (2), 485–496.
12. Sun, X. Q.; Luo, H. M.; Dai, S. Ionic Liquids-Based Extraction: A Promising Strategy for the Advanced Nuclear Fuel Cycle. *Chem. Rev.* **2012**, *112* (4), 2100–2128.
13. Sun, X. Q.; Ji, Y.; Guo, L.; Chen, J.; Li, D. A Novel Ammonium Ionic Liquid Based Extraction Strategy for Separating Scandium from Yttrium and Lanthanides. *Sep. Purif. Technol.* **2011**, *81* (1), 25–30.
14. Kumari, A.; Sinha, M. K.; Sahu, S. K.; Pandey, B. D. Solvent Extraction and Separation of Trivalent Lanthanides Using Cyphos IL 104, a Novel Phosphonium Ionic Liquid as Extractant. *Solvent Extr. Ion Exch.* **2016**, *34* (5), 469–484.
15. Rama, R.; Rout, A.; Venkatesan, K. A.; Antony, M. P.; Vasudeva Rao, P. R. Extraction Behavior of Americium(III) in Benzoylpyrazolone Dissolved in Pyrrolidinium Based Ionic Liquid. *Sep. Sci. Technol.* **2015**, *50* (14), 2164–2169.
16. Papaiconomou, N.; Lee, J.-M.; Salminen, J.; von Stosch, M.; Prausnitz, J. M. Selective Extraction of Copper, Mercury, Silver, and Palladium Ions from Water Using Hydrophobic Ionic Liquids. *Ind. Eng. Chem. Res.* **2008**, *47* (15), 5080–5086.
17. Tokuda, H.; Hayamizu, K.; Ishii, K.; Susan, M. A. B. H.; Watanabe, M. Physicochemical Properties and Structures of Room Temperature Ionic Liquids. 2. Variation of Alkyl Chain Length in Imidazolium Cation. *J. Phys. Chem. B* **2005**, *109* (13), 6103–6110.
18. Sowmiah, S.; Srinivasadesikan, V.; Tseng, M.-C.; Chu, Y.-H. On the Chemical Stabilities of Ionic Liquids. *Molecules* **2009**, *14* (9), 3780.
19. Freire, M. G.; Neves, C. M. S. S.; Marrucho, I. M.; Coutinho, J. A. P.; Fernandes, A. M. Hydrolysis of Tetrafluoroborate and Hexafluorophosphate Counter Ions in Imidazolium-Based Ionic Liquids. *J. Phys. Chem. A* **2010**, *114* (11), 3744–3749.
20. Wankowski, J. L.; Dietz, M. L. Ionic Liquid (IL) Cation and Anion Structural Effects on Metal Ion Extraction into Quaternary Ammonium-based ILs. *Solvent Extr. Ion Exch.* **2016**, *34* (1), 48–59.

21. Freire, M. G.; Santos, L. M. N. B. F.; Fernandes, A. M.; Coutinho, J. A. P.; Marrucho, I. M. An Overview of the Mutual Solubilities of Water–Imidazolium-Based Ionic Liquids Systems. *Fluid Phase Equilibr.* **2007**, *261* (1–2), 449–454.
22. Zhou, T.; Chen, L.; Ye, Y.; Chen, L.; Qi, Z.; Freund, H.; Sundmacher, K. An Overview of Mutual Solubility of Ionic Liquids and Water Predicted by COSMO-RS. *Ind. Eng. Chem. Res.* **2012**, *51* (17), 6256–6264.
23. Freire, M. G.; Neves, C. M. S. S.; Ventura, S. P. M.; Pratas, M. J.; Marrucho, I. M.; Oliveira, J.; Coutinho, J. A. P.; Fernandes, A. M. Solubility of Non-Aromatic Ionic Liquids in Water and Correlation Using a QSPR Approach. *Fluid Phase Equilibr.* **2010**, *294* (1–2), 234–240.
24. Kakiuchi, T. Mutual Solubility of Hydrophobic Ionic Liquids and Water in Liquid-Liquid Two-phase Systems for Analytical Chemistry. *Anal. Sci.* **2008**, *24* (10), 1221–1230.
25. Samek, Z.; Jan Langmaier, J.; Kakiuchi, T. Charge-Transfer Processes at the Interface between Hydrophobic Ionic Liquid and Water. *Pure Appl. Chem.* **2009**, *81* (8), 1473–1488.
26. Stockmann, T. J.; Guterman, R.; Ragogna, P. J.; Ding, Z. Trends in Hydrophilicity/Lipophilicity of Phosphonium Ionic Liquids as Determined by Ion-Transfer Electrochemistry. *Langmuir* **2016**, *32* (49), 12966–12974.
27. Trindade, J. R.; Visak, Z. P.; Blesic, M.; Marrucho, I. M.; Coutinho, J. A. P.; Canongia Lopes, J. N.; Rebelo, L. P. N. Salting-Out Effects in Aqueous Ionic Liquid Solutions: Cloud-Point Temperature Shifts. *J. Phys. Chem. B* **2007**, *111* (18), 4737–4741.
28. Gutowski, K. E.; Broker, G. A.; Willauer, H. D.; Huddleston, J. G.; Swatloski, R. P.; Holbrey, J. D.; Rogers, R. D. Controlling the Aqueous Miscibility of Ionic Liquids: Aqueous Biphasic Systems of Water-Miscible Ionic Liquids and Water-Structuring Salts for Recycle, Metathesis, and Separations. *J. Am. Chem. Soc.* **2003**, *125* (22), 6632–6633.
29. Najdanovic-Visak, V.; Lopes, J.; Visak, Z.; Trindade, J.; Rebelo, L. Salting-Out in Aqueous Solutions of Ionic Liquids and K_3PO_4: Aqueous Biphasic Systems and Salt Precipitation. *Int. J. Mol. Sci.* **2007**, *8* (8), 736.
30. Bonhôte, P.; Dias, A.-P.; Papageorgiou, N.; Kalyanasundaram, K.; Grätzel, M. Hydrophobic, Highly Conductive Ambient-Temperature Molten Salts. *Inorg. Chem.* **1996**, *35* (5), 1168–1178.
31. Liu, Y.; DiFoggio, R.; Sanderlin, K.; Perez, L.; Zhao, J. Measurement of Density and Viscosity of Dodecane and Decane with a Piezoelectric Tuning Fork over 298–448 K and 0.1–137.9 MPa. *Sens. Actuators, A* **2011**, *167* (2), 347–353.
32. Dymond, J. H.; O/ye, H. A. Viscosity of Selected Liquid n-Alkanes. *J. Phys. Chem. Ref. Data* **1994**, *23* (1), 41–53.
33. Xie, Z.-B.; Kang, H.-T.; Chen, Z.-S.; Zhang, S.-H.; Le, Z.-G. Liquid–Liquid Extraction of U(VI) using Malonamide in Room Temperature Ionic Liquid. *J. Radioanal. Nucl. Chem.* **2016**, *308* (2), 573–578.
34. Ghatee, M. H.; Zare, M.; Moosavi, F.; Zolghadr, A. R. Temperature-Dependent Density and Viscosity of the Ionic Liquids 1-Alkyl-3-

methylimidazolium Iodides: Experiment and Molecular Dynamics Simulation. *J. Chem. Eng. Data* **2010**, *55* (9), 3084–3088.

35. Hoogerstraete, T. V.; Onghena, B.; Binnemans, K. Homogeneous Liquid–Liquid Extraction of Metal Ions with a Functionalized Ionic Liquid. *J. Phys. Chem. Lett.* **2013**, *4* (10), 1659–1663.
36. Jacquemin, J.; Husson, P.; Padua, A. A. H.; Majer, V. Density and Viscosity of Several Pure and Water-Saturated Ionic Liquids. *Green Chem.* **2006**, *8* (2), 172–180.
37. Rodríguez, H.; Brennecke, J. F. Temperature and Composition Dependence of the Density and Viscosity of Binary Mixtures of Water + Ionic Liquid. *J. Chem. Eng. Data* **2006**, *51* (6), 2145–2155.
38. Wu, J.-Y.; Chen, Y.-P.; Su, C.-S. Density and Viscosity of Ionic Liquid Binary Mixtures of 1-n-Butyl-3-Methylimidazolium Tetrafluoroborate with Acetonitrile, *N,N*-Dimethylacetamide, Methanol, and *N*-Methyl-2-Pyrrolidone. *J. Solution Chem.* **2015**, *44* (3), 395–412.
39. Gong, Y.-h.; Shen, C.; Lu, Y.-z.; Meng, H.; Li, C.-x. Viscosity and Density Measurements for Six Binary Mixtures of Water (Methanol or Ethanol) with an Ionic Liquid ([BMIM][DMP] or [EMIM][DMP]) at Atmospheric Pressure in the Temperature Range of (293.15 to 333.15) K. *J. Chem. Eng. Data* **2012**, *57* (1), 33–39.
40. Li, W.; Zhang, Z.; Han, B.; Hu, S.; Xie, Y.; Yang, G. Effect of Water and Organic Solvents on the Ionic Dissociation of Ionic Liquids. *J. Phys. Chem. B* **2007**, *111* (23), 6452–6456.
41. Widegren, J. A.; Laesecke, A.; Magee, J. W. The Effect of Dissolved Water on the Viscosities of Hydrophobic Room-Temperature Ionic Liquids. *Chem. Commun.* **2005** (12), 1610–1612.
42. Siqueira, L. J. A.; Ribeiro, M. C. C. Alkoxy Chain Effect on the Viscosity of a Quaternary Ammonium Ionic Liquid: Molecular Dynamics Simulations. *J. Phys. Chem. B* **2009**, *113* (4), 1074–1079.
43. Zhou, Z.-B.; Matsumoto, H.; Tatsumi, K. Low-Melting, Low-Viscous, Hydrophobic Ionic Liquids: 1-Alkyl(Alkyl Ether)-3-Methylimidazolium Perfluoroalkyltrifluoroborate. *Chem. Eur. J.* **2004**, *10* (24), 6581–6591.
44. Holbrey, J. D.; Visser, A. E.; Spear, S. K.; Reichert, W. M.; Swatloski, R. P.; Broker, G. A.; Rogers, R. D. Mercury(II) Partitioning from Aqueous Solutions with a New, Hydrophobic Ethylene-Glycol Functionalized Bis-Imidazolium Ionic Liquid. *Green Chem.* **2003**, *5* (2), 129–135.
45. Chen, Z. J.; Xue, T.; Lee, J.-M. What Causes the Low Viscosity of Ether-Functionalized Ionic Liquids? Its Dependence on the Increase of Free Volume. *RSC Adv.* **2012**, *2* (28), 10564–10574.
46. Yu, G.; Zhao, D.; Wen, L.; Yang, S.; Chen, X. Viscosity of Ionic Liquids: Database, Observation, and Quantitative Structure-Property Relationship Analysis. *AIChE J.* **2012**, *58* (9), 2885–2899.
47. Alcalde, R.; García, G.; Atilhan, M.; Aparicio, S. Systematic Study on the Viscosity of Ionic Liquids: Measurement and Prediction. *Ind. Eng. Chem. Res.* **2015**, *54* (43), 10918–10924.

48. Ikeda, H.; Suzuki, A. Radiolysis of n-Dodecane and Its Physical Property Change Based on the Dose in One Pass through a Reference HA Column. *J. Nucl. Sci. Technol.* **1998**, *35* (10), 697–704.
49. Mincher, B. J.; Modolo, G.; Mezyk, S. P. Review Article: The Effects of Radiation Chemistry on Solvent Extraction 3: A Review of Actinide and Lanthanide Extraction. *Solvent Extr. Ion Exch.* **2009**, *27* (5–6), 579–606.
50. Allen, D.; Baston, G.; Bradley, A. E.; Gorman, T.; Haile, A.; Hamblett, I.; Hatter, J. E.; Healey, M. J. F.; Hodgson, B.; Lewin, R.; Lovell, K. V.; Newton, B.; Pitner, W. R.; Rooney, D. W.; Sanders, D.; Seddon, K. R.; Sims, H. E.; Thied, R. C. An Investigation of the Radiochemical Stability of Ionic Liquids. *Green Chem.* **2002**, *4* (2), 152–158.
51. Le Rouzo, G.; Lamouroux, C.; Dauvois, V.; Dannoux, A.; Legand, S.; Durand, D.; Moisy, P.; Moutiers, G. Anion Effect on Radiochemical Stability of Room-temperature Ionic Liquids under Gamma Irradiation. *Dalton Trans.* **2009** (31), 6175–6184.
52. Bosse, E.; Berthon, L.; Zorz, N.; Monget, J.; Berthon, C.; Bisel, I.; Legand, S.; Moisy, P. Stability of [$MeBu_3N$][Tf_2N] under Gamma Irradiation. *Dalton Trans.* **2008** (7), 924–931.
53. Berthon, L.; Nikitenko, S. I.; Bisel, I.; Berthon, C.; Faucon, M.; Saucerotte, B.; Zorz, N.; Moisy, P. Influence of Gamma Irradiation on Hydrophobic Room-Temperature Ionic Liquids [BuMeIm]PF_6 and [BuMeIm]$(CF_3SO_2)_2N$. *Dalton Trans.* **2006** (21), 2526–2534.
54. Jagadeeswara Rao, C.; Venkatesan, K. A.; Tata, B. V. R.; Nagarajan, K.; Srinivasan, T. G.; Vasudeva Rao, P. R. Radiation Stability of Some Room Temperature Ionic Liquids. *Radiat. Phys. Chem.* **2011**, *80* (5), 643–649.
55. Ao, Y.; Zhou, H.; Yuan, W.; Wang, S.; Peng, J.; Zhai, M.; Wang, J.; Zhao, Z.; Zhao, L.; Wei, Y. α-Radiolysis of Ionic Liquid Irradiated with Helium Ion Beam and the Influence of Radiolytic Products on Dy^{3+} Extraction. *Dalton Trans.* **2014**, *43* (14), 5580–5585.
56. Sugo, Y.; Taguchi, M.; Sasaki, Y.; Hirota, K.; Kimura, T. Radiolysis Study of Actinide Complexing Agent by Irradiation with Helium Ion Beam. *Radiat. Phys. Chem.* **2009**, *78* (12), 1140–1144.
57. Pearson, J.; Jan, O.; Miller, G.; Nilsson, M. A Comparison of Low and High LET (Linear Energy Transfer) Induced Radiolysis of Solvent Extraction Processes. *Procedia Chem.* **2012**, *7*, 334–340.
58. Cotton, S. A.; Harrowfield, J. M., Lanthanides: Solvation. In *Encyclopedia of Inorganic and Bioinorganic Chemistry*; John Wiley & Sons, Ltd.: 2011.
59. Harrowfield, J. M.; Skelton, B. W.; White, A. H.; Wilner, F. R. Lanthanide(III) Solvation: *N,N'*-Dimethylformamide as a Unidentate *O*-Donor. *Inorg. Chim. Acta* **2004**, *357* (8), 2358–2364.
60. B. Deacon, G.; Gortler, B.; C. Junk, P.; Lork, E.; Mews, R.; Petersen, J.; Zemva, B. Syntheses and Structures of Some Homoleptic Acetonitrile Lanthanoid(III) Complexes. *J. Chem. Soc., Dalton Trans.* **1998** (22), 3887–3892.
61. Chaumont, A.; Wipff, G. Solvation of M^{3+} Lanthanide Cations in Room-Temperature Ionic Liquids. A Molecular Dynamics Investigation. *Phys. Chem. Chem. Phys.* **2003**, *5* (16), 3481–3488.

62. Chaumont, A.; Wipff, G. Solvation of Ln(III) Lanthanide Cations in the [BMI][SCN], [$MeBu_3N$][SCN], and [BMI]$_5$[Ln(NCS)$_8$] Ionic Liquids: A Molecular Dynamics Study. *Inorg. Chem.* **2009**, *48* (10), 4277–4289.
63. Zuo, Y.; Liu, Y.; Chen, J.; Li, D. Q. The Separation of Cerium(IV) from Nitric Acid Solutions Containing Thorium(IV) and Lanthanides(III) Using Pure [C_8mim]PF_6 as Extracting Phase. *Ind. Eng. Chem. Res.* **2008**, *47* (7), 2349–2355.
64. Giridhar, P.; Venkatesan, A. K.; Srinivasan, G. T.; Vasudeva Rao, R. P. Extraction of Uranium(VI) from Nitric Acid Medium by 1.1 M Tri-n-Butylphosphate in Ionic Liquid Diluent. *J. Radioanal. Nucl. Chem.* **2005**, *265* (1), 31–38.
65. Mohapatra, P. K.; Sengupta, A.; Iqbal, M.; Huskens, J.; Verboom, W. Highly Efficient Diglycolamide-Based Task-Specific Ionic Liquids: Synthesis, Unusual Extraction Behaviour, Irradiation, and Fluorescence Studies. *Chem. Eur. J.* **2013**, *19* (9), 3230–3238.
66. Ouadi, A.; Gadenne, B.; Hesemann, P.; Moreau, J. J. E.; Billard, I.; Gaillard, C.; Mekki, S.; Moutiers, G. Task-Specific Ionic Liquids Bearing 2-Hydroxybenzylamine Units: Synthesis and Americium-Extraction Studies. *Chem. Eur. J.* **2006**, *12* (11), 3074–3081.
67. Ansari, S. A.; Pathak, P. N.; Manchanda, V. K.; Husain, M.; Prasad, A. K.; Parmar, V. S. *N,N,N',N'*-Tetraoctyl Diglycolamide (TODGA): A Promising Extractant for Actinide-Partitioning from High-Level Waste (HLW). *Solvent Extr. Ion Exch.* **2005**, *23* (4), 463–479.
68. Narita, H.; Yaita, T.; Tamura, K.; Tachimori, S. Study on the Extraction of Trivalent Lanthanide Ions with *N,N'*-Dimethyl-*N,N'*-Diphenyl-Malonamide and Diglycolamide. *J. Radioanal. Nucl. Chem.* **1999**, *239* (2), 381–384.
69. Philip Horwitz, E.; Kalina, D. C.; Diamond, H.; Vandegrift, G. F.; Schulz, W. W. The TRUEX Process - A Process for the Extraction of the Transuranic Elements from Nitric Acid in Wastes Utilizing Modified PUREX. *Solvent Extr. Ion Exch.* **1985**, *3* (1–2), 75–109.
70. Mincher, B. J. A Slope Analysis Investigation of Trivalent f-Series Metal CMPO Complex. *Solvent Extr. Ion Exch.* **1992**, *10* (4), 615–622.
71. Braatz, A. D.; Antonio, M. R.; Nilsson, M. Structural Study of Complexes Formed by Acidic and Neutral Organophosphorus Reagents. *Dalton Trans.* **2017**, *46* (4), 1194–1206.
72. Williams, N. J.; Dehaudt, J.; Bryantsev, V. S.; Luo, H. M.; Abney, C. W.; Dai, S. Selective Separation of Americium from Europium Using 2,9-Bis(triazine)-1,10-Phenanthrolines in Ionic Liquids: A New Twist on an Old Story. *Chem. Commun.* **2017**, *53* (18), 2744–2747.
73. Visser, A. E.; Jensen, M. P.; Laszak, I.; Nash, K. L.; Choppin, G. R.; Rogers, R. D. Uranyl Coordination Environment in Hydrophobic Ionic Liquids: An *in Situ* Investigation. *Inorg. Chem.* **2003**, *42* (7), 2197–2199.
74. Gaillard, C.; Boltoeva, M.; Billard, I.; Georg, S.; Mazan, V.; Ouadi, A.; Ternova, D.; Hennig, C. Insights into the Mechanism of Extraction of Uranium (VI) from Nitric Acid Solution into an Ionic Liquid by using Tri-n-butyl phosphate. *ChemPhysChem* **2015**, *16* (12), 2653–2662.

75. Shimojo, K.; Kurahashi, K.; Naganawa, H. Extraction Behavior of Lanthanides Using a Diglycolamide Derivative TODGA in Ionic Liquids. *Dalton Trans.* **2008** (37), 5083–5088.
76. Panja, S.; Mohapatra, P. K.; Tripathi, S. C.; Gandhi, P. M.; Janardan, P. A highly Efficient Solvent System Containing TODGA in Room Temperature Ionic Liquids for Actinide Extraction. *Sep. Purif. Technol.* **2012**, *96*, 289–295.
77. Fu, J.; Yang, Y. I.; Zhang, J.; Chen, Q.; Shen, X.; Gao, Y. Q. Structural Characteristics of Homogeneous Hydrophobic Ionic Liquid–HNO_3–H_2O Ternary System: Experimental Studies and Molecular Dynamics Simulations. *J. Phys. Chem. B* **2016**, *120* (23), 5194–5202.
78. Billard, I.; Ouadi, A.; Jobin, E.; Champion, J.; Gaillard, C.; Georg, S. Understanding the Extraction Mechanism in Ionic Liquids: UO_2^{2+}/HNO_3/ TBP/C_4mimTf$_2$N as a Case Study. *Solvent Extr. Ion Exch.* **2011**, *29* (4), 577–601.
79. Garvey, S. L.; Dietz, M. L. Ionic Liquid Anion Effects in the Extraction of Metal Ions by Macrocyclic Polyethers. *Sep. Purif. Technol.* **2014**, *123*, 145–152.
80. Luo, H. M.; Dai, S.; Bonnesen, P. V.; Haverlock, T. J.; Moyer, B. A.; Buchanan, A. C. A Striking Effect of Ionic-Liquid Anions in the Extraction of Sr^{2+} and Cs^{+} by Dicyclohexano-18-Crown-6. *Solvent Extr. Ion Exch.* **2006**, *24* (1), 19–31.
81. Ma, J.; Dong, X.; Fang, Q.; Li, X.; Wang, J. Toxicity of Imidazolium-Based Ionic Liquids on Physa Acuta and the Snail Antioxidant Stress Response. *J. Biochem. Mol. Toxicol.* **2014**, *28* (2), 69–75.
82. Romero, A.; Santos, A.; Tojo, J.; Rodríguez, A. Toxicity and Biodegradability of Imidazolium Ionic Liquids. *J. Hazard. Mater.* **2008**, *151* (1), 268–273.
83. Luo, H. M.; Dai, S.; Bonnesen, P. V.; Buchanan, A. C.; Holbrey, J. D.; Bridges, N. J.; Rogers, R. D. Extraction of Cesium Ions from Aqueous Solutions Using Calix[4]arene-bis(tert-octylbenzo-crown-6) in Ionic Liquids. *Anal. Chem.* **2004**, *76* (11), 3078–3083.
84. Jensen, M. P.; Neuefeind, J.; Beitz, J. V.; Skanthakumar, S., Soderholm, L. Mechanisms of Metal Ion Transfer into Room-Temperature Ionic Liquids: The Role of Anion Exchange. *J. Am. Chem. Soc.* **2003**, *125* (50), 15466–15473.
85. Rout, A.; Venkatesan, K. A.; Srinivasan, T. G.; Vasudeva Rao, P. R. Unusual Extraction of Plutonium(IV) from Uranium(VI) and Americium(III) Using Phosphonate Based Task Specific Ionic Liquid. *Radiochim. Acta* **2010**, *98* (8), 459.
86. Ansari, S. A.; Mohapatra, P. K.; Mazan, V.; Billard, I. Extraction of Actinides by Tertiary Amines in Room Temperature Ionic Liquids: Evidence for Anion Exchange as a Major Process at High Acidity and Impact of Acid Nature. *RSC Adv.* **2015**, *5* (45), 35821–35829.
87. Murali, M. S.; Bonville, N.; Choppin, G. R. Uranyl Ion Extraction into Room Temperature Ionic Liquids: Species Determination by ESI and MALDI-MS. *Solvent Extr. Ion Exch.* **2010**, *28* (4), 495–509.

88. Cocalia, V. A.; Jensen, M. P.; Holbrey, J. D.; Spear, S. K.; Stepinski, D. C.; Rogers, R. D. Identical Extraction Behavior and Coordination of Trivalent or Hexavalent f-Element Cations Using Ionic Liquid and Molecular Solvents. *Dalton Trans.* **2005** (11), 1966–1971.
89. Dietz, M. L.; Stepinski, D. C. Anion Concentration-Dependent Partitioning Mechanism in the Extraction of Uranium into Room-Temperature Ionic Liquids. *Talanta* **2008**, *75* (2), 598–603.
90. Bonnaffé-Moity, M.; Ouadi, A.; Mazan, V.; Miroshnichenko, S.; Ternova, D.; Georg, S.; Sypula, M.; Gaillard, C.; Billard, I. Comparison of Uranyl Extraction Mechanisms in an Ionic Liquid by Use of Malonamide or Malonamide-Functionalized Ionic Liquid. *Dalton Trans.* **2012**, *41* (25), 7526–7536.
91. Gaillard, C.; Boltoeva, M.; Billard, I.; Georg, S.; Mazan, V.; Ouadi, A. Ionic Liquid-Based Uranium(VI) Extraction with Malonamide Extractant: Cation Exchange vs. Neutral Extraction. *RSC Adv.* **2016**, *6* (74), 70141–70151.
92. Giernoth, R. Task-Specific Ionic Liquids. *Angew. Chem., Int. Ed.* **2010**, *49* (16), 2834–2839.
93. Rahman, M.; Sarkar, A.; Ghosh, M.; Majee, A.; Hajra, A. Catalytic Application of Task Specific Ionic Liquid on the Synthesis of Benzoquinazolinone Derivatives by a Multicomponent Reaction. *Tetrahedron Lett.* **2014**, *55* (1), 235–239.
94. Sawant, A. D.; Raut, D. G.; Darvatkar, N. B.; Salunkhe, M. M. Recent Developments of Task-Specific Ionic Liquids in Organic Synthesis. *Green Chem. Lett. Rev.* **2011**, *4* (1), 41–54.
95. Bates, E. D.; Mayton, R. D.; Ntai, I.; Davis, J. H. CO_2 Capture by a Task-Specific Ionic Liquid. *J. Am. Chem. Soc.* **2002**, *124* (6), 926–927.
96. Li, Q.-F.; Yue, D.; Ge, G.-W.; Du, X.; Gong, Y.; Wang, Z.; Hao, J. Water-Soluble Tb^{3+} and Eu^{3+} Complexes Based on Task-specific Ionic Liquid Ligands and their Application in Luminescent Poly(vinyl alcohol) Films. *Dalton Trans.* **2015**, *44* (38), 16810–16817.
97. Visser, A. E.; Swatloski, R. P.; Reichert, W. M.; Mayton, R.; Sheff, S.; Wierzbicki, A.; Davis, J. J. H.; Rogers, R. D. Task-Specific Ionic Liquids for the Extraction of Metal Ions from Aqueous Solutions. *Chem. Commun.* **2001** (1), 135–136.
98. Ouadi, A.; Klimchuk, O.; Gaillard, C.; Billard, I. Solvent Extraction of U(VI) by Task Specific Ionic Liquids Bearing Phosphoryl Groups. *Green Chem.* **2007**, *9* (11), 1160–1162.
99. Li, H.; Wang, B.; Liu, S. Synthesis of Pyridine-Based Task-specific Ionic Liquid with Alkyl Phosphate Cation and Extraction Performance for Uranyl Ion. *Ionics* **2015**, *21* (9), 2551–2556.
100. Rout, A.; Venkatesan, K. A.; Srinivasan, T. G.; Vasudeva Rao, P. R. Separation of Plutonium(IV) from Uranium(VI) Using Phosphonate-based Task-Specific Ionic Liquid. *Desalin. Water Treat.* **2012**, *38* (1–3), 179–183.
101. Vicente, J. A.; Mlonka, A.; Gunaratne, H. Q. N.; Swadzba-Kwasny, M.; Nockemann, P. Phosphine Oxide Functionalised Imidazolium Ionic Liquids as Tuneable Ligands for Lanthanide Complexation. *Chem. Commun.* **2012**, *48* (49), 6115–6117.

102. Mohapatra, P. K.; Kandwal, P.; Iqbal, M.; Huskens, J.; Murali, M. S.; Verboom, W. A Novel CMPO-Functionalized Task Specific Ionic Liquid: Synthesis, Extraction and Spectroscopic Investigations of Actinide and Lanthanide Complexes. *Dalton Trans.* **2013**, *42* (13), 4343–4347.
103. Luo, H. M.; Dai, S.; Bonnesen, P. V.; Buchanan Iii, A. C. Separation of Fission Products Based on Ionic Liquids: Task-Specific Ionic Liquids Containing an Aza-Crown Ether Fragment. *J. Alloys Compd.* **2006**, *418* (1–2), 195–199.
104. Sengupta, A.; Mohapatra, P. K.; Kadam, R. M.; Manna, D.; Ghanty, T. K.; Iqbal, M.; Huskens, J.; Verboom, W. Diglycolamide-Functionalized Task Specific Ionic Liquids for Nuclear Waste Remediation: Extraction, Luminescence, Theoretical and EPR Investigations. *RSC Adv.* **2014**, *4* (87), 46613–46623.
105. Jensen, M. P.; Yaita, T.; Chiarizia, R. Reverse-Micelle Formation in the Partitioning of Trivalent f-Element Cations by Biphasic Systems Containing a Tetraalkyldiglycolamide. *Langmuir* **2007**, *23* (9), 4765–4774.
106. Wu, Y.; Zhang, Y.; Fan, F.; Luo, H. M.; Hu, P.; Shen, Y. Synthesis of Task-Specific Ionic Liquids with Grafted Diglycolamide Moiety. Complexation and Stripping of Lanthanides. *J. Radioanal. Nucl. Chem.* **2014**, *299* (3), 1213–1218.
107. Mehdi, H.; Binnemans, K.; Van Hecke, K.; Van Meervelt, L.; Nockemann, P. Hydrophobic Ionic Liquids with Strongly Coordinating Anions. *Chem. Commun.* **2010**, *46* (2), 234–236.
108. Rout, A.; Venkatesan, K. A.; Srinivasan, T. G.; Vasudeva Rao, P. R. Ionic Liquid Extractants in Molecular Diluents: Extraction Behavior of Plutonium (IV) in 1,3-Diketonate Ionic Liquids. *Solvent Extr. Ion Exch.* **2011**, *29* (4), 602–618.
109. Rout, A.; Binnemans, K. Solvent Extraction of Neodymium(III) by Functionalized Ionic Liquid Trioctylmethylammonium Dioctyl Diglycolamate in Fluorine-Free Ionic Liquid Diluent. *Ind. Eng. Chem. Res.* **2014**, *53* (15), 6500–6508.
110. Sun, X. Q.; Ji, Y.; Hu, F.; He, B.; Chen, J.; Li, D. The Inner Synergistic Effect of Bifunctional Ionic Liquid Extractant for Solvent Extraction. *Talanta* **2010**, *81* (4–5), 1877–1883.
111. Rout, A.; Kotlarska, J.; Dehaen, W.; Binnemans, K. Liquid-liquid Extraction of Neodymium(III) by Dialkylphosphate Ionic Liquids from Acidic Medium: The Importance of the Ionic Liquid Cation. *Phys. Chem. Chem. Phys.* **2013**, *15* (39), 16533–16541.
112. Weaver, B.; Kappelmann, F. A. *TALSPEAK: A New Method Of Separating Americium and Curium From The Lanthanides By Extraction From An Aqueous Solution Of An Aminopolyacetic Acid Complex With A Monoacidic Organophosphate Or Phosphate*; C. T. Division Report ORNL 3559; Oak Ridge National Laboratory: Oak Ridge, TN, 1964.
113. Weaver, B.; Kappelmann, F. A. Preferential Extraction of Lanthanides over Trivalent Actinides by Monoacidic Organophosphates from Carboxylic Acids and from Mixtures of Carboxylic and Aminopolyacetic Acids. *J. Inorg. Nucl. Chem.* **1968**, *30* (1), 263–272.

114. Nash, K. L. The Chemistry of TALSPEAK: A Review of the Science. *Solvent Extr. Ion Exch.* **2015**, *33* (1), 1–55.
115. Williams, N. J.; Do-Thanh, C.-L.; Stankovich, J. J.; Luo, H. M.; Dai, S. Extraction of Lanthanides Using 1-Hydroxy-6-*N*-Octylcarboxamido-2(1H)-Pyridinone as an Extractant *via* Competitive Ligand Complexations between Aqueous and Organic Phases. *RSC Adv.* **2015**, *5* (129), 107054–107057.
116. Sun, X. Q.; Bell, J. R.; Luo, H. M.; Dai, S. Extraction Separation of Rare-Earth Ions *via* Competitive Ligand Complexations between Aqueous and Ionic-Liquid Phases. *Dalton Trans.* **2011**, *40* (31), 8019–8023.
117. Sun, X. Q.; Luo, H. M.; Dai, S. Solvent Extraction of Rare-earth Ions Based on Functionalized Ionic Liquids. *Talanta* **2012**, *90*, 132–137.
118. Sun, X. Q.; Luo, H. M.; Dai, S. Mechanistic Investigation of Solvent Extraction Based on Anion-Functionalized Ionic Liquids for Selective Separation of Rare-Earth Ions. *Dalton Trans.* **2013**, *42* (23), 8270–8275.
119. Sun, X. Q.; Do-Thanh, C.-L.; Luo, H. M.; Dai, S. The Optimization of an Ionic Liquid-based TALSPEAK-Like Process for Rare Earth Ions Separation. *Chem. Eng. J.* **2014**, *239*, 392–398.
120. Rout, A.; Karmakar, S.; Venkatesan, K. A.; Srinivasan, T. G.; Vasudeva Rao, P. R. Room Temperature Ionic Liquid Diluent for the Mutual Separation of Europium(III) from Americium(III). *Sep. Purif. Technol.* **2011**, *81* (2), 109–115.
121. Shkrob, I. A.; Marin, T. W.; Jensen, M. P. Ionic Liquid Based Separations of Trivalent Lanthanide and Actinide Ions. *Ind. Eng. Chem. Res.* **2014**, *53* (9), 3641–3653.
122. Hudson, M. J.; Harwood, L. M.; Laventine, D. M.; Lewis, F. W. Use of Soft Heterocyclic *N*-Donor Ligands To Separate Actinides and Lanthanides. *Inorg. Chem.* **2013**, *52* (7), 3414–3428.
123. Panak, P. J.; Geist, A. Complexation and Extraction of Trivalent Actinides and Lanthanides by Triazinylpyridine *N*-Donor Ligands. *Chem. Rev.* **2013**, *113* (2), 1199–1236.
124. Bhattacharyya, A.; Ansari, S. A.; Gadly, T.; Ghosh, S. K.; Mohapatra, M.; Mohapatra, P. K. A Remarkable Enhancement in Am^{3+}/Eu^{3+} Selectivity by an Ionic Liquid Based Solvent Containing Bis-1,2,4-Triazinyl Pyridine Derivatives: DFT Validation of Experimental Results. *Dalton Trans.* **2015**, *44* (13), 6193–6201.
125. Kolarik, Z. Complexation and Separation of Lanthanides(III) and Actinides(III) by Heterocyclic *N*-Donors in Solutions. *Chem. Rev.* **2008**, *108* (10), 4208–4252.
126. Bhattacharyya, A.; Mohapatra, P. K.; Roy, A.; Gadly, T.; Ghosh, S. K.; Manchanda, V. K. Ethyl-Bis-Triazinylpyridine (Et-BTP) for the Separation of Americium(III) from Trivalent Lanthanides Using Solvent Extraction and Supported Liquid Membrane Methods. *Hydrometallurgy* **2009**, *99* (1–2), 18–24.
127. Kobayashi, T.; Yaita, T.; Suzuki, S.; Shiwaku, H.; Okamoto, Y.; Akutsu, K.; Nakano, Y.; Fujii, Y. Effect of the Introduction of Amide Oxygen into 1,10-Phenanthroline on the Extraction and Complexation of Trivalent Lanthanide in Acidic Condition. *Sep. Sci. Technol.* **2010**, *45* (16), 2431–2436.

128. Marie, C.; Miguirditchian, M.; Guillaneux, D.; Bisson, J.; Pipelier, M.; Dubreuil, D. New Bitopic Ligands for the Group Actinide Separation by Solvent Extraction. *Solvent Extr. Ion Exch.* **2011**, *29* (2), 292–315.
129. Galletta, M.; Scaravaggi, S.; Macerata, E.; Famulari, A.; Mele, A.; Panzeri, W.; Sansone, F.; Casnati, A.; Mariani, M. 2,9-Dicarbonyl-1,10-Phenanthroline Derivatives with an Unprecedented Am(III)/Eu(III) Selectivity under Highly Acidic Conditions. *Dalton Trans.* **2013**, *42* (48), 16930–16938.
130. Dehaudt, J.; Williams, N. J.; Shkrob, I. A.; Luo, H. M.; Dai, S. Selective Separation of Trivalent f-Ions Using 1,10-Phenanthroline-2,9-Dicarboxamide Ligands in Ionic Liquids. *Dalton Trans.* **2016**, *45* (29), 11624–11627.

Biomass Processing

Chapter 8

Viscosity and Rheology of Ionic Liquid Mixtures Containing Cellulose and Cosolvents for Advanced Processing

David L. Minnick, Raul A. Flores,[1] and Aaron M. Scurto*

Department of Chemical & Petroleum Engineering and Center for Environmentally Beneficial Catalysis (CEBC), University of Kansas, Lawrence, Kansas 66045, United States
***Phone: +1 (785) 864-4947; Fax: +1 (785) 864-4967; E-mail: ascurto@ku.edu**
[1]Current Address: Department of Chemical Engineering, Stanford University, Stanford, California 94305, United States

Select ionic liquids (ILs) are capable of dissolving significant quantities of biomass (cellulose, hemicellulose, and lignin) but are limited by prohibitively high viscosities even for relatively small biomass loadings. This may be rectified by using mixtures of ionic liquids with certain polar aprotic solvents, which can improve thermodynamic solubility, reduce transport properties, and decrease solvent cost compared to using pure ionic liquids. Here, the viscosity and rheology of microcrystalline cellulose dissolved in IL/cosolvent mixtures were measured at 40°C and 60°C using 1-ethyl-3-methylimidazolium diethyl phosphate [EMIm][DEP] and dimethyl sulfoxide (DMSO) as a model ionic liquid (IL) and cosolvent. Increasing the DMSO cosolvent composition significantly reduced the apparent viscosity of IL/cellulose mixtures. The zero-shear viscosity and Power Law (Ostwald-de Waele) parameters are reported for the various mixtures. The cellulose solution exhibits increasingly Newtonian behavior with higher cosolvent loadings. The cost of a mixed solvent system for processing cellulose was analyzed. Optimal ratios of cosolvent to ionic liquid based on cost and solubility were found to be between 25-50 mass% of cosolvent, which corresponds to viscosities that were lower than the pure IL system by up to 80%.

Introduction

Cellulose is an abundant and renewable biopolymer which may represent a future feedstock for the widespread production of chemicals, fuels, and materials. However, utilizing biomass remains a challenge as cellulose and hemicellulose are difficult to process using conventional techniques and solvents. Cellulose is insoluble in nearly all aqueous and organic solvents due to the complex hydrogen bonding network of cellulose hydroxyl groups. However, some ionic liquids (ILs) can dissolve large quantities of cellulose driven primarily by the anion's ability to solvate cellulose and disrupt cellulose-cellulose interactions (*1*). The cation is believed to contribute, albeit to a lesser extent, to the dissolution mechanism (*2*). Many different ionic liquids have been screened for biomass solubility (*1*, *3*) with the most widely studied ILs being 1-butyl-3-methylimidazolium chloride ([BMIm][Cl]) (*4–6*) and 1-ethyl-3-methylimidazolium acetate ([EMIm][OAc]) (*7–9*). Ionic liquids with acetate ([OAc]) anions have been shown to react or degrade in the presence of cellulose, high temperatures, and CO_2 (*10–16*). There is some indication that cellulose also degrades in [BMIm][Cl] above 95°C (*17*). Furthermore, many of the most promising ILs for biomass processing have relatively high viscosities in their pure form. Moreover, upon dissolution of even small cellulose quantities, ionic liquid (IL)-cellulose mixtures demonstrate significantly high viscosities and some ILs, such as [BMIm][Cl] form gels and undergo other solid-transitions upon dissolution of larger quantities of cellulose (*18*). Alternatively, we have recently identified the ionic liquid, 1-ethyl-3-methylimidazolium diethyl phosphate [EMIm][DEP] (Figure 1) as a potential model IL for biomass dissolution as it has a relatively low pure-component viscosity (60 cP at 60°C), high cellulose solubility, and does not form gels with large amounts of dissolved cellulose (*19*, *20*). In addition, [EMIm][DEP] is not known to react with cellulose and does not contain halides that can lead to corrosion issues with metals (*21*, *22*).

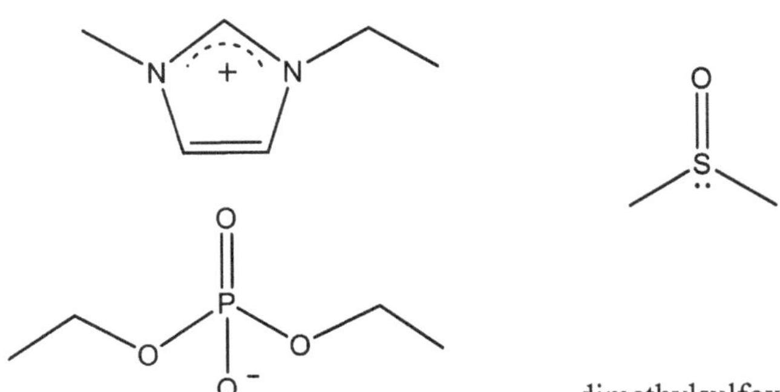

[EMIm][DEP]

dimethylsulfoxide
DMSO

Figure 1. 1-Ethyl-3-methylimidazolium diethyl phosphate [EMIm][DEP] and DMSO.

A significant disadvantage of using pure ILs for biomass processing is their high viscosities which can range from 20-2000+ cP depending on temperature.

Table 1 displays the pure component viscosities for several ionic liquids capable of dissolving cellulosic biomass at 40°C and 60°C. Whereas [BMIm][Cl] has a significant biomass dissolution capacity, (estimated at 19+ mass% as it forms gels), its pure component viscosity (no cellulose) at 40°C is 3800 cP, which, from a processing standpoint, presents severe mass transfer limitations to biomass dissolution and chemical conversion. Furthermore, mixtures of [BMIm][Cl] containing high cellulose loadings are known to form gels thus reducing their processability. As will be shown here, in addition to previous literature discussions, the viscosity of an increases exponentially with cellulose composition; for instance the viscosity of [EMIm][DEP] containing 3 mass% cellulose at 40°C is greater than 3000 cP. These high mixture viscosities present significant transport limitations for industrial biomass and cellulose processing applications. The elevated viscosity will most directly affect pumping and mixing unit operations of IL-biomass processes. In addition, high viscosities will reduce the rate of mass transfer of biomass or reagent dissolution into the IL. Importantly, any chemical processing steps such as hydrolysis may be diffusionally, not kinetically, limited by the significant mixture viscosities of IL-cellulose solutions. In addition, the rate of heat transfer into or out of the mixture will be significantly diminished. A final disadvantage of using pure ILs is their relative expense, which poses an economic challenge to the implementation of IL technologies/processes at industrial scales, especially for commodity chemicals.

Table 1. Viscosity of select ionic liquids that dissolve cellulosic biomass.

Ionic Liquid	*Viscosity at 40°C [cP]*	*Viscosity at 60°C [cP]*
[BMIm][Cl][a]	3800	700
[BMIm][OAc][b]	112	43
[EMIm][OAc][c]	61	26
[EMIm][DEP][d]	142	59

[a] From Ref. (*23*) where [BMIm][Cl] is technically a subcooled liquid below its pure component melting point. [b] Ref. (*24*). [c] Ref. (*25*). [d] Experimental Data here.

One potential solution to overcome IL/cellulose mixture viscosity limitations is to use an IL/cosolvent mixture instead of a pure IL. However, most aqueous and organic solvents act as *anti-solvents* when added to IL/cellulose mixtures thereby decreasing the solubility of cellulose in IL mixtures (*8*). At first glance, using liquid antisolvents like water, alcohols, *etc.* provide what appears to be a relatively convenient separation method using benign (or relatively benign) reagents. However, the overall energy and cost of removing and recycling the antisolvent from the IL must be carefully considered. In addition, any IL/antisolvent system will inevitably absorb water from even "dried" biomass, which also must be removed prior to recycle. Alternatively, there are a few polar aprotic solvents which behave as cosolvents for cellulose in certain composition and temperature regimes. We have recently reported that solvents including dimethyl sulfoxide (DMSO), dimethylformamide (DMF), and 1,3-dimethyl-2-imidazolidinone (DMI) can actually increase the solubility of cellulose in ionic liquids versus pure ILs despite these solvents having no cellulose solubility of their own (*20*). Aprotic cosolvents do not seem to interfere with the hydrogen bonding between the cellulose and primarily the anion. Figure 2 illustrates the enhancement in equilibrium solubility of cellulose in IL mixtures with various cosolvents (*20*). As shown, there is a very wide composition range where the cosolvent does not negatively affect the solubility and may actually increase the solubility with maxima in the 70-80 mole% cosolvent range (approximately 50% cosolvent loading on a mass scale). However, beyond certain cosolvent compositions the solubility of cellulose in the mixture rapidly decreases. Thus, considerable quantities of ionic liquid may be replaced with a less expensive and lower viscosity cosolvent. Costa Gomes and coworkers have demonstrated that dissolution time, temperature, and mixture viscosity of cellulose in IL solutions can be reduced by inclusion of a cosolvent (*26*, *27*). Therefore, ideal cosolvents should lower the mixture viscosity but not decrease the solubility of cellulose in the mixture. Additionally, the cosolvent species should be mutually soluble in the IL at a wide composition range so that a liquid-liquid phase split does not occur. Finally, chemical stability is also imperative: the cosolvent should not react with either the IL or cellulose. The use of comparatively cheaper cosolvents will have the additional effect of lowering the cost of the bulk IL/cosolvent mixture because less IL will be needed.

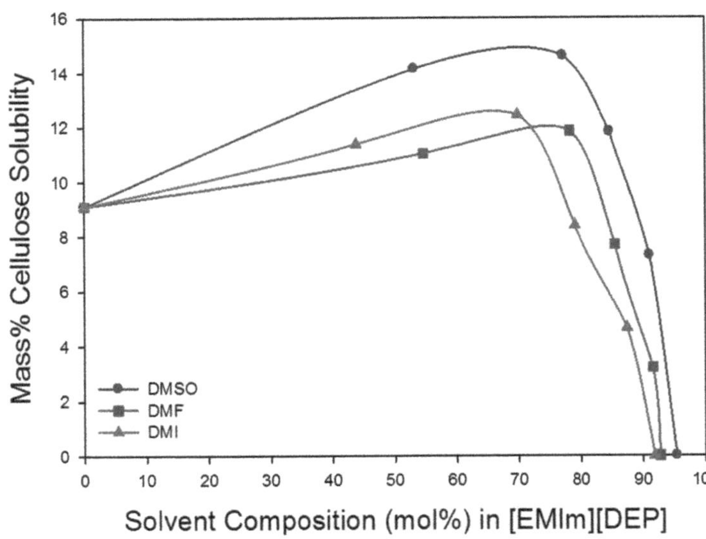

Figure 2. Cellulose solubility in [EMIm][DEP]-cosolvent mixtures at 40°C where cosolvent loading is represented on a molar percent basis. Lines are smoothed data. Data from Ref (20).

In this contribution, we present viscosity and rheology data for cellulose in mixtures of the model ionic liquid 1-ethyl-3-methylimidazolium diethyl phosphate [EMIm][DEP] with DMSO as a model aprotic cosolvent. Temperatures of 40°C and 60°C were investigated for the pure IL and IL/cosolvent mixtures containing cellulose loadings from 1 to 5 mass%. A brief analysis discussing the solvent feedstock cost reduction resulting from utilizing ionic liquid/cosolvents mixtures for cellulose processing is also provided.

Materials and Methods

Materials

1-Methylimidazole (99% purity) was obtained from Acros Organics. Triethyl phosphate (99.9% purity), dimethyl sulfoxide (99.9% purity, 63ppm H_2O), and microcrystalline cellulose (MCC) were obtained from Sigma Aldrich, Inc. The microcrystalline cellulose sample used in this study was previously characterized by our group and had a measured crystallinity value of 61%, determined by a solid-state NMR technique (*28*). Furthermore, the MCC sample had an average MW of 152,789 g/mol ± 3,000 g/mol corresponding to a degree of polymerization of 937 anhydroglucose units (± 19 AGU) determined by a previously described technique using an Ubbelohde viscometer (*20*).

Ionic Liquid Synthesis

1-Ethyl-3-methylimidazolium diethyl phosphate [EMIm][DEP] (MW 264.26 g/mol) was synthesized from 1-methylimidazole and triethyl phosphate using a Biotage Initiator microwave reactor according to a previously described procedure (*29*). Unreacted precursors and impurities were removed from the IL by liquid-liquid extraction with ethyl acetate. [EMIm][DEP] was dried on a high vacuum line ($<10^{-4}$ Torr at 100 °C) for 48 hours prior to use and contained less than 150 ppm H_2O. Ionic liquid purity was determined to be greater than 99% by quantitative NMR. ^{1}H NMR [EMIm][DEP]: (500 MHz, $CDCl_3$), $\delta \times 10^{-6}$: 9.92 (1H, s, NC*H*N), 7.22 (1H, s, NC*H*CHN), 7.18 (1H, s, NCHC*H*N), 3.77 (2H, q, $NC\mathit{H_2}CH_3$), 3.46 (3H, s, $NC\mathit{H_3}$), 3.31 (4H, m, $P(OC\mathit{H_2}CH_3)_2$), 0.95 (3H, t, $NCH_2C\mathit{H_3}$), 0.57 (6H, t, $P(OCH_2C\mathit{H_3})_2$).

Experimental Methodology

Sample Preparation

Mixtures of ionic liquid [EMIm][DEP] and cosolvent DMSO were gravimetrically prepared at compositions of 0.0, 25.0, 50.0, and 75.0 mass% dimethyl sulfoxide on a binary (IL-DMSO) basis. The mixtures were subsequently heated to 80°C and a measured quantity of microcrystalline cellulose (MCC) was added to each to achieve cellulose concentrations of 0.0, 1.0, 3.0, and 5.0 mass% MCC on a ternary (IL-DMSO-cellulose) basis. Cellulose dissolution was achieved in less than 30 minutes for all samples after which they were cooled to room temperature, purged with argon, and sealed.

Rheology and Viscosity Measurements

Mixture viscosities of [EMIm][DEP]-DMSO-cellulose solutions were measured using an AR-2000 rheometer (TA Instruments) equipped with a 40mm smooth plate geometry. Samples were loaded onto a smooth Peltier plate which maintained constant experimental temperatures of 40°C and 60°C (± 0.1°C) for the studies presented herein. After sample application to the Peltier plate, the geometry was quickly lowered to the desired gap distance (500 μm) and the sample was trimmed to ensure uniform coverage without overflow. Temperature equilibration was achieved in approximately five minutes and the experiments were conducted over a period of time no greater than 20 minutes. Care was taken to limit exposure of the mixture to the atmosphere minimizing moisture uptake. Mixture viscosity was obtained in triplicate as a function of shear rate which was ramped typically from 100 – 1000 (1/sec) with 0.3 – 5100 (1/sec) for select systems.

Results and Discussion

Rheological Behavior of Mixtures of Cellulose, [EMIm][DEP], and Cosolvent

The individual IL ([EMIm][DEP]), cosolvent (DMSO), and all cellulose-free IL/cosolvent mixtures exhibited Newtonian behavior as anticipated. However, all of the mixtures with cellulose were non-Newtonian to various degrees. The apparent viscosity as a function of shear rate was measured at 40°C and 60°C for various compositions of cellulose and various ratios of ionic liquid and cosolvent. The composition of DMSO is given on a cellulose-free basis, *i.e.* initial composition of IL and DMSO whereas the cellulose composition is given on a total mixture basis. Figure 3 illustrates a representative example at 3 mass% cellulose with increasing composition of DMSO. The lines are smoothed data from three individual trials at uniform conditions. The average standard deviations between the trials at the different shear rates at 40°C were 0.052, 0.023, 0.005, and 0.002 Pa·s for IL-cosolvent mixtures of 0%, 25%, 50%, and 75% DMSO respectively. These are above the inherent uncertainty of the instrument and thus should approximate maximum error in the measurements. The rheological properties were not found to be time or temperature dependent and did not display hysteresis. Gericke report that viscosity of cellulose in [EMIm][OAc] at 20°C after cycling for several hours to 100°C was approximately 20-30% different (*30*). However, [EMIm][OAc] is known to react with cellulose and the researchers attribute the change in viscosity to the degradation of the cellulose at 100°C.

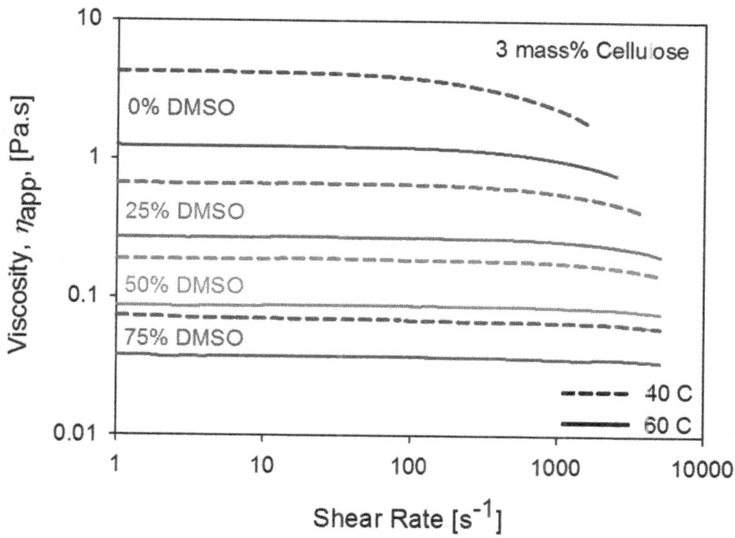

Figure 3. Apparent viscosity versus shear rate for various [EMIm][DEP]/DMSO mixtures (solute-free basis) with 3 mass% cellulose at 40°C (dashed line) and 60°C (solid lined). Lines are smoothed data. DMSO composition given on cellulose-free basis; cellulose on total solution basis

In Figure 3, the top two curves display the viscosity of solutions containing 3 mass% cellulose in pure [EMIm][DEP], *i.e.* cosolvent free. Newtonian behavior is observed over 2-3 orders of magnitude in shear rate followed by shear-thinning behavior at higher shear rates. This is similar for a number of studies of cellulose in ionic liquids in the literature (*17, 30–33*). In our measurements, shear-thinning behavior is observed after approximately 100 s^{-1} in the pure IL. These are relatively viscous solutions even at 3% cellulose as the zero-shear viscosities (η_0) are between 1 and 5 Pa·s versus the pure ionic liquid's viscosity of 0.142 Pa·s and 0.059 Pa·s at 40°C and 60°C respectively. At 25% DMSO at the same 3% cellulose loading, the mixture displays shear thinning behavior but at noticeably higher shear rates. At 50% and 75%, only slight shear thinning behavior is observed. As the temperature is increased, the apparent viscosity decreased for all systems as expected. While the data in Figure 3 maintained a constant cellulose loading of 3 mass%, other cellulose compositions (1% and 5%) have similar qualitative behavior with shear rate, temperature, and cosolvent composition. Most of the trends discussed below will refer to the zero-shear rate viscosity, η_0 of the mixture. A power law analysis is additionally provided to quantify the shear-thinning regime.

Despite the comparatively low pure [EMIm][DEP] viscosity of 0.142 Pa·s at 40°C and 0.059 Pa·s at 60°C, even small quantities of cellulose, when dissolved in the pure IL, result in exponential increases in the mixture viscosity. For instance, Figure 4 demonstrates that the dissolution of 3 mass% cellulose in [EMIm][DEP] at 40°C and 60°C results in an increase in the mixture zero-shear viscosity by more than an order of magnitude. When 5 mass% cellulose is dissolved in the pure IL the mixture zero-shear viscosities at 40°C and 60°C are 7.73 Pa·s and 2.466 Pa·s respectively. Considering most IL-biomass processes will operate at high biomass loadings (*i.e.* 10-20 mass%) and that increased cellulose capacity has nearly an exponential effect on the mixture viscosity, transport limitations of these solutions must be carefully understood and quantified.

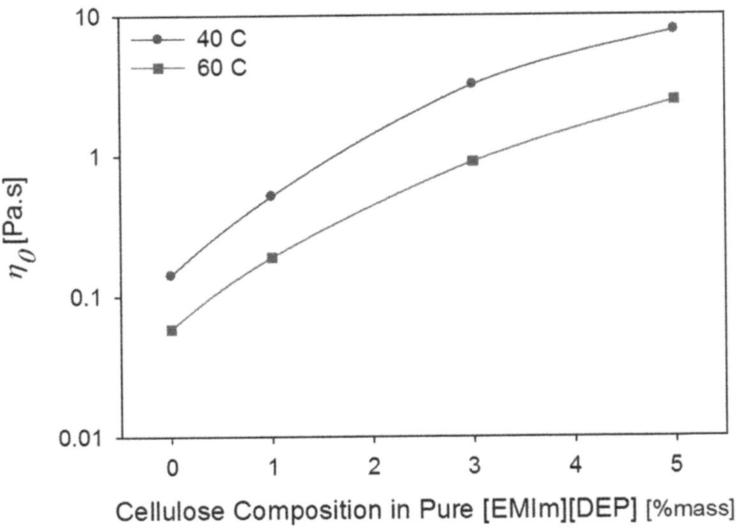

Figure 4. Experimental zero-shear viscosities of [EMIm][DEP]-cellulose mixtures at 40 and 60°C. Lines are smoothed data.

Zero-shear viscosity was also analyzed as a function of cellulose composition. For the 1-5 mass% cellulose composition range, the power law relationship between the viscosity and composition (in mass%) at both 40°C and 60°C was approximately 1.7, $\eta_0 \propto C^{1.7}$. Using specific viscosity instead, η_{sp}, ($\eta_{sp} = (\eta_0 - \eta_{solvent}) / \eta_{solvent}$), the dependence was approximately to the 2.0 power. This is very similar to the results of Haward *et al.*, who also found the power law dependence of 2.0 for specific viscosity of cellulose (DP ~860) in [EMIm][OAc] (*32*). For their data, this dependence was valid for compositions between 0.5% to 8 mass% cellulose. Above and below this composition range, they found the dependence to be 1 or 14/3 respectively. However, other researchers have found higher order dependencies, although with different molecular weight cellulose samples (*30*, *34*).

Effect of Cosolvent and Temperature on the Viscosity of Cellulose Solutions

As previously discussed, at select temperatures and mass loadings, aprotic cosolvents increase the solubility of cellulose in ionic liquid mixtures. The beneficial cosolvent properties also extend to the mass transport properties of IL-cellulose mixtures. For instance, aprotic cosolvents significantly reduce the mixture viscosity of IL-biomass mixtures, thus improving the rheological behavior of cellulose containing solutions. The comparatively small cosolvent molecules (demonstrated by DMSO and applicable to additional aprotic cosolvent systems) act as a buffer between cellulose chains and IL molecules in solution thus reducing frictional interactions (*27*, *33*). The result is a significant reduction in mixture viscosity as shown in Figure 5 and Figure 6 for [EMIm][DEP]-cellulose solutions with varying DMSO cosolvent compositions at 40°C and 60°C respectively.

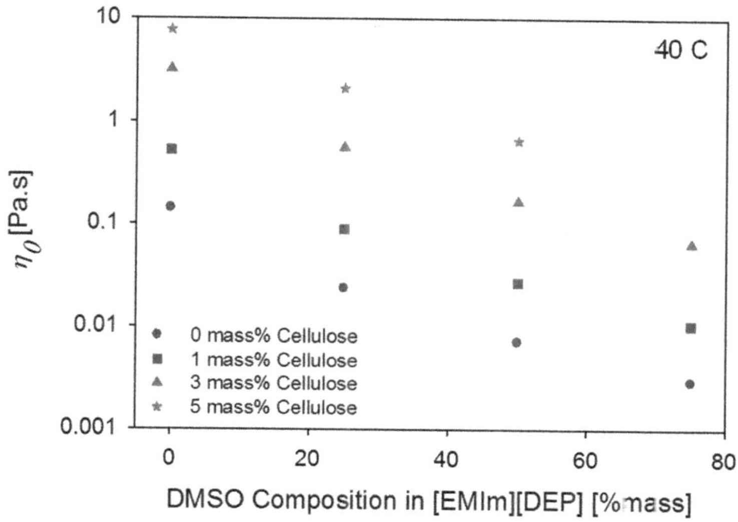

Figure 5. Zero-shear viscosity of cellulose-IL mixtures as a function of cosolvent loading at 40°C. DMSO composition given on cellulose-free basis; cellulose on total basis.

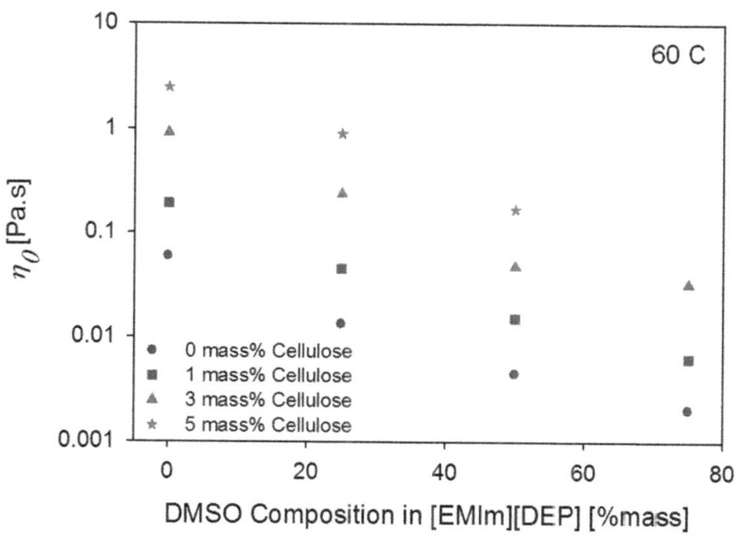

Figure 6. Zero-shear viscosity of cellulose-IL mixtures as a function of cosolvent loading at 60°C. DMSO composition given on cellulose-free basis.

At 40°C a sample of pure [EMIm][DEP] containing 5 mass% cellulose (total solution basis) has a mixture zero-shear viscosity of 7.73 Pa·s. At the same 5 mass% cellulose loading an IL-DMSO mixture containing 50 mass% DMSO (solute free basis) has a mixture viscosity of 0.651 Pa·s. Again, these samples have identical cellulose compositions indicating that the significant reduction in mixture viscosity (one order of magnitude, ~92%) is directly due to the presence of cosolvent in solution. Similar results are observed for the pure IL system as well as [EMIm][DEP]-DMSO mixtures with 1 and 3 mass% cellulose loadings. Interestingly, plots displaying the log of mixture zero-shear viscosity against DMSO composition result in fairly linear trends until about 50% cosolvent loading irrespective of cellulose loading. Therefore, the measured viscosity data presented here could potentially be interpolated or extrapolated to predict the viscosity effects as a function of cosolvent composition, cellulose composition, *etc.*

Temperature also has a moderate impact on mixture viscosity as displayed in Figure 7. For instance, the mixture viscosity of 1 mass% cellulose dissolved in pure [EMIm][DEP] is reduced from 0.520 Pa·s to 0.190 Pa·s by increasing the temperature from 40°C to 60°C. This corresponds to a ~63% reduction in mixture viscosity by a 20°C temperature elevation. Alternatively, at a fixed temperature of 40°C and a constant 1 mass% cellulose loading, increasing the DMSO cosolvent composition from 0 to 25 mass% (cellulose free basis) results in an ~80% decrease in the mixture viscosity. Therefore, while both increased temperature (at a fixed cosolvent loading) and increased cosolvent composition (at a fixed temperature) have positive effects on mixture viscosity reduction, ultimately the cosolvent effects are more significant and have a larger impact on the mixture properties than the effects of temperature.

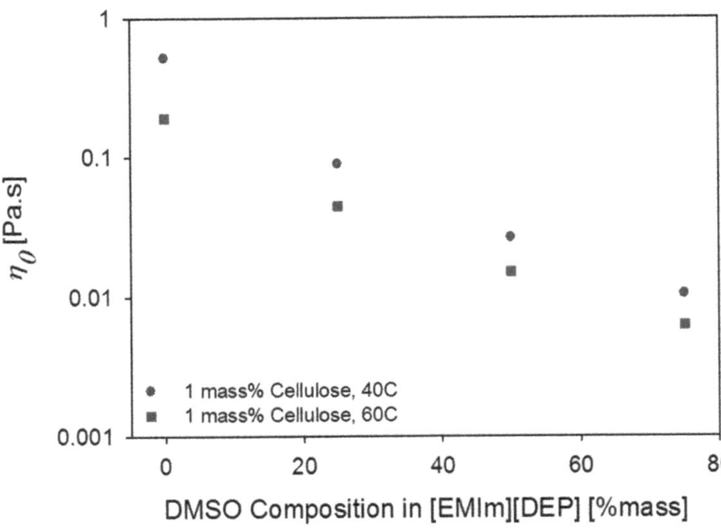

Figure 7. Temperature effect of mixture viscosity of 1% mass cellulose solutions at various cosolvent (DMSO) compositions with [EMIm][DEP]. DMSO composition given on cellulose-free basis.

The composition dependence of cellulose on the on the zero-shear viscosity was also analyzed at the different cosolvent compositions. The power law relationship ($\eta_0 \propto C^A$) between the viscosity and composition (in mass%) at both 40° and 60°C was approximately 1.9 and 1.6 respectively. The dependence on the specific viscosity, η_{sp}, was approximately to the 1.8 and 1.7 power respectively. Compared to the pure IL, IL-cosolvent mixtures seem to have a lower viscosity dependence on cellulose composition.

There are only a few studies in the literature that attempt to quantify the viscosity of cellulose/ionic liquid/cosolvent mixtures. Le *et al.* calculate the intrinsic viscosity of cellulose in [EMIm][OAc] and DMSO solutions applying a novel methodology which uses extrapolated data in the dilute regime (~2% cellulose) to determine the intrinsic viscosity (*7*). They found that the intrinsic viscosity is not a function of the DMSO cosolvent composition and, thus, hypothesize that the hydrodynamic size of the dissolved cellulose remains constant. In the closest study to compare with our data, Wang *et al.* measured the rheology of cellulose in [EMIm][Cl] and DMSO mixtures (*27*). While pure [EMIm][Cl] has a relatively high melting point (T_m~ 80°C), only 10% DMSO allows for a liquid mixture at 25°C. Their cellulose sample had a similar molecular weight (DP=850 vs. 937 here). They find that the power law functionality of specific viscosity with composition to be approximately to the power of ~2.1 for compositions between 0.5 and 2 mass% cellulose (vs. 1.6-1.9 for this study albeit at much higher temperatures). Additionally, they found the power law dependence of specific viscosity to composition to be ~4.7 for compositions of 2 to 10 mass%, where we find in the 3%-5% composition range the power to be ~2.0 for our system. [EMIm][Cl] is a higher-melting solid with stronger cation-anion interactions which may help explain these trends.

Power Law Analysis with Shear Rate of Mixed IL-Cosolvent Systems for Cellulose Dissolution

The Power-Law regime of the rheology data was fit to the Ostwald-de Waele equation

$$\eta_{app} = K\dot{\gamma}^{n-1} \qquad \text{(Eqn. 1)}$$

where η_{app} is the apparent viscosity in Pa·s, K is the flow consistency index, $\dot{\gamma}$ is the shear rate in s^{-1}, and n is the flow behavior index. Flow behavior indices, n, close to one indicate Newtonian behavior; while less than one indicate non-Newtonian shear thinning behavior. The regressed indices are reported in Table 2.

Table 2. Zero-shear viscosity and regressed Power Law parameters

			η_0[a]		*Power Law*	
T *[°C]*	*Cellulose* *[mass%]*	*DMSO*[b] *[mass%]*	*[Pa·s]*	*± SD*[c]	*K* *[Pa·sn]*	*n*
40	0	0	0.142	0.004	-	-
		25	0.0232	0.0001	-	-
		50	0.0072	0.0001	-	-
		75	0.0029	0.0001	-	-
60	0	0	0.0588	0.0002	-	-
		25	0.0133	0.0001	-	-
		50	0.0044	0.0001	-	-
		75	0.0020	0.0001	-	-
40	1.0	0	0.520	0.003	0.7135	0.938
		25	0.0892	0.004	0.1195	0.956
		50	0.0266	0.0002	0.0293	0.985
		75	0.0105	0.0001	0.0111	0.991
60	1.0	0	0.1901	0.0006	0.1888	0.9934
		25	0.0451	0.0001	0.0459	0.9987
		50	0.0152	0.0001	0.0156	0.9995
		75	0.0063	0.0001	0.0063	1.0000
40	3.0	0	3.21	0.09	18.7675	0.685
		25	0.556	0.003	1.0526	0.889
		50	0.1656	0.0010	0.2431	0.938
		75	0.0652	0.0008	0.0857	0.948
60	3.0	0	0.907	0.007	1.7877	0.882
		25	0.2354	0.0006	0.3092	0.954
		50	0.0476	0.0003	0.0586	0.965
		75	0.0325	0.0001	0.0385	0.973
40	5.0	0	7.73	0.43	78.0096	0.568
		25	2.10	0.04	9.9506	0.728
		50	0.651	0.004	1.5864	0.850
60	5.0	0	2.466	0.041	9.864	0.759

Continued on next page.

Table 2. (Continued). Zero-shear viscosity and regressed Power Law parameters

			η_0[a]		Power Law	
T *[°C]*	*Cellulose* *[mass%]*	*DMSO*[b] *[mass%]*	*[Pa·s]*	*± SD*[c]	*K* *[Pa sn]*	*n*
		25	0.896	0.005	1.857	0.876
		50	0.169	0.003	0.264	0.931

[a] For mixtures with no cellulose, this is the actual dynamic viscosity. [b] Cellulose-free basis. [c] Standard deviation of at least three measurements.

Mixtures with 1% cellulose demonstrated only slight shear-thinning behavior as demonstrated by the flow indices, *n*, being very close to one in Table 2. However as the composition of cellulose was increased to 3 and 5 mass%, the solutions became increasingly non-Newtonian. For example the flow index was 0.568 and 0.759 at 40°C and 60°C respectively for the sample containing 5 mass% cellulose in pure [EMIm][DEP]. However, when the composition of DMSO was increased to 50% for a IL-cosolvent mixture containing an identical cellulose loading, these indices became 0.85 and 0.931. Thus despite the same composition of cellulose, the IL-cosolvent mixtures were much more Newtonian than the pure IL system at the same temperature.

Figure 8 illustrates the change in the flow parameter, *n*, with composition of DMSO. Increasing the composition of DMSO significantly increases the flow parameter and, thus, indicates less shear thinning behavior at higher shear rates. Comparing the trends at 40°C and 60°C indicates that the effect is more pronounced at lower temperature.

The effect of cellulose loading on the flow behavior index is shown in Figure 9 at 40°C. The index drastically decreases with increasing cellulose composition in the pure IL. However, with increasing composition of the cosolvent (as shown with 25% and 50% DMSO), the decrease is more subtle. Similar results are observed at 60°C, albeit at higher values of *n*. Overall, cosolvents not only significantly lower the absolute viscosity, zero-shear viscosity, *etc.*, but they also produce mixtures that behave increasingly Newtonian at the same composition of cellulose.

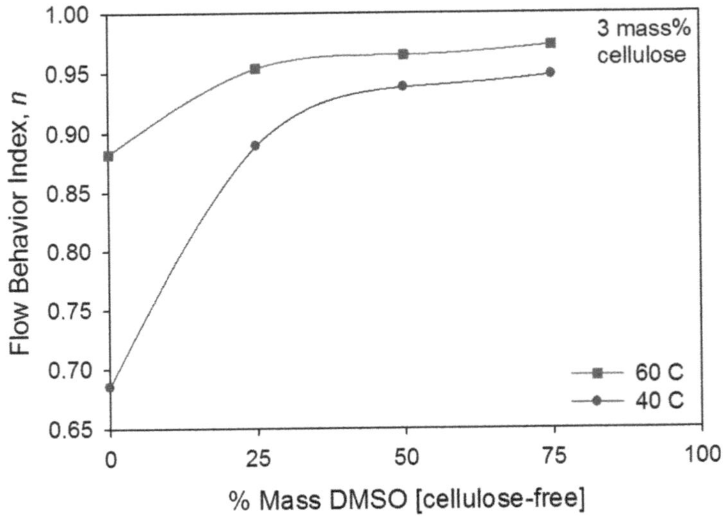

Figure 8. Flow behavior index, n, with composition of cosolvent with [EMIm][DEP] for a 3% cellulose solution.

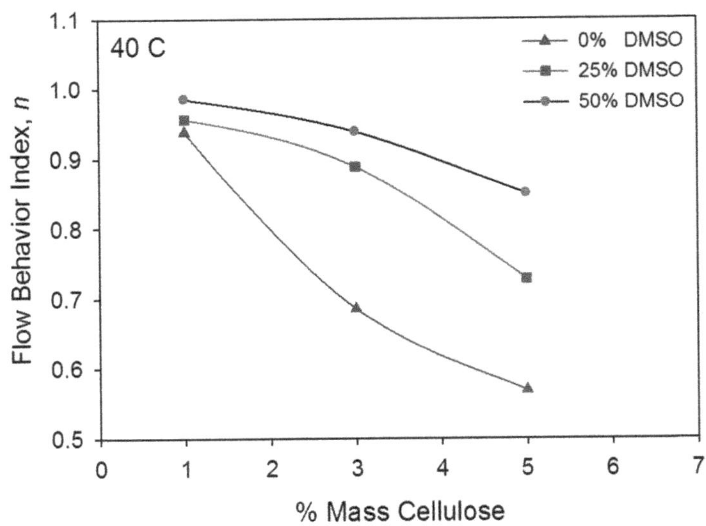

Figure 9. Flow behavior index with composition of cellulose at 40°C. Lines are smoothed data.

Economics of Mixed IL-Cosolvent Systems for Cellulose Dissolution

From an economic perspective, the inclusion of cosolvents in mixtures with ILs for biomass processing significantly reduces the capital costs involved with purchasing the IL feedstock. For instance, an economic study on biomass processing with ionic liquids indicated that ~93% of the operating costs were associated with purchasing the IL feedstock (*35*). The findings further indicated that the IL purchase price (studied on [EMIm][Cl]) made the process economically prohibitive. Previous attempts at improving IL based processes have primarily focused on developing less expensive ionic liquids. While these efforts are indeed necessary the majority of inexpensive protic ILs developed to this point do not have the characteristic high hydrogen bond accepting abilities found in imidazolium ILs are therefore not suitable for biomass applications.

Through the utilization of mixed IL-cosolvent systems, an alternative approach for reducing the economic burden of the IL feedstock is presented. For instance, cosolvents can displace up to 75 mass% of the ionic liquid while maintaining high cellulose solubilities. The cost of bulk [EMIm][Cl] production (and presumed to be similar to the production costs of [EMIm][DEP]) was estimated in a techno-economic study to be ~$5.75 per kg (*35*) which is significantly more expensive than the bulk price of dimethyl sulfoxide $1.30 per kg, and other potential cosolvents including dimethylformamide $0.89 per kg and dimethyl imidazolidinone $4.88 per kg (solvent price based on the purchase of 1 metric ton) (*36*). A hypothetical process which requires 1000 kg of solvent to process lignocellulosic biomass would require an estimated $5750 to purchase a pure IL feedstock. Alternatively, if 50% of the IL were displaced with DMSO, with all other conditions being the same, the mixed solvent feedstock cost would be $3525, a ~39% reduction compared to the pure IL scenario. Additionally, blended IL-cosolvent mixtures of 25 mass% and 75 mass% DMSO would provide cost reductions of 19% and 58% respectively.

Cost minimization can also be analyzed by investigating the economics of making a mixed solvent system required to process (dissolve) 1.0 kg of dry cellulosic biomass. In conjunction with the pure solvent prices previously discussed, this analysis was developed on the basis of previously published ternary cellulose solubility results in IL-cosolvent mixtures (*20*). Figure 10 displays the optimized solvent cost results for the [EMIm][DEP]-DMSO system at 40, 60, and 80°C as a function of cosolvent loading in the mixture. The minima occur between 25% and 50% cosolvent composition with an approximate average over the temperatures of ~35% cosolvent loading. Interpolating the viscosity data in Table 2 for a 35% DMSO mixture with the IL, this would correspond to a 5% cellulose solution of a zero-shear viscosity of approximately 1.4 and 0.4 Pa·s at 40° and 60°C respectively. This would be a viscosity reduction of ~82% over using the pure IL and, as shown in Figure 10, up to a 65% reduction in cost over a pure IL system. It should be pointed out that cost reduces at the margin with increasing temperatures. In terms of rheology, a 35% cosolvent mixture would have a flow index of approximately 0.9 compared with 0.76 for the pure IL case.

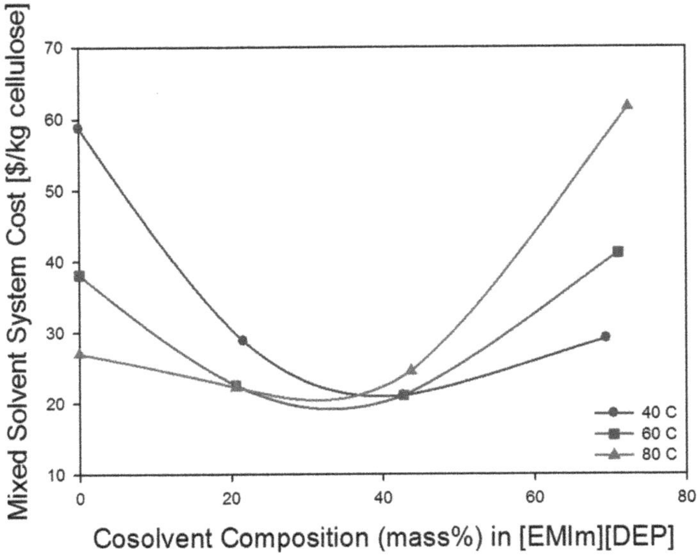

Figure 10. Cost of mixed [EMIm][DEP]-DMSO solvent systems for the dissolution of 1 kg of cellulosic biomass. Results presented based on optimizing the cellulose dissolution capacity of the mixed solvent systems with solvent cost.

The cost of dissolving 1 kg of cellulose for other cosolvents were also calculated using previously published solubility data (*20*). As displayed in Table 3, at 40°C the mixed IL-cosolvent system cost for DMSO, DMF, and DMI were all approximately $19.50 per kg. However, the quantity of cosolvent which optimized the cellulose capacity and solvent cost varied significantly. For instance, the optimized DMSO system at 40°C consisted of a mixture of 46.3 mass% cosolvent and 39.6 mass% IL on a ternary basis which has an approximate cellulose capacity of 14.1 mass%. Alternatively, the DMF solvent system at 40°C was composed of a ternary mixture of 34.8 mass% cosolvent, 53.4 mass% [EMIm][DEP], and 11.8 mass% cellulose.

The mixed solvent system of DMSO is slightly less expensive on a per kg basis compared to DMF likely due to the higher cellulose dissolution capacity of the [EMIm][DEP]-DMSO system at 40°C compared to mixtures of the IL with DMF. However, the purchase price of pure DMF ($0.89 per kg) is less expensive than DMSO ($1.30 per kg). Therefore, at 60 and 80°C when the solubilities of the various cosolvent systems become increasingly similar, the lower DMF cost renders this mixed solvent less expensive. Furthermore, while DMI demonstrates the second best cellulose dissolution capacity of the three IL-cosolvent mixtures, the high pure-component DMI cost ($4.88 per kg) becomes comparatively expensive at 60 and 80°C relative to DMSO and DMF. Interestingly, the least expensive mixed solvent costs required to solubilize 1 kg of cellulose are observed at 40°C and are comparatively less than any solvent system at 60 and 80°C. The data presented in Table 3 represents the raw material cost only and does not account for the cost of thermal energy required to heat the mixtures to their respective temperatures. Therefore, operating at lower temperatures (*i.e.* 40°C) will have additional operating cost savings compared to the higher temperature systems. While a techno-economic evaluation is required to assess the quantitative savings of mixed IL-cosolvent systems, preliminary indicators demonstrate the potential for significant cost reduction relative to utilizing pure ILs.

Table 3. Cost of mixed IL-cosolvent systems required to dissolve 1 kg of dry cellulosic biomass.

Temperature [°C]	*Cosolvent mixed with [EMIm][DEP]*	*Optimized Cosolvent Loading (mass%)*	*Optimized Solvent Cost ($/kg)*
40	DMSO	46.0	19.48
40	DMF	34.8	19.75
40	DMI	41.1	19.38
60	DMSO	34.2	28.44
60	DMF	30.5	22.62
60	DMI	33.4	23.63
80	DMSO	22.7	30.98
80	DMF	15.0	20.56
80	DMI	23.0	24.30

Summary

The rheology and viscosity of mixed [EMIm][DEP]/DMSO/cellulose systems were investigated at 40°C and 60°C at cellulose compositions ranging between 1-5 mass% and DMSO compositions at 25%, 50%, and 75 mass%. The viscosity of the mixture had a power law functionality with the composition of cellulose (to the power of 1.5-2.0) which is similar to other systems in the literature. The presence of the cosolvent, DMSO, significantly decreased the apparent viscosity and resulted in increasingly Newtonian mixtures. The zero-shear viscosity and Power Law (Ostwald-de Waele) parameters in the shear-thinning regime are reported for the various mixtures. The costs of a mixed solvent system for processing cellulose significantly decreases for ratios of cosolvent to ionic liquid between 25%-50 mass%. Importantly, this cosolvent loading region also reduces the viscosity of the solution by about 80%. The economic results also indicate that using other cosolvents including dimethylformamide (DMF) and dimethyl imidazolidinone (DMI) also improve the process economics.

Acknowledgments

We would like to acknowledge support from the Madison and Lila Self Graduate Fellowship (DLM) and the KU Center for Environmentally Beneficial Catalysis (CEBC).

References

1. Wang, H.; Gurau, G.; Rogers, R. D. Ionic Liquid Processing of Cellulose. *Chem. Soc. Rev.* **2012**, *41* (4), 1519–1537.
2. Lu, B. L.; Xu, A. R.; Wang, J. J. Cation Does Matter: How Cationic Structure Affects the Dissolution of Cellulose in Ionic Liquids. *Green Chem.* **2014**, *16* (3), 1326–1335.
3. Pinkert, A.; Marsh, K. N.; Pang, S. S.; Staiger, M. P. Ionic Liquids and Their Interaction with Cellulose. *Chem. Rev.* **2009**, *109* (12), 6712–6728.
4. Rinaldi, R. Instantaneous Dissolution of Cellulose in Organic Electrolyte Solutions. *Chem. Commun.* **2011**, *47* (1), 511–513.
5. Gross, A. S.; Bell, A. T.; Chu, J. W. Thermodynamics of Cellulose Solvation in Water and the Ionic Liquid 1-Butyl-3-Methylimidazolim Chloride. *J. Phys. Chem. B* **2011**, *115* (46), 13433–13440.
6. Li, Y.; Liu, X. M.; Zhang, S. J.; Yao, Y. Y.; Yao, X. Q.; Xu, J. L.; Lu, X. M. Dissolving Process of a Cellulose Bunch in Ionic Liquids: A Molecular Dynamics Study. *Phys. Chem. Chem. Phys.* **2015**, *17* (27), 17894–17905.
7. Le, K. A.; Rudaz, C.; Budtova, T. Phase Diagram, Solubility Limit and Hydrodynamic Properties of Cellulose in Binary Solvents with Ionic Liquid. *Carbohydr. Polym.* **2014**, *105*, 237–243.
8. Olsson, C.; Idstrom, A.; Nordstierna, L.; Westman, G. Influence of Water on Swelling and Dissolution of Cellulose in 1-Ethyl-3-Methylimidazolium Acetate. *Carbohydr. Polym.* **2014**, *99*, 438–446.
9. Xu, A. R.; Zhang, Y. J.; Zhao, Y.; Wang, J. J. Cellulose Dissolution at Ambient Temperature: Role of Preferential Solvation of Cations of Ionic Liquids by a Cosolvent. *Carbohydr. Polym.* **2013**, *92* (1), 540–544.
10. Barber, P. S.; Griggs, C. S.; Gurau, G.; Liu, Z.; Li, S.; Li, Z. X.; Lu, X. M.; Zhang, S. J.; Rogers, R. D. Coagulation of Chitin and Cellulose from 1-Ethyl-3-Methylimidazolium Acetate Ionic-Liquid Solutions Using Carbon Dioxide. *Angew. Chem., Int. Ed.* **2013**, *52* (47), 12350–12353.
11. Besnard, M.; Cabaco, M. I.; Chavez, F. V.; Pinaud, N.; Sebastiao, P. J.; Coutinho, J. A. P.; Mascetti, J.; Danten, Y. CO_2 in 1-Butyl-3-Methylimidazolium Acetate. 2. NMR Investigation of Chemical Reactions. *J. Phys. Chem. A* **2012**, *116* (20), 4890–4901.
12. Denning, D. M.; Falvey, D. E. Solvent-Dependent Decarboxylation of 1,3-Dimethylimdazolium-2-Carboxylate. *J. Org. Chem.* **2014**, *79* (10), 4293–4299.
13. Sun, X. F.; Chi, Y. L.; Mu, T. C. Studies on Staged Precipitation of Cellulose from an Ionic Liquid by Compressed Carbon Dioxide. *Green Chem.* **2014**, *16* (5), 2736–2744.

14. Clough, M. T.; Geyer, K.; Hunt, P. A.; Son, S.; Vagt, U.; Welton, T. Ionic Liquids: Not Always Innocent Solvents for Cellulose. *Green Chem.* **2015**, *17* (1), 231–243.
15. Clough, M. T.; Geyer, K.; Hunt, P. A.; Mertes, J.; Welton, T. Thermal Decomposition of Carboxylate Ionic Liquids: Trends and Mechanisms. *Phys. Chem. Chem. Phys.* **2013**, *15* (47), 20480–20495.
16. Du, H. B.; Qian, X. H. The Effects of Acetate Anion on Cellulose Dissolution and Reaction in Imidazolium Ionic Liquids. *Carbohydr. Res.* **2011**, *346* (13), 1985–1990.
17. Chen, X.; Zhang, Y. M.; Cheng, L. Y.; Wang, H. P. Rheology of Concentrated Cellulose Solutions in 1-Butyl-3-Methylimidazolium Chloride. *J. Polym. Environ.* **2009**, *17* (4), 273–279.
18. Gericke, M.; Fardim, P.; Heinze, T. Ionic Liquids - Promising but Challenging Solvents for Homogeneous Derivatization of Cellulose. *Molecules* **2012**, *17* (6), 7458–7502.
19. Normazlan, W. M. D. W.; Sairi, N. A.; Alias, Y.; Udaiyappan, A. F.; Jouyban, A.; Khoubnasabjafari, M. Composition and Temperature Dependence of Density, Surface Tension, and Viscosity of Emim Dep/Mmim Dmp + Water+1-Propanol/2-Propanol Ternary Mixtures and Their Mathematical Representation Using the Jouyban Acree Model. *J. Chem. Eng. Data* **2014**, *59* (8), 2337–2348.
20. Minnick, D. L.; Flores, R. A.; DeStefano, M. R.; Scurto, A. M. Cellulose Solubility in Ionic Liquid Mixtures: Temperature, Cosolvent, and Antisolvent Effects. *J. Phys. Chem. B* **2016**, *120* (32), 7906–7919.
21. Deetlefs, M.; Seddon, K. R. Assessing the Greenness of Some Typical Laboratory Ionic Liquid Preparations. *Green Chem.* **2010**, *12* (1), 17–30.
22. Francis, R. The Effect of Chlorine Additions to Cooling Water on Corrosion of Copper Alloy Condenser Tubes. *Mater. Perform.* **1982**, *21* (8), 44–49.
23. Seddon, K. R.; Stark, A.; Torres, M. J. Viscosity and Density of 1-Alkyl-3-Methylimidazolium Ionic Liquids. *Clean Solvents* **2002**, *819*, 34–49.
24. Almeida, H. F. D.; Passos, H.; Lopes-da-Silva, J. A.; Fernandes, A. M.; Freire, M. G.; Coutinho, J. A. P. Thermophysical Properties of Five Acetate-Based Ionic Liquids. *J. Chem. Eng. Data* **2012**, *57* (11), 3005–3013.
25. Freire, M. G.; Teles, A. R. R.; Rocha, M. A. A.; Schroder, B.; Neves, C. M. S. S.; Carvalho, P. J.; Evtuguin, D. V.; Santos, L. M. N. B. F.; Coutinho, J. A. P. Thermophysical Characterization of Ionic Liquids Able to Dissolve Biomass. *J. Chem. Eng. Data* **2011**, *56* (12), 4813–4822.
26. Andanson, J. M.; Bordes, E.; Devemy, J.; Leroux, F.; Padua, A. A. H.; Gomes, M. F. C. Understanding the Role of Co-Solvents in the Dissolution of Cellulose in Ionic Liquids. *Green Chem.* **2014**, *16* (5), 2528–2538.
27. Wang, L. J.; Gao, L.; Cheng, B. W.; Ji, X. J.; Song, J.; Lu, F. Rheological Behaviors of Cellulose in 1-Ethyl-3-Methylimidazolium Chloride/Dimethylsulfoxide. *Carbohydr. Polym.* **2014**, *110*, 292–297.
28. Minnick, D. L.; Scurto, A. M. Reversible and Non-Reactive Cellulose Separations from Ionic Liquid Mixtures with Compressed Carbon Dioxide. *Chem. Commun.* **2015**, *51* (63), 12649–12652.

29. Thomas, M. F.; Li, L. L.; Handley-Pendleton, J. M.; van der Lelie, D.; Dunn, J. J.; Wishart, J. F. Enzyme Activity in Dialkyl Phosphate Ionic Liquids. *Bioresour. Technol.* **2011**, *102* (24), 11200–11203.
30. Gericke, M.; Schlufter, K.; Liebert, T.; Heinze, T.; Budtova, T. Rheological Properties of Cellulose/Ionic Liquid Solutions: From Dilute to Concentrated States. *Biomacromolecules* **2009**, *10* (5), 1188–1194.
31. Collier, J. R.; Watson, J. L.; Collier, B. J.; Petrovan, S. Rheology of 1-Butyl-3-Methylimidazolium Chloride Cellulose Solutions. Ii. Solution Character and Preparation. *J. Appl. Polym. Sci.* **2009**, *111* (2), 1019–1027.
32. Haward, S. J.; Sharma, V.; Butts, C. P.; McKinley, G. H.; Rahatekar, S. S. Shear and Extensional Rheology of Cellulose/Ionic Liquid Solutions. *Biomacromolecules* **2012**, *13* (5), 1688–1699.
33. Sammons, R. J.; Collier, J. R.; Rials, T. G.; Petrovan, S. Rheology of 1-Butyl-3-Methylimidazolium Chloride Cellulose Solutions. I. Shear Rheology. *J. Appl. Polym. Sci.* **2008**, *110* (2), 1175–1181.
34. Sescousse, R.; Le, K. A.; Ries, M. E.; Budtova, T. Viscosity of Cellulose-Imidazolium-Based Ionic Liquid Solutions. *J. Phys. Chem. B* **2010**, *114* (21), 7222–7228.
35. Sen, S. M.; Binder, J. B.; Raines, R. T.; Maravelias, C. T. Conversion of Biomass to Sugars Via Ionic Liquid Hydrolysis: Process Synthesis and Economic Evaluation. *Biofuels, Bioprod. Biorefin.* **2012**, *6* (4), 444–452.
36. alibaba.com (Bulk DMSO Purchase Price December 2016).

Chapter 9

Ultra-Low Cost Ionic Liquids for the Delignification of Biomass

Florence J. V. Gschwend, Agnieszka Brandt-Talbot, Clementine L. Chambon, and Jason P. Hallett*

Department of Chemical Engineering, Imperial College London, South Kensington, London SW11 2AZ, United Kingdom
***E-mail: j.hallett@imperial.ac.uk**

Low-cost pretreatment of lignocellulosic biomass is an essential next step toward large-scale deployment as renewable liquid fuels, materials or chemicals. Ionic liquids (ILs) are highly effective at pretreatment, but high IL cost has hindered commercial viability. We have recently developed low-cost (ca. $1/kg) ILs, such as triethylammonium hydrogen sulphate, for pretreatment. In this chapter we discuss the fractionation of the grass *Miscanthus x giganteus*, wherein we deconstruct the lignocellulosic matrix into a cellulose-rich pulp, a recovered lignin fraction and an organic distillate. More than 80% of the lignin and quantitative hemicelluloses are removed during extraction. This results in 70-90% glucose release during enzymatic saccharification. The IL can also be successfully recovered and reused, with >99% IL recovery and minimal effects on efficiency of extraction. A detailed mass balance of all components and subsequent economic analysis revealed this efficient pretreatment with an ultra-low cost IL could result in an economically viable pretreatment process.

Introduction

The transportation sector and chemical industry currently rely on fossil fuels for their major products; mainly these are derived from petroleum. The continued use of petroleum is not sustainable, and this use constitutes approximately one-third of anthropogenic CO_2 emissions, making it a major factor behind climate change. Lignocellulosic (woody) biomass is an abundant (billon ton plus), renewable and carbon-neutral alternative to petroleum-based production (*1*). While biomass accounts for 12% of current energy, this is mainly within low-grade heat applications (*2*). To replace fossil fuels with renewable alternatives, biomass processing must advance beyond its current state (*3*, *4*). While separation of the biopolymers within woody biomass is desirable, the isolation and valorization of these components remains a major research challenge.

Lignocellulose is a natural composite material, and in general its composition consists of 65% polysaccharides (40% cellulose and 25% hemicellulose), 25% lignin and *ca.* 10% minor components (extractives or ash, on a dry basis). Woody biomass is difficult to process because it is recalcitrant towards mild chemical or biological degradation; therefore a pretreatment (deconstruction or separation) step is required before the inherent sugars contained within the woody matrix can be accessed for biological or chemical conversion into useful fuels, chemicals or materials. There are several pretreatment methods available, including the use of steam (*5*), ammonia (*6*), dilute acid (*7*), organosolv (*8*), and, most recently, ionic liquids (*9*) - liquid organic salts with tuneable solvent properties (*10*).

Thermochemical pretreatment methods (e.g. dilute acid, steam explosion, ammonia, AFEX) have the primary aim of increasing sugar release, rather than on delignification of the biomass. Removal of lignin during pretreatment (fractionation) imparts additional benefits, including lower enzyme loading during polysaccharide hydrolysis due to elimination of non-productive enzyme binding (*11*). While this can improve overall fermentation yields due to lower inhibitor levels (*12*), it importantly reduces the size of process units, resulting in an intensified process with higher sugar throughput, and may have additional benefits such as production of higher grade lignin for subsequent processing (*13*).

Ionic liquid pretreatment began with the report of cellulose-dissolving ILs (*14*), opening up new possibilities for cellulose processing without derivatization (e.g. the Ionocell-F process) (*15*) a and chemical transformation of sugars into platform chemicals, where ionic liquids display superior yields and selectivities compared to aqueous systems (*16*). For biomass pretreatment, two distinct options are currently under development (*9*): the Dissolution Pretreatment involves the use of cellulose dissolving ionic liquids to decrystallize cellulose within biomass, increasing surface area available for enzymatic attack (*17–20*). Meanwhile, ionoSolv pretreatment dissolves lignin and hemicellulose out of biomass (similar to Organosolv processing) and leaves a highly crystalline, but relatively pure, cellulose pulp for further processing (*21–23*). Key advantages of ionic liquid pretreatments are low process pressures (*24*), reduced friction/abrasion (*25*) and novel product separations (*26*).

The Dissolution Pretreatment utilizes highly hydrogen bonding basic ionic liquids to dissolve cellulose, most prominently 1-ethyl-3-methylimidazolium acetate, [C_2C_1im][OAc]. This, and similar ionic liquids, dissolve the lignocellulose without substantial delignification or hemicellulose removal, disrupt hydrogen bonds within the cellulose fibrils, and regenerate an amorphous pulp after post-pretreatment addition of water. The amorphous cellulose exhibits *ca.* 50 times higher enzymatic hydrolysis rate (*26*, *27*) than untreated biomass due to increased surface area and low cellulose crystallinity (*28*). Unfortunately, these ionic liquids possess a few drawbacks, including very high solvent price (estimates for $[C_2C_1im][OAc]$ range from \$20-101/kg) (*29*), low thermal stability (*30*) and low tolerance to moisture (*31*). Low water content is difficult to achieve due to the high moisture content of freshly harvested biomass (up to 50%) and these ionic liquids' high affinity for water (*32, 33*).

In contrast, the ionoSolv pretreatment only dissolves lignin and hemicellulose, even with high water content, leaving a cellulose-rich pulp (*23*). Despite the high crystallinity of the cellulose, enzymatic saccharification yields are high (70-90% of the theoretical maximum) due to increased accessibility of the fibril surface. Our previous studies demonstrated that 10-40% water content in ionoSolv ILs is effective and even necessary for pretreatment (*21, 23*). This is has been attributed to several effects, from a curb on sugar degradation pathways to a reduction in medium viscosity and acidity. We first started our explorations with the ionic liquid 1-butyl-3-methylimidazolium hydrogen sulfate, $[C_4C_1im][HSO_4]$ (*21*), an effective but relatively expensive solvent choice. We sought to reduce the cost and improve the flexibility of the media by using its protic analogue 1-butylimidazolium hydrogen sulfate, $[HC_4im][HSO_4]$ (*23*), taking advantage of the continuum of acid strengths available in these systems. Use of 1-butylimidazolium hydrogen sulfate resulted in 90% fermentable glucose after enzymatic saccharification of *Miscanthus* pulps, and the use of a protic ionic liquid can lead to a significant increase in the economic viability of the ionoSolv pretreatment, as these ionic liquids are inevitably cheaper than their peralkylated analogues, due the simplification of the synthesis (alkylation followed by ion exchange for the fully alkylated solvents; simple mixing of an acid and an amine for protic ILs). Our techno-economic analysis of the bulk-scale synthesis of 1-methylimidazolium hydrogen sulfate demonstrated that these ionic liquids can be produced for \$2.96–5.88/kg (*34*), well below estimates of future bulk prices of dialkylimidazolium ionic liquids(\$40-81/kg (*35*) or 5 to 20 times the price of common organic solvents (*36*)).

It has now been widely demonstrated that the anion of an ionic liquid often determines the solvation chemistry taking place, with the cation having a smaller effect (*9*). With this in mind, we sought to further decrease the cost of ionic liquid production by using a less expensive, and more widely available at bulk scale, alkylamines (*22*). In that study we concluded that triethylammonium hydrogen sulfate, $[N_{0222}][HSO_4]$ (Figure 1), exhibited the best performance under the conditions of our screening. We also demonstrated that triethylammonium hydrogen sulfate water mixtures can be produced at bulk scale for as little as \$1.24/kg (*34*), a cost similar to that of common organic solvents such as acetone and toluene.

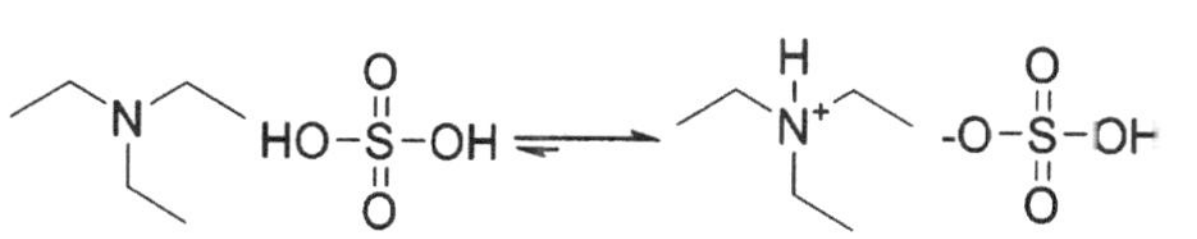

Figure 1. One-step synthesis of triethylammonium hydrogen sulfate ($[N_{0222}][HSO_4]$) from triethylamine and sulfuric acid.

Results and Discussion

Pretreatment of Miscanthus with Triethylammonium Hydrogensulfate, $[N_{0222}][HSO_4]$

Miscanthus x giganteus is a popular perennial grass proposed as a bioenergy crop (*9*) and is representative of common grasses such as switchgrass or agricultural byproducts such as corn stover. We demonstrated the successful use of the easily synthesized, low-cost ionic liquid triethylammnoium hydrogensulfate $[N_{0222}][HSO_4]$ for the pretreatment of miscanthus, giving rise to high saccharification yields (*37*). In Figure 2, we show the recovery of solid pulp from the biomass after ionic liquid pretreatment and the mass of lignin precipitated from the ionic liquid liquor. Recovery of the ionic liquid after the pretreatment was in all cases close to 100% (see Figure 3); values above 100% were likely due to biomass fragments dissolved in the ionic liquid consisting mainly of hemicellulose degradation products.

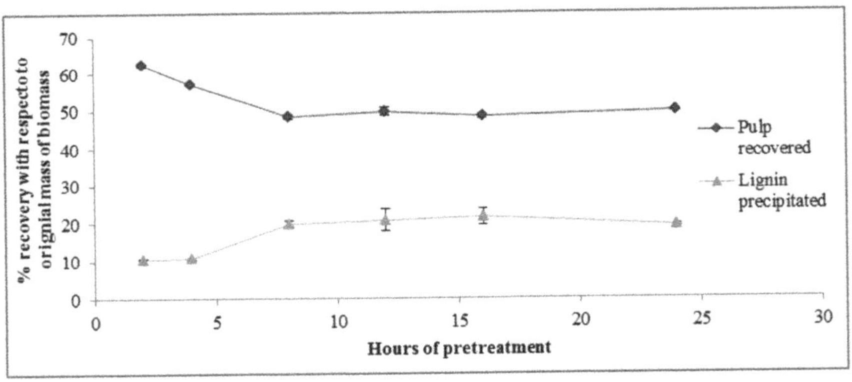

Figure 2. Pulp recovered and lignin precipitated after pretreatment and washing. Recovery based on oven-dried weight. Data from reference (37)

Enzymatic Saccharification and Pulp Composition

Saccharification of the recovered pulp was used to assay the effectiveness of the pretreatment. Comparing the saccharification yields with the mass loss during pretreatment (Figure 4), a high mass loss corresponds roughly to a higher glucan yield. This observation has been reported in literature before and is attributed to the fact that inhibitory compounds, such as lignin and hemicellulose and decomposition products of those biopolymers, are removed during pretreatment, leading to less nonproductive enzyme binding during the saccharification (*24*, *38*, *39*).

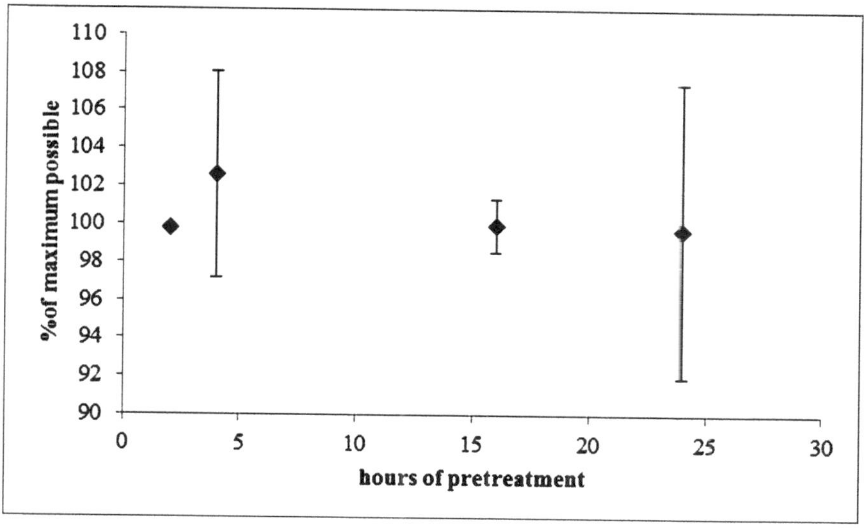

Figure 3. Recovery of $[N_{0222}][HSO_4]$ after the pretreatment of miscanthus at 120°C and precipitation of lignin. Data from reference (37)

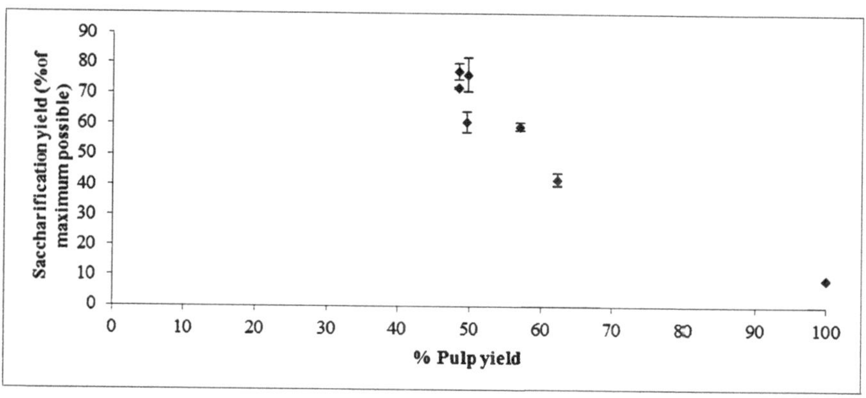

Figure 4. Pulp yield after pretreatment and glucose yield after 7 days of enzymatic saccharification. Data from reference (37)

Figure 5 shows the saccharification yield at different pretreatment times. The maximum xylan yield is reached after less than 2 hours of pretreatment while the maximum glucan yield is achieved only after ca. 10 hours of pretreatment. The fact that xylan and glucan yields cannot be optimised simultaneously has been reported previously (*21*, *24*, *40*). Compositional analysis of the recovered pulp (Table 1 and Figure 6) confirms the removal of hemicelluloses with prolonged pretreatment.

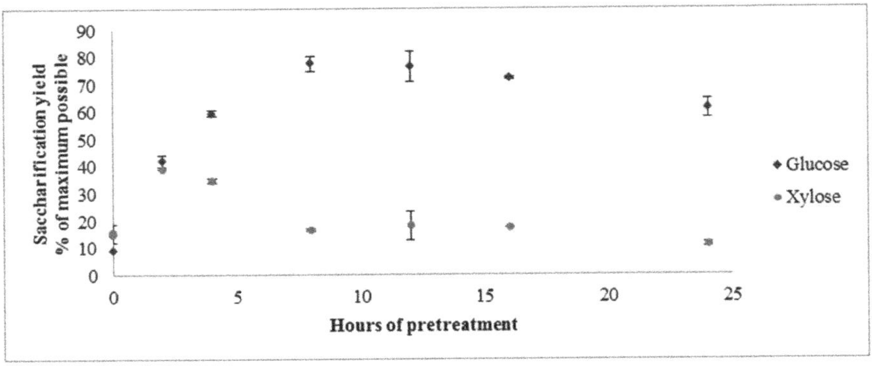

Figure 5. Sugar yield after 7 days of saccharification of pretreated micanthus pulp. Data from reference (37)

Table 1. Pulp recovery and lignin precipitation after pretreatment. Data from reference (37)

Time			*Pulp*				*Lignin*[b]
	Glucose[a]	*Xylose*[a]	*AIL*[a]	*ASL*[a]	*Ash*[a]	*Total*[b]	
0 h [c]	47.7	24.5	23.4	1.1	n/a	100	-
2 h	43.4 ±0.0	11.1 ±0.1	7.7 ±0.6	0.4 ±0.0	0.0	62.3 ±0.5	10.5 ±0.2
4 h	42.8 ±0.2	8.7 ±0.1	5.2 ±0.1	0.3 ±0.0	0.0	57.1 ±0.3	10.8 ±0.2
8 h	43.5 ±0.6	1.7 ±0.0	2.9 ±0.3	0.2 ±0.0	0.0	48.4 ±0.7	19.8 ±0.7
12 h	43.9 ±0.2	2.3 ±0.6	3.9 ±0.2	0.2 ±0.0	0.0	49.7 ±1.0	20.8 ±2.9
16 h	43.9 ±0.4	3.5 ±0.5	3.2 ±0.7	0.2 ±0.0	0.0	48.4 ±0.5	21.5 ±2.0
24 h	41.5 ±0.4	0.0 ±0.0	7.5 ±0.0	0.1 ±0.0	0.0	49.6 ±0.4	19.1 ±0.2
Verdia *et al.* (*24*)	35.2	1.5	1.7	0.6	1.0	40.1	24.3

[a] Pulp analysed by compositional analysis in duplicates. [b] Total recovery and lignin precipitation from pretreatment experiments in triplicates. [c] 24 hours at 120°C in $[HC_4im][HSO_4]$ with 20wt% water.

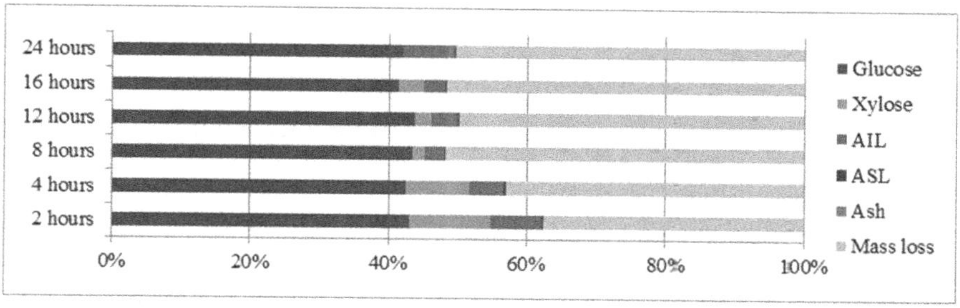

Figure 6. Compositional analysis of the miscanthus pulp recovered after pretreatment with [N_{0222}][HSO_4] at 120°C. Data from reference (37)

Although saccharification yields decrease after longer pretreatment times, the glucan content of the recovered biomass is relatively stable throughout the pretreatment. Hemicellulose content decreased continuously over time. While the saccharification indicates that at least 10% of the hemicellulose originally present in the biomass is still present after 24 hours of pretreatment, the compositional analysis shows complete removal, likely due to artifacts introduced by the concentrated sulfuric acid used in compositional analysis. The structure of biomass which has already been pretreated with ionic liquids is most likely already broken up to a certain degree and the concentrated sulfuric acid is therefore likely to hydrolyse and degrade part of the polysaccharides, giving rise to discrepancies between results obtained from saccharification experiments and compositional analysis.

The apparent acid-insoluble lignin (AIL) content in the recovered biomass decreases initially, then rises again after longer pretreatment. An increase in acid-insoluble solids content was also observed during pretreatment with [HC_4im][HSO_4] with high acid contents (1.5:1 acid to base ratio) and is attributed to the formation of insoluble carbohydrate degradation products (pseudolignin). Another possible explanation is the occurrence of recondensation reactions within the lignin polymer, leading to higher molecular weight lignin with decreased solubility in the ionic liquid. One drawback associated with compositional analysis of pulps is that acid insoluble lignin cannot be distinguished from other acid-insoluble byproducts.

Technoeconomic Considerations

There are a number of technical and economic factors that underpin the viability of using a solvent in a given application, particularly an ionic liquid. Most academic studies have naturally focused on technical considerations, and these are rightly viewed as go/no go criteria for research purposes. A key example is the high viscosity of most ionic liquids, which can limit transport in applications involving multiphase processing (especially gas/liquid transport). Transport considerations have not yet limited our development of the ioncSolv process, though the scales involved are not yet large enough to make a final determination (we have only used the process at up to the 1 L scale, though preliminary results with wood chips are promising). The high thermal stability of the ILs under study here was previously published by us (they can be operated up to 270 C) and can limit the use of ILs at large scale as solvent losses to degradation can quickly become crippling. However, other potential issues (toxicity, disposal, corrosion) have not yet been examined.

Economic considerations are also of import if translation of IL-based technologies is to become a reality. We discussed the solvent cost earlier, and in the context of bioethanol it is important to note that if the ionic liquid itself is 100 times more valuable than the proposed product, solvent recovery will quickly dominate the economics of the process. We have focused our efforts on 1) minimizing solvent cost; 2) minimizing solvent losses; 3) maximizing biomass loading (to minimize solvent use); and we next identify key energy factors as potentially limiting. While there are sustainability considerations to take into account when a solvent is proposed for use, the attending energy costs regarding solvent recovery are likewise of key economic importance. Only a full process technoeconomic model can identify these key issues, and we have not completed this to date.

Conclusions

In this chapter we described lignocellulose fractionation with a low-cost ionic liquid, and some of the key parameters affecting cost and pretreatment effectiveness. Pretreatment with triethylammonium hydrogen sulfate $[N_{0222}][HSO_4]$ at a mild temperature (120 °C) resulted in ca. 80% yield of glucose after 8 h. More than 85% lignin was removed and 80% of the lignin could be recovered, with up to 80% of its ether bonds cleaved. With harsher conditions (> 16 h pretreatment time) condensed lignin and pseudo-lignin became associated with the pulp, inhibiting saccharification. Evidence for lignin condensation was seen in the HSQC NMR spectra and in the molecular weight data. Lignin re-precipitation at prolonged pretreatment times also decreased lignin recovery.

A high-level technoeconomic assessment predicts that capital and operating costs will be lower than for the bench-mark dilute acid pretreatment. Areas of future process development work include optimization of pretreatment temperature, biomass loading, hemicellulose recovery, up-scaling and heat integration. The ionoSolv pretreatment system shows clear potential for industrial scale-up due to high efficiency and very low solvent cost compared to other ionic liquid or organic solvent based pretreatment systems.

Experimental

Ionic Liquid Synthesis and Characterisation

Starting materials for ionic liquid synthesis were purchased from Sigma Aldrich and, unless stated otherwise, used as received. ^{1}H, ^{13}C, HSQC, HMQC and HMBC NMR were recorded on a Bruker 400 MHz spectrometer. Chemical shifts (δ) are reported in ppm, the DMSO signal at 2.500 (^{1}H dimension) and 39.520 (^{13}C dimension). Mass spectrometry was measure by Dr. Lisa Haigh

(Imperial College London, Chemistry Department) on a Micromass Premier spectrometer.

Synthesis of Triethylammonium Hydrogensulfate [N_{0222}][HSO_4]

Triethylamine (76.1 g, 750 mmol) was cooled with an acetone dry ice mix. Under stirring, 150 ml of 5M H_2SO_4 (1.25 mol) were added dropwise. The water was removed under reduced pressure and the product dried in vacuum at 70°C overnight.

^{1}H NMR: δH (400 MHz, DMSO-d^6)/ppm: 3.39 (s (br), [HSO_4]$^-$, N-H^+), 3.10 (q, J = 7.3 Hz, 6H, N-CH_2), 1.20 (t, J=7.3 Hz, 9H, N-CH_2-CH_3).

^{13}C NMR: δC (101 MHz, DMSO-d^6)/ppm: 46.21 (N-CH_2), 9.15 (N-CH_2-CH_3).

MS (Magnet FB$^+$) m/z: 102 ([N_{0222}]$^+$, 100%), (Magnet FB$^-$) m/z: 79 ([HSO_4]$^-$, 100%).

Fractionation of Biomass

During pretreatment experiments, all weights were recorded using an A&D GH-252 with an accuracy of ±0.1 mg. Pretreatments were run in a Thermo Scientific HERA THERM convection oven, which was also used for the determination of oven dried weights. Lignin was dried in a Binder VD 23 vacuum oven. A C-28 centrifuge from Boeco, Germany was used.

Miscanthus x giganteus was obtained from Silwood park campus (Imperial College London, UK). It was air-dried, ground and sieved (180-850 μm, 20 + 80 US mesh scale). For the untreated biomass and recovered pulp the moisture content was determined by weighing out approximately 100 mg of biomass/pulp onto a preweighed piece of aluminium foil and recording the weight. The foil with the biomass/pulp was folded and oven dried (T=105°C) overnight. The next day, the hot packets were taken out of the oven and placed in a desiccator to allow cooling to room temperature. The new weight was recorded immediately afterwards and the moisture content calculated. This was done in triplicates for untreated biomass and once per sample for recovered pulp.

For the pretreatments, an ionic liquid/water master-mix was prepared by adding the required amount of water to the dried ionic liquid. The water content was confirmed by Karl-Fischer titration in triplicates. 10±0.05 g of ionic liquid/water master-mix is weighed into a glass pressure tube and the exact weight recorded. Between 1.04 and 1.09 g of miscanthus was added, the vials capped and the content mixed with a vortex shaker. They were then placed into a preheated convection oven. After the pretreatment period, they were taken out and allowed to cool to room temperature. Experiments were carried out in triplicate.

After the pretreatment, 40 mL of ethanol was added to the pretreatment mixture and the suspension transferred into a 50 mL Falcon tube. The tube was shaken for one minute and the mixture then left at room temperature for at least 1 hour. The tube was mixed again for 30 seconds and then centrifuged at 4000 rpm for 50 minutes. The supernatant was decanted carefully into a round bottom flask. The washing step was repeated three more times. The remaining pulp was then transferred into a cellulose thimble and further washed by Soxhlet extraction with refluxing ethanol (150 mL) for 22 hours. The thimbles were then left on the bench overnight to dry. The ethanol used for the Soxhlet extraction was combined with the previous washes and evaporated under reduced pressure at 40°C, leaving the dried ionic liquid/lignin mixture. To the dried ionic liquid/lignin mixture, 30 mL of water was added in order to precipitate the lignin. The suspension was transferred into a 50 mL falcon tube, shaken for one minute and then left at room temperature for at least 1 hour. The tube was centrifuged and the supernatant decanted and collected in a round bottom flask. This washing step was repeated twice more.

The air-dried pulp yield was determined by weighing the recovered biomass from the cellulose thimbles. The oven-dried yield was determined as described for the untreated biomass. The lid of the Falcon tube containing the lignin was pierced and the tube put into a vacuum oven overnight to dry at 40°C under vacuum. The dried lignin was weighed the next day.

Compositional Analysis

Determination of Structural Carbohydrates and Lignin in Biomass. 200-300 mg of air-dry biomass or recovered biomass was weighed out into a pressure tube and the weight recorded (Sartoriaum CPA 1003 S balance, ±0.001 g). 3 mL of 72% sulfuric acid was added, the samples stirred with a Teflon stir rod and the pressure tubes placed into a preheated water bath at 30°C. The samples were stirred again every 15 min. for one hour, they were then diluted with 84 mL distilled water and the lids closed. The samples were autoclaved (Sanyo Labo Autoclave ML5 3020 U) for 1 hour at 121°C and left to cool to close to ambient temperature. The samples were then filtered through filtering ceramic crucibles of a known weight. The filtrate was filled in two Falcon tubes and the remaining black solid washed with distilled water. The crucibles were placed into a convection oven (VWR Venti-Line 115) at 105°C for 24±2 hours. They were then taken out and placed in a desiccator for 15 min before they were weighed and the weight recorded. They were then placed into a muffle oven (Nabertherm + controller P 330) and ashed to constant weight at 575°C. The weight after ashing was recorded. The content of acid insoluble lignin (AIL) was determined according to Equation 1. The content of one of the Falcon tubes was used for the determination of acid soluble lignin content (ASL) by UV analysis at 240 nm (Equation 2) (Perkin Elmer Lambda 650 UV/Vis spectrometer).

$$\%AIL = \frac{Weight_{crucible\ plus\ AIR} - Weight_{crucible\ plus\ ash}}{ODW_{sample}} \cdot 100 \qquad \text{(eq. 1)}$$

$$\%ASL = \frac{A}{l \cdot \varepsilon \cdot c} \cdot 100 = \frac{A \cdot V_{filtrate}}{l \cdot \varepsilon \cdot ODW_{sample}} \cdot 100 \qquad \text{(eq. 2)}$$

where Weightcrucibles plus AIR is the weight of the oven-dried crucibles plus the acid insoluble residue, Weightcrucibles plus ash is the weight of the crucibles after ashing to constant temperature at 575°C, A is the absorbance at 240 nm, l is the pathlength of the cuvette in cm (1 cm in this case), ε is the extinction coefficient (12 L/g cm), c is the concentration in mg/mL, ODW is the oven-dried weight of the sample in mg and Vfiltrate is the volume of the filtrate in mL and equal to 86.73 mL.

To the contents of the other Falcon tube calcium carbonate was added until the pH reached 5. The liquid was passed through a 0.2 µm PTFE syringe filter and subsequently submitted to HPLC analysis (Shimadzu, Aminex HPX-97P from Bio rad, 300 x 7.8 mm, purified water as mobile phase at 0.6 ml/min, column temperature 85°C) for the determination of total sugar content. Calibration standards with concentrations of 0.1, 1, 2 and 4 mg/mL of glucose, xylose, mannose, arabinose and galactose were used. Sugar recovery standards were made as 10 mL aqueous solutions close to the expected sugar concentration of the samples and transferred to pressure tubes. 278 µL 72% sulfuric acid was added, the pressure tube closed and autoclaved and the sugar content determined as described above. The sugar recovery coefficient (SRC) was determined according to equation 3 and the sugar content of the analysed sample using equation 4:

$$SRC = \frac{c_{HPLC} \cdot V}{initial\ weight} \qquad \text{(eq. 3)}$$

$$\%Sugar = \frac{c_{HPLC} \cdot V \cdot corr_{anhydro}}{SRC \cdot ODW_{sample}} \cdot 100 \qquad \text{(eq. 4)}$$

where cHPLC is the sugar concentration detected by HPLC, V is the initial volume of the solution in mL (10.00 mL for the sugar recovery standards and 86.73 mL for the samples), initial weight is the mass of the sugars weighed in, corranhydro is the correction for the mass increase during hydrolysis of polymeric sugars (0.90 for C6 sugars glucose, galactose and mannose and 0.88 for C5 sugars xylose and arabinose) and ODW is the oven-dried weight of the sample in mg.

Saccharification Assay

Saccharification assays were carried out in triplicates with blanks (also triplicates). All reagents and enzymes were purchased from Sigma Aldrich.

100 mg of air-dried biomass was placed into a Sterilin tube and the weight recorded. Three blanks were run with 100 µL of purified water in order to correct for sugar residues present in the enzyme solutions. 9.9 mL solution consisting of 5 mL 1M sodium citrate buffer at pH 4.8, 40 µL Tetracyline antibiotic solution (10 mg/mL in 70% ethanol), 30 µL Cycloheximide antibiotic solution (10 mg/mL in purified water), 4.71 mL purified water, 60 µL Cellulase from Trichoderma reesei ATCC 26921 solution and 60 µL Cellobiase from Aspergillus niger solution was added, the tubes closed and placed into an Stuart Orbital Incubator (S1500) at 50°C and 250 rpm.

Time point samples were taken after 4, 18, 48, and 96 hours and an end point sample after 168 hours. For time point samples, 500 μL of the saccharification mixture was taken out (representative amount of solids and liquids) and transferred to a microcentrifuge tube. The samples were centrifuged in a VWR MICRO STAR 17R centrifuge at 4°C and 13.3 G for 10 min. The supernatant was pipetted off into another microcentrifuge tube and frozen until analysis. Prior to analysis, they were shaken with a vortex shaker and centrifuged again at 4°C and 13.3 G for 5 min. End point samples were obtained by filtering 1 mL of the saccharification mixture though a PTFE syringe filter. Samples were run on Shimadzu HPLC with an AMINEX HPX-97P column (Bio rad, 300 x 7.8 mm) with purified water as mobile phase (0.6 mL/min). The column temperature was 85°C and acquisition was run for 40 min. Calibration standards with concentrations of 0.1, 1, 2 and 4 mg/mL of glucose, xylose, mannose, arabinose and galactose and 8 mg/mL of glucose were used.

Acknowledgments

The authors wish to acknowledge the Engineering and Physical Sciences Research Council (EP/K014676/1) for funding for AB, the Grantham Institute for Climate Change and the Environment for a studentship for FJVG and Imperial College London for a studentship for CLC. Additional funding was provided by Climate-KIC for CLC and FJVG.

References

1. Stöcker, M. Biofuels and Biomass-to-Liquid Fuels in the Biorefinery: Catalytic Conversion of Lignocellulosic Biomass Using Porous Materials. *Angew. Chem., Int. Ed.* **2008**, *47* (48), 9200–9211.
2. Muhammad, N.; Omar, W. N.; Man, Z.; Bustam, M. A.; Rafiq, S.; Uemura, Y. Effect of Ionic Liquid Treatment on Pyrolysis Products from Bamboo. *Ind. Eng. Chem. Res.* **2012**, *51* (5), 2280–2289.
3. Shah, N.; Vallejo, L.; Cockerill, T.; Gambhir, A.; Heyes, A.; Hills, T.; Jennings, M.; Jones, O.; Kalas, N.; Keirstead, J.; Khor, C.; Mazur, C.; Napp, T.; Strapasson, A.; Tong, D.; Woods, J. *Halving Global CO_2 by 2050: Technologies and Costs*; Energy Futures Lab and Grantham Institute for Climate Change, Imperial College London: 2013.
4. FitzPatrick, M.; Champagne, P.; Cunningham, M. F.; Whitney, R. A. A Biorefinery Processing Perspective: Treatment of Lignocellulosic Materials for the Production of Value-Added Products. *Bioresour. Technol.* **2010**, *101* (23), 8915–8922.
5. Liu, C.; Wyman, C. E. Impact of Fluid Velocity on Hot Water Only Pretreatment of Corn Stover in a Flowthrough Reactor; In *Proceedings of the Twenty-Fifth Symposium on Biotechnology for Fuels and Chemicals*; May 4–7, 2003, Breckenridge, CO; Springer: 2004; pp 977−987.
6. Balan, V.; Sousa, L. d. C.; Chundawat, S. P.; Marshall, D.; Sharma, L. N.; Chambliss, C. K.; Dale, B. E. Enzymatic Digestibility and Pretreatment

Degradation Products of AFEX-Treated Hardwoods (Populus Nigra). *Biotechnol. Prog.* **2009**, *25* (2), 365–375.

7. Lloyd, T. A.; Wyman, C. E. Combined Sugar Yields for Dilute Sulfuric Acid Pretreatment of Corn Stover Followed by Enzymatic Hydrolysis of the Remaining Ssolids. *Bioresour. Technol.* **2005**, *96* (18), 1967–1977.
8. Zhao, X.; Cheng, K.; Liu, D. Organosolv Pretreatment of Lignocellulosic Biomass for Enzymatic Hydrolysis. *Appl. Microbiol. Biotechnol.* **2009**, *82* (5), 815–827.
9. Brandt, A.; Gräsvik, J.; Hallett, J. P.; Welton, T. Deconstruction of Lignocellulosic Biomass with Ionic Liquids. *Green Chem.* **2013**, *15* (3), 550–583.
10. Hallett, J. P.; Welton, T. Room-Temperature Ionic Liquids: Solvents for Synthesis and Catalysis. 2. *Chem. Rev.* **2011**, *111* (5), 3508–3576.
11. Gao, D.; Haarmeyer, C.; Balan, V.; Whitehead, T. A.; Dale, B. E.; Chundawat, S. P. Lignin Triggers Irreversible Cellulase Loss During Pretreated Lignocellulosic Biomass Saccharification. *Biotechnol. Biofuels* **2014**, *7* (1), 1–13.
12. Palmqvist, E.; Hahn-Hägerdal, B. Fermentation of Lignocellulosic Hydrolysates. II: Inhibitors and Mechanisms of Inhibition. *Bioresour. Technol.* **2000**, *74* (1), 25–33.
13. Ragauskas, A. J.; Beckham, G. T.; Biddy, M. J.; Chandra, R.; Chen, F.; Davis, M. F.; Davison, B. H.; Dixon, R. A.; Gilna, P.; Keller, M.; Langan, P.; Naskar, A. K.; Saddler, J. N.; Tschaplinski, T. J.; Tuskan, G. A.; Wyman, C. E. Lignin Valorization: Improving Lignin Processing in the Biorefinery. *Science* **2014**, *344* (6185), 1246843.
14. Swatloski, R. P.; Spear, S. K.; Holbrey, J. D.; Rogers, R. D. Dissolution of Cellulose with Ionic Liquids. *J. Am. Chem. Soc.* **2002**, *124* (18), 4974–4975.
15. Michud, A.; Tanttu, M.; Asaadi, S.; Ma, Y.; Netti, E.; Kääriainen, P.; Persson, A.; Berntsson, A.; Hummel, M.; Sixta, H. Ioncell-F: Ionic Liquid-Based Cellulosic Textile Fibers as an Alternative to Viscose and Lyocell. *Textile Res. J.* **2016**, *86* (5), 543–552.
16. Eminov, S.; Wilton-Ely, J. D.; Hallett, J. P. Highly Selective and Near-Quantitative Conversion of Fructose to 5-Hydroxymethylfurfural Using Mildly Acidic Ionic Liquids. *ACS Sustainable Chem. Eng.* **2014**, *2* (4), 978–981.
17. Sathitsuksanoh, N.; Holtman, K. M.; Yelle, D. J.; Morgan, T.; Stavila, V.; Pelton, J.; Blanch, H.; Simmons, B. A.; George, A. Lignin Fate and Characterization During Ionic Liquid Biomass Pretreatment for Renewable Chemicals and Fuels Production. *Green Chem.* **2014**, *16* (3), 1236–1247.
18. Wang, H.; Maxim, M. L.; Gurau, G.; Rogers, R. D. Microwave-Assisted Dissolution and Delignification of Wood in 1-Ethyl-3-methylimidazolium Acetate. *Bioresour. Technol.* **2013**, *136*, 739–742.
19. da Costa Lopes, A. M.; Joao, K. G.; Bogel-Łukasik, E.; Roseiro, L. B.; Bogel-Łukasik, R. Pretreatment and Fractionation of Wheat Straw Using Various Ionic Liquids. *J. Agric. Food Chem.* **2013**, *61* (33), 7874–7882.

20. Groff, D.; George, A.; Sun, N.; Sathitsuksanoh, N.; Bokinsky, G.; Simmons, B. A.; Holmes, B. M.; Keasling, J. D. Acid Enhanced Ionic Liquid Pretreatment of Biomass. *Green Chem.* **2013**, *15* (5), 1264–1267.
21. Brandt, A.; Ray, M. J.; To, T. Q.; Leak, D. J.; Murphy, R. J.; Welton, T. Ionic Liquid Pretreatment of Lignocellulosic Biomass with Ionic Liquid-Water Mixtures. *Green Chem.* **2011**, *13* (9), 2489–2499.
22. George, A.; Brandt, A.; Tran, K.; Zahari, S. M. S. N. S.; Klein-Marcuschamer, D.; Sun, N.; Sathitsuksanoh, N.; Shi, J.; Stavila, V.; Parthasarathi, R.; Singh, S.; Holmes, B. M.; Welton, T.; Simmons, B. A.; Hallett, J. P. Design of Low-Cost Ionic Liquids for Lignocellulosic Biomass Pretreatment. *Green Chem.* **2015**, *17* (3), 1728–1734.
23. Verdía, P.; Brandt, A.; Hallett, J. P.; Ray, M. J.; Welton, T. Fractionation of Lignocellulosic Biomass with the Ionic Liquid 1-Butylimidazolium Hydrogen Sulfate. *Green Chem.* **2014**, *16* (3), 1617–1627.
24. Diedericks, D.; Rensburg, E. v.; del Prado García-Aparicio, M.; Görgens, J. F. Enhancing the Enzymatic Digestibility of Sugarcane Bagasse Through the Application of an Ionic Liquid in Combination with an Acid Catalyst. *Biotechnol. Prog.* **2012**, *28* (1), 76–84.
25. Brandt, A.; Erickson, J. K.; Hallett, J. P.; Murphy, R. J.; Potthast, A.; Ray, M. J.; Rosenau, T.; Schrems, M.; Welton, T. Soaking of Pine Wood Chips with Ionic Liquids for Reduced Energy Input during Grinding. *Green Chem.* **2012**, *14* (4), 1079–1085.
26. Liu, C.-Z.; Wang, F.; Stiles, A. R.; Guo, C. Ionic liquids for Biofuel Production: Opportunities and Challenges. *Appl. Energy* **2012**, *92*, 406–414.
27. Labbé, N.; Kline, L. M.; Moens, L.; Kim, K.; Kim, P. C.; Hayes, D. G. Activation of Lignocellulosic Biomass by Ionic Liquid for Biorefinery Fractionation. *Bioresour. Technol.* **2012**, *104*, 701–707.
28. Cheng, G.; Varanasi, P.; Arora, R.; Stavila, V.; Simmons, B. A.; Kent, M. S.; Singh, S. Impact of Ionic Liquid Pretreatment Conditions on Cellulose Crystalline Structure Using 1-Ethyl-3-methylimidazolium Acetate. *J. Phys. Chem. B* **2012**, *116* (33), 10049–10054.
29. Viell, J. *A Pretreatment Process for Wood Based on Ionic Liquids*; RWTH Aachen: Düsseldorf, 2014; p 212.
30. Clough, M. T.; Geyer, K.; Hunt, P. A.; Mertes, J.; Welton, T. Thermal Decomposition of Carboxylate Ionic Liquids: Trends and Mechanisms. *Phys. Chem. Chem. Phys.* **2013**, *15* (47), 20480–20495.
31. Doherty, T. V.; Mora-Pale, M.; Foley, S. E.; Linhardt, R. J.; Dordick, J. S. Ionic Liquid Solvent Properties as Predictors of Lignocellulose Pretreatment Efficacy. *Green Chem.* **2010**, *12* (11), 1967–1975.
32. Hall, C. A.; Le, K. A.; Rudaz, C.; Radhi, A.; Lovell, C. S.; Damion, R. A.; Budtova, T.; Ries, M. E. Macroscopic and Microscopic Study of 1-Ethyl-3-methyl-imidazolium Acetate–Water Mixtures. *J. Phys. Chem. B* **2012**, *116* (42), 12810–12818.
33. Brandt, A. *Pretreatment of Lignocellulosic Biomass with Ionic Liquids*; Imperial College London: London, 2012.

34. Chen, L.; Sharifzadeh, M.; Mac Dowell, N.; Welton, T.; Shah, N.; Hallett, J. P. Inexpensive Ionic Liquids: $[HSO_4]^-$-Based Solvent Production at Bulk Scale. *Green Chem.* **2014**, *16* (6), 3098–3106.
35. Wasserscheid, P.; Welton, T. *Ionic Liquids in Synthesis*; Wiley-VCH: 2002.
36. Plechkova, N. V.; Seddon, K. R. Applications of Ionic Liquids in the Chemical Industry. *Chem. Soc. Rev.* **2008**, *37* (1), 123–150.
37. Brandt, A.; Gschwend, F. J. V.; Fennell, P. S.; Lammens, T. M.; Tan, B.; Weale, J.; Hallett, J. P. An Economically Viable Ionic Liquid for the Pretreatment of Lignocellulosic Biomass. *Green Chem.* **2017**, *19*, 3078–3102.
38. Cox, B. J.; Ekerdt, J. G. Pretreatment of Yellow Pine in an Acidic Ionic Liquid: Extraction of Hemicellulose and Lignin to Facilitate Enzymatic Digestion. *Bioresour. Technol.* **2013**, *134*, 59–65.
39. Li, H.; Pu, Y.; Kumar, R.; Ragauskas, A. J.; Wyman, C. E. Investigation of Lignin Deposition on Cellulose during Hydrothermal Pretreatment, its Effect on Cellulose Hydrolysis, and Underlying Mechanisms. *Biotechnol. Bioeng.* **2014**, *111* (3), 485–492.
40. Sun, N.; Liu, H.; Sathitsuksanoh, N.; Stavila, V.; Sawant, M.; Bonito, A.; Tran, K.; George, A.; Sale, K. L.; Singh, S. Production and Extraction of Sugars from Switchgrass Hydrolyzed in Ionic Liquids. *Biotechnol. Biofuels* **2013**, *6* (1), 1–15.

Fundamentals

Chapter 10

Water at Ionic Liquid Interfaces

Alicia Broderick and John T. Newberg*

Department of Chemistry and Biochemistry, University of Delaware, Newark, Delaware 19716, United States
***E-mail: jnewberg@udel.edu**

Water is known to affect bulk properties of ionic liquids (ILs) including density, viscosity, conductivity and gas absorption. It is also becoming increasingly recognized that water gives rise to significant changes at IL interfaces. In this chapter we review the surface sensitive analytical techniques and molecular dynamic (MD) simulations utilized to probe the IL-vacuum, IL-gas and IL-solid interfaces in the presence of water. An overview is first given from the perspective of surface science experiments, followed by a focus on results from MD simulations. Experimental studies in most cases examined the IL in the melted state, while a few studies examined the IL-vacuum interface while the IL transitioned between frozen and melted states. Over the past two years there has been a significant increase in atomic force microscopy (AFM) studies probing the effects of water on the IL-solid interface of both neutral surfaces and electrified surfaces. Experimental and MD simulation studies that vary the amount of water often reveal water mole fraction (x_w) dependent structural changes and IL layering at IL-gas and IL-solid interfaces. Under low x_w conditions the concentration of water at the interface can be significantly different than in the bulk. While water is often viewed as a ubiquitous contaminant, it is also possible to envisage utilizing water as a chemical knob to influence x_w dependent interfacial IL chemistry and structure with significant implications in gas absorption, electrochemical and surface catalytic studies.

Introduction

The structural versatility of ILs allows for the potential to synthesize and tune various properties including polarity, conductivity, viscosity, density, melting point, acidity, surface tension, and/or electrochemical window (*1*). Understanding the interaction of ILs with small gas phase molecules (i.e., sorbates) has important implications in gas separation, sequestration, energy storage, fuels, lubrication, sensors and catalysis (*2–8*). Figure 1 illustrates sorbate interactions for three example processes of gas separation, heterogeneous catalysis and electrocatalysis which involve: i) sorbate collisions with the IL-gas interface, ii) capture of sorbate at the IL-gas interface, iii) diffusion of sorbate into the bulk, iv) adsorption and/or reaction at the IL-solid interface, v) diffusion of sorbate back to the IL-gas interface, and vi) desorption from IL-gas interface. The overall process of uptake and release is determined by sorbate interactions in the bulk and at IL interfaces (including IL-gas and/or IL-solid). Understanding the interactions occurring in the bulk and at interfaces will provide molecular level knowledge for the development of task specific ILs for capture and conversion technologies.

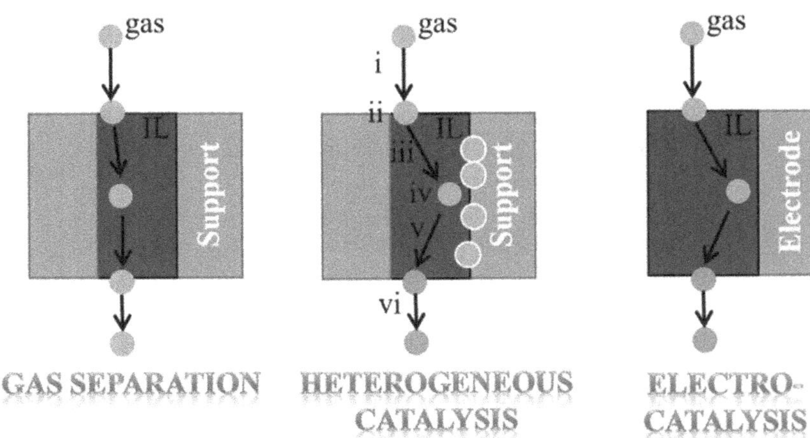

Figure 1. Processes involving IL-gas and IL-solid interactions, including i) sorbate collisions with the IL-gas interface, ii) capture of sorbate at the IL-gas interface, iii) diffusion of sorbate into the bulk, iv) adsorption and/or reaction at the IL-solid interface, v) diffusion of sorbate back to the IL-gas interface, and vi) desorption from IL-gas interface

Water is often viewed as an impurity for ILs exposed to the ambient environment, where even hydrophobic ILs are known to absorb small amounts of water over time (*9*). Water affects a number of bulk IL properties including density, viscosity, conductivity and gas absorption capacity (*10–14*). Water also affects bulk IL orientational dynamics which in turn can affect reaction rates (*15*). Given the ubiquity of water it is important to understand the interplay water has with ILs at IL-gas and IL-solid interfaces in order to assess the potential beneficial and/or deleterious role water may have on the aforementioned processes (Figure 1).

An overview of IL studies involving interfaces was recently published in two book chapters discussing experimental and theoretical examinations of the IL-vacuum, IL-gas, IL-liquid and IL-solid interfaces (*16*). Studies have been predominantly on imidazolium based ILs. The molecular level picture of ILs present at interfaces is often described on a layer-by-layer basis even though the liquid phase ions are dynamic in nature. The IL-vacuum and IL-gas interface can be described as having three distinct regions (*16*). The outermost region is commonly enriched with imidazolium cations with the hydrophobic alkyl chain pointing towards the vacuum and the charged imidazolium ring interacting with the corresponding anion. Below this outer most interfacial region is a second transition region which consists of both cations and anions and is typically several ion-pair diameters in length. Beyond this transition region is a third region which continues into the IL bulk and can be either homogeneous or bicontinuous. The IL-solid interface can be described as having oscillating IL density profiles within the first few nm adjacent to the solid surface, although depending on the nature of the IL and the solid interface this oscillating nature can also be absent. The IL structure is significantly impacted when an electric potential is applied to the interface which has significant implications in electrochemistry and electrocatalysis.

There are a number of reviews that have highlighted studies involving the interaction between water and ILs, emphasizing the importance of solute-solvent interactions (*17*, *18*), water induced changes in surface structure (*19*, *20*) and the use of computational studies to shed light on molecular level details (*21*). The goal of this Chapter is to give an up to date overview of the surface sensitive analytical techniques and theoretical studies utilized to probe the IL-vacuum, IL-gas and IL-solid interfaces in the presence of water. The focus herein will be on the use of microscopy, spectroscopy, mass spectrometry and scattering techniques (Table 1), as well as molecular dynamic simulations (Table 2). The most significant contribution over the past two years has been from microscopy studies shedding new light on the influence of water at the IL-solid interface. As will be shown in this chapter, some experiments are performed by depositing water onto an IL that is initially frozen at the start of experiments, while others perform studies with the IL in the liquid state throughout the entire experiment. We will refer to the former condition as water "adsorption" onto a solid surface. In some cases, water is found to be enhanced at the IL interface relative to the bulk while it is in a liquid state. Under these circumstances we will refrain from the use of the term "adsorbed" to describe water in the interfacial region of an IL in a liquid state since there is freedom for water to diffuse into the IL bulk.

Table 1. Experimental Studies Probing Ionic Liquid Interfaces in the Presence of Water

Interface	*Technique*	*H_2O[a]*	*IL Anions[b]*	*P_{max} (Torr)[c]*	*Findings*	*Refs*
IL-Vac	NICISS	l	[Cl]	1.5×10^{-4}	Cation enhanced at interface, $[Cl]^-$ migrates towards bulk upon introduction of water.	(*22*)
	NICISS	l	[Cl]	7.5×10^{-6}	For low x_w water induces $[Cl]^-$ enhancement at interface.	(*23*)
	UHV XPS	g	$[BF_4]$	3×10^{-6}	-76 kJ mol^{-1} heat of adsorption for water while IL is transitioning from solid to liquid.	(*24*)
	LOSMS	g	$[BF_4]$,$[NTf_2]$	4×10^{-8}	Water desorption energies -43 to -44 kJ mol^{-1} at 1 ML coverage.	(*25*)
IL-Gas	SFG	g	[imide]	200	Above 10^{-4} Torr cation ring tipped up from the surface. No water observed up to 200 Torr.	(*29*)
	SFG	g	$[BF_4]$,[imide]	20	Cation reorients in presence of water for hydrophobic [imide]- but not for the hydrophilic $[BF_4]^-$.	(*30*)
	SFG	l	$[BF_4]$	Air	Terminal methyl in butyl chain normal to surface and unaffected by changes in x_w.	(*31*)
	SFG	g,l	$[BF_4]$	24 Torr	Cation ring nearly parallel to water surface and unaffected by changes in x_w.	(*32*)
	SFG	l	$[BF_4]$	Air	Peak shifts in terminal methyl suggest there may be reorientation for high x_w.	(*33*)
	SFG	g	$[PF_6]$	Air	Pure IL in lab RH = 40 %, cation ring parallel to surface with butyl chain sticking out.	(*34*)
	SFG	l	[FAP]	Air	Comparing neat IL and dilute IL in water, cation reorients in presence of water.	(*35*)
	SFG	l	$[O_3SOC_1]$	Air	Cation changes orientation with x_w.	(*36*)
	XR	l	$[NTf_2]$	N_2	Cations and water present near interface.	(*37*)
	NR	l	[Br]	Air	IL depleted at surface above critical micelle concentrations of 0.15 mol dm^{-3}.	(*38*)

Interface	*Technique*	H_2O^a	*IL Anions*[b]	P_{max} *(Torr)*[c]	*Findings*	*Refs*
	FJSB-KW	g	[NTf_2]	250	Determined D_2O dissolution $\Delta H = -53\ kJ\ mol^{-1}$ and $\Delta S = -210\ J\ mol^{-1}\ K^{-1}$.	*(39)*
	APXPS	g	[Ace]	5	Direct measure of interfacial water versus pressure, ~5 waters per IL pair at 5 Torr.	*(40)*
				Interface		
IL-Solid	AFM	g,l	[NTf_2]	Mica	Interfacial layering depends on alkyl chain length and presence of sufficient water.	*(44)*
	AFM	g	[O_3SOC_2]	Mica	Dry vs. 45 % RH shows water disrupts surface layering at IL-mica interface.	*(45)*
	AFM	g	[FAP],[C_2SO_4]		Dry vs. 37 % RH shows water influences ion-pair orientation and slip conditions.	*(46)*
	AFM	g,l	[NTf_2]	Mica	Dry vs. water saturated ILs shows water disruptions layering at IL-mica interface	*(47)*
	AFM	l	[BF_4],[NTf_2]	Mica/ Silica	Water disturbs solvation layers depending on IL used and/or silica vs. mica substrate.	*(48)*
	AFM(V)[d]	l	[dca]	HOPG	Interfacial layering depends on presence of water, potential applied and IL examined.	*(49)*
	AFM(V)	l	[NTf_2]	Mica, Au	Interfacial layering on Au significantly affected under positive potential.	*(50)*
	AFM(V)	l	[OTf]	Au(111)	Structure of interfacial layering depends on applied potential and the water amount.	*(51)*
	SEIRAS	g	[NTf_2]	Au	Water is in first ionic layer and bonds to anion more strongly than cation	*(52)*

[a] IL exposed to gas phase water (g) and/or mixed with liquid water (l). [b] Lists anions used, most studies (except ref. *(37)*) used [$C_n$mim] imidazolium cations. References *(37)* and *(49)* used [C_4mpyr]. [c] Approximate maximum pressures (P_{max}) during in situ probing of IL interface. Experiments under atmospheric pressure either exposed to ambient air or N_2. [d] AFM(V) studies examined surfaces under an applied electric potential.

Table 2. Molecular Dynamics Studies of Ionic Liquid Interfaces in the Presence of Water.

Interface	*Other Species*	*Ionic Liquids*	*Findings*	*Refs*
IL-Vac	-	[C_1mim][Cl]	Water enhanced at IL-vacuum interface.	(*55, 56*)
	-	[C_4mim][BF_4],[PF_6]	Cation alkyl enhanced at IL-vacuum interface, water present under cation layer along with anion.	(*57*)
	-	[C_4mim][BF_4]	Presents interfaces but little emphasis placed on understanding water vs. IL density profile details.	(*58*)
	CO_2	[C_4mim][BF_4],[NTf_2], [PF_6]	PMF shows monotonic decrease in free energy of water molecule crossing IL-vacuum interface.	(*59, 60*)
	CO_2	[C_4mim][NTf_2]	Cation alkyl enhanced at IL-vacuum interface, water present under cation layer along with anion.	(*61*)
	CO_2, N_2, O_2	[C_2mim][Gly]	Water present at inner layer of IL blocking CO_2 absorption.	(*62*)
	CO_2, N_2, O_2	[MP][lac],[MP][PR],[EP][lac]	Water present at inner layer through strong interactions with anions.	(*63*)
IL-Solid	-	[C_4mim][PF_6],[NTf_2]	Water is within sub-nm of electrode and water adsorption increases with voltage. Water manifests near positive electrodes where anions are present.	(*64*)
	-	[C_4mim][BF_4]	Water depleted at neutral/negative graphene interface, enhanced at positive interface	(*65*)

Ionic Liquid-Vacuum Interface

The low vapor pressures of ILs make them amenable to probing by vacuum based surface spectroscopy techniques. Techniques utilized to examine the IL-vacuum interface include Neutral impact collision ion scattering spectroscopy (NICISS) (*22, 23*), ultra-high vacuum X-ray photoelectron spectroscopy (UHV XPS) (*24*) and temperature program desorption (TPD) coupled to line of sight mass spectroscopy (LOSMS) (*25*). All three of these surface techniques are traditionally vacuum based because the inbound probes aimed at the sample (ions for NICISS) and/or outbound probes coming off the sample (ions for LOSMS, electrons for XPS, and neutrals for NICISS) are significantly scattered in the presence of a gas phase. Experiments performed with these techniques were done at pressures of 150 μTorr or lower. In the next section we consider the IL-gas interface which probes the IL interface in the presence of Torr level pressures all the way up to 1 atmosphere.

NICISS uses inert ions (typically He^+) which collide with the sample interface and scatter neutral He atoms for detection. NICISS is atomic specific and highly surface sensitive with the ability to do concentration dependent depth profiling with angstrom level resolution. NICISS studies (Table 1) investigated the IL-vacuum interface by preparing IL-water mixtures, which were then introduced into the NICISS probing chamber and pumped on during analysis. Under such conditions water was evaporating from the surface leading to a background chamber pressure reported in Table 1 for each experiment. Both NICISS studies examined [C_6mim][Cl]/water mixtures at high (*22*) and low (*23*) water mole fractions (x_w) in the range of 0.71 to 0.0025.

NICISS depth profile results are shown in Figure 2, investigating the neat IL (x_w = 0) and x_w from 0.43 to 0.71 (*22*). For neat [C_6mim][Cl] the carbon profile (black data; imidazolium cation) is enhanced in the interfacial region at 0.4 and 1.4 nm, where 0 nm represents the IL-vacuum interface. The $[Cl]^-$ (green data) is enhanced at 1 nm, suggesting a layering of the interfacial region with the cation closest to the vacuum interface and the chloride anion present as an adjacent layer beneath the cation. When water is introduced to the IL (red data; oxygen profile), water is enhanced at ~1 nm from the interface present beneath the top cation layer. As x_w increases the top cation layer and water layer remain enhanced near 0.4 and 1 nm, respectively, while the $[Cl]^-$ enhancement moves deeper into the bulk.

In a separate study investigating the same IL-water mixture using NICISS at much lower water mole fractions (0.0025 to 0.025) it was again shown that the cation has a strong propensity for the surface with an underlayer of $[Cl]^-$ (*23*). Similar to the previous study, it was concluded that water influences the composition and charge distribution at the IL-vacuum interface. However, for these lower x_w regimes the $[Cl]^-$ was reported to have a higher propensity for the IL-vacuum interface with increasing x_w. These seemingly contradictory results highlight the necessity to further investigate the interfacial regime of imidazolium chloride ILs in the presence of water in order to determine whether this difference in behavior for different x_w regimes is real.

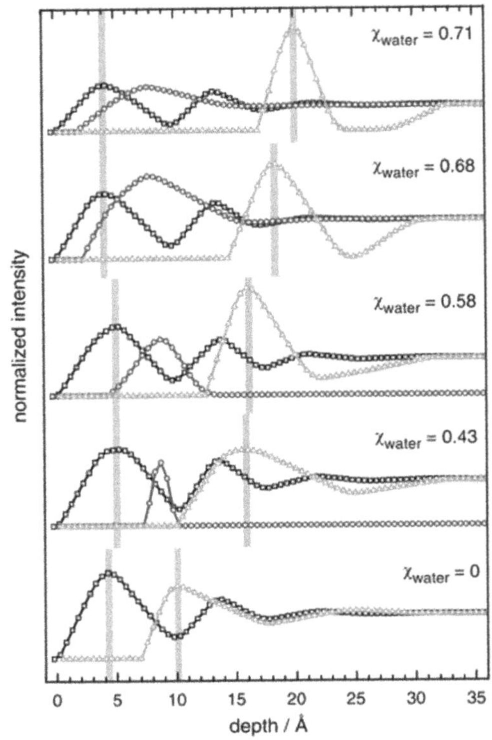

Figure 2. NICISS depth profiles of carbon (black), oxygen (red), and chlorine (green) for pure [C_6mim][Cl] and its mixture with water. Vertical gray lines represent maximum positions for carbon and chlorine. (Used with permission from Reference (22)).

UHV XPS is a surface sensitive vacuum based spectroscopy that has been used fairly extensively to investigate the IL-vacuum interface of neat ILs (*26*, *27*). X-ray photons (Al or Mg K_α) penetrate the surface on the order of μm, which eject core level electrons that are emitted from the interface with a probing depth on the order of ~10 nm or less depending on the angle at which the photoelectrons are collected (*28*). UHV XPS provides both elemental and chemical information of the IL-vacuum interface. Because traditional UHV XPS setups require samples to be under vacuum, in-situ probing of the IL interface needs to be at cryogenic temperatures for water to adsorb under typical conditions of Langmuir (μTorr s) exposures.

UHV XPS has been used to study the desorption of multilayer water from [C_8mim][BF_4], where water vapor was initially exposed to solid [C_8mim][BF_4] in-vacuo at 175 K (*24*). Figure 3a shows XPS survey spectra before and after dosing where a large O 1s peak due to water adsorption is evident after dosing. Figure 3b shows high resolution F 1s, N 1s, C 1s, and B 1s spectra before and after dosing. The IL peaks (F 1s, N 1s, C 1s, B 1s) show significant attenuation after depositing a thin solid water film on top of solid [C_8mim][BF_4]. The sample was then heated leading to the desorption of water from the IL surface, and XPS spectra was recorded over a temperature range of 175 – 300 K as [C_8mim][BF_4] transitioned from a solid to a liquid. By monitoring the decrease in O 1s intensity and assuming first order desorption kinetics and a pre-exponential factor of 10^{13} s^{-1}, the heat of adsorption of water onto [C_8mim][BF_4] was estimated to be -76 kJ mol^{-1}.

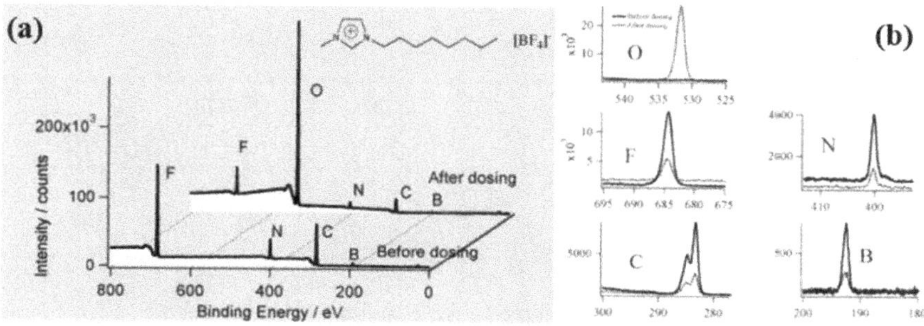

Figure 3. UHV XPS before and after dosing [C_8mim][BF_4] with water. (a) Survey spectra. (b) High resolution O 1s, F 1s, N 1s, C 1s, and B 1s spectra. (Used with permission from Reference (24)).

LOSMS probes the mass to charge ratio of different ionized fragments at rapid speeds. Line of sight analysis allows for the collection of desorbing species from a small focal point on the sample surface, reducing the background signal from other sources in the analysis chamber. Due to pressure limitations in the ionization region, this technique requires a vacuum. LOSMS was used to study the interaction of water with ILs by performing sticking probability and TPD experiments with [C_8mim][BF_4] and [C_2mim][NTf_2] (*25*). The sticking probability measurements were carried out by exposing the IL initially in the liquid state at 295 K to 4 x 10^{-8} Torr water vapor while monitoring the $[OH]^+$ fragment. Isothermal studies were then performed over the range of 295 K down to 113 K, determining the sticking probability as a function of temperature as the IL transition from liquid to solid. TPD experiments were performed by depositing water vapor onto solid IL at 100K, followed by heating the IL from 100 K to 300 K while monitoring the $[OH]^+$ fragment as the IL transitioned from solid to liquid. TPD spectra were then captured as a function of initial water coverage. For TPD experiments water completely desorbed below the glass transition temperature (while the IL was still solid). Figure 4 shows the generated potential energy diagram from these studies of enthalpy versus distance across the [C_6mim][BF_4] IL-vacuum interface for water. The TPD experiments determined the physisorption energy of water from the solid IL surface to be -41 kJ mol^{-1} assuming first order desorption kinetics and a pre-exponential factor of 10^{13} s^{-1}. The graph also shows the enthalpy of bulk absorption for water (-34 kJ mol^{-1}) and enthalpy of adsorption of -76 kJ mol^{-1} into an ionic underlayer previously determined with UHV XPS (*24*). These results indicate that water prefers to reside in the ionic underlayer below the imidazolium cation, consistent with NICISS studies discussed previously (*22*, *23*).

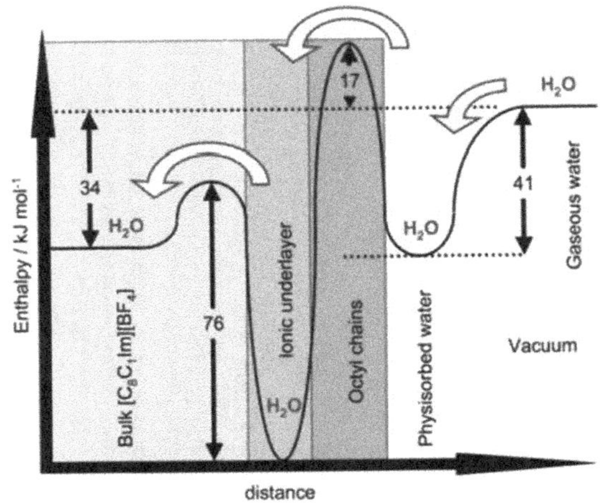

Figure 4. Potential energy diagram of water adsorption enthalpy versus distance relative to the surface for [C₆mim][BF₄] obtained from LOSMS and UHV XPS studies. (Used with permission from Reference (25)).

Ionic Liquid-Gas Interface

The influence of water at the IL-gas interface has been investigated by sum frequency generation (SFG) (*29–36*), X-ray reflectivity (XR) (*37*), neutron reflectometry (NR) (*38*), flowing jet sheet beam King Wells (FJSB-KW) (*39*) and ambient pressure XPS (APXPS) (*40*). In ambient air, water vapor is present at Torr level pressures. For these IL-gas studies the total pressures ranged from 5 Torr up to atmospheric pressure. In some cases the IL was exposed to gas phase water (or D_2O) within a vacuum chamber, while in other cases IL mixtures with liquid water were examined in lab air or N_2.

SFG is a nonlinear vibrational spectroscopy technique which uses two beams (visible laser at a fixed frequency and variable infrared laser) that overlap at an interface to generate a third beam that is the sum of the two incident beams (*28*). SFG detects anisotropic vibrations, making it a highly surface sensitive technique to probe molecular vibrations at interfaces. As seen from Table 1, SFG has been the predominant surface science probe utilized to examine the IL under elevated gas pressures. One of the main advantages of SFG is the ability to probe surfaces in-situ in the presence of isotropic gases and liquids. For IL studies using SFG, polarization dependent vibrational modes of the $[C_n mim]^+$ cation are the primary probe.

SFG was used to examine the cation orientation for hydrophilic [C_4mim][BF_4] and hydrophobic [C_4mim][imide] in the presence of water vapor ranging from 5 x 10^{-5} Torr up to 20 Torr (*30*). Figure 5 shows the SFG spectra (ssp polarization) for [C_4mim][imide] (a,b) and [C_4mim][BF_4] (c,d) at two water vapor pressures. At 5 x 10^{-5} Torr both the hydrophobic and hydrophilic IL (Figure 5 a,c respectively) spectra show notable features of aliphatic C-H modes of a butyl chain indicating the imidazolium ring is parallel to the surface plane. When the water vapor is increased to 20 Torr, the spectra for the hydrophilic [C_4mim][BF_4] does not change indicating no orientation change from the addition of water (Figure 5d). However, the hydrophobic IL (Figure 5b) shows two new distinct vibrations assigned to the anti-symmetric and symmetric stretch of the H-C(4)C(5)-H within the imidazolium ring. It is suggested that the cation reorients for hydrophobic ILs to help solvate the water molecules. A common observation in SFG studies is [C_4mim]$^+$ cation reorientation influenced by the presence of water for various different anions (*29*, *30*, *35*, *36*). It has also been suggested that for very dilute concentrations of [C_4mim][BF_4] in water ($0.95 < x_w < 1$) the [C_4mim]$^+$ cation may reorient as evidenced from butyl CH_3 peak shifts (*33*).

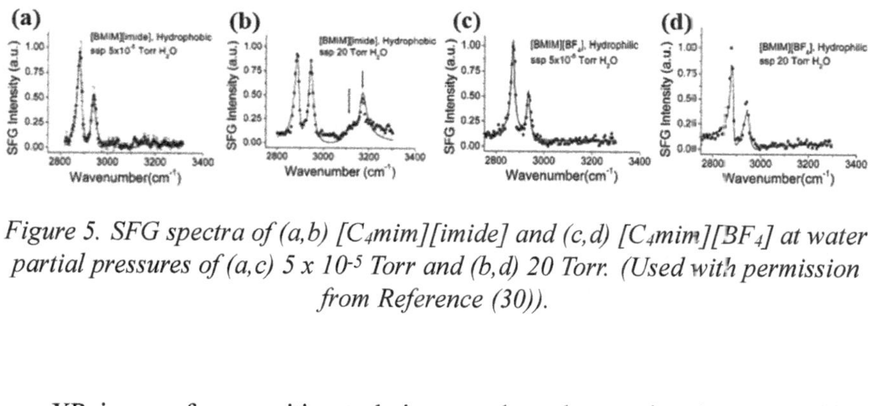

Figure 5. SFG spectra of (a,b) [C_4mim][imide] and (c,d) [C_4mim][BF_4] at water partial pressures of (a,c) 5 x 10^{-5} Torr and (b,d) 20 Torr. (Used with permission from Reference (30)).

XR is a surface sensitive technique used to characterize the composition of films and interface structures (*28*). This technique measures the surface reflected X-ray intensity at a grazing angle. It is a powerful method to investigate liquid surfaces on the tens of nanometers scale with sub-nanometer resolution. However, XR does have a loss of phase information (which can lead to data misinterpretations), so it should be complemented with simulations to reach conclusions about the X-ray reflectivity curves (*41*). XR was used to study the IL-gas interface of [C_4mpyr][NTf_2]/water mixtures in a N_2 environment (*37*). Figure 6 shows the scattering length density (SLD) profile of the IL-gas interface for $x_W = 0.15$ as a function of depth. In the presence of water, the interface consists of an outer layer composed of the cation alkyl chain pointing towards the gas phase (region A) with the underlying imidazolium ring cation interacting with water (region B). Underneath are anions followed by a mixture of cations, anions and water (region C) which extends into the bulk. In this same study complementary MD simulations of the SLD profile show a clear interfacial enhancement of water in region B.

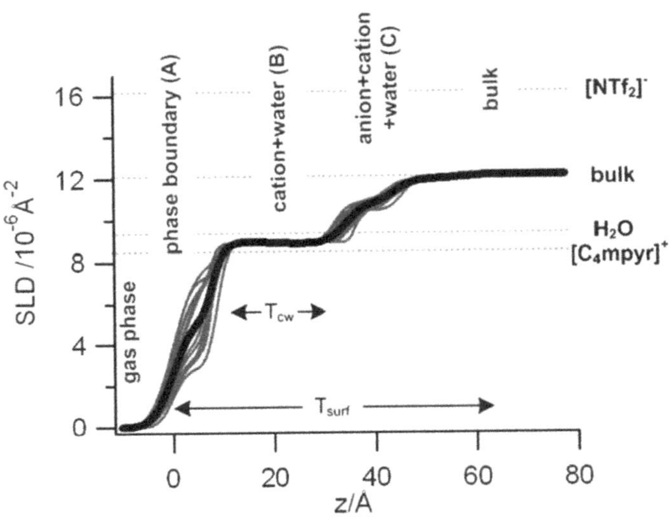

Figure 6. XR data of scattering length density (SLD) profile for a [C_4mpyr][NTf_2]/water mixture at x_w = 0.15. (Used with permission from Reference (37)).

NR is a diffraction technique that shines a beam of neutrons onto a flat surface and measures the intensity of the reflected beam. Neutron scattering amplitudes vary randomly between elements and are sensitive to lighter elements (*42*). Surface contamination is a concern for water studies leading to a rise in incoherent background during an experiment. NR was used to examine [C_8mim][Br]/water mixtures in air as a function of increasing IL concentration up to a maximum of 0.43 mol dm^{-3} (*38*). Surface tension data in this same concentration regime reveal a minimum at 0.15 mol dm^{-3}, which is attributed to the critical micelle concentration in bulk solution. NR results reveal a change in the surface structure above the critical micelle concentration, which was attributed to a depletion of [C_8mim][Br] at the interface. These results suggest there is a connection between surface structure and bulk solution aggregation.

FJSB-KW was recently used to study [C_4mim][NTf_2] interacting with gas phase D_2O (*39*). For these experiments a FJSB of liquid [C_4mim][NTf_2] is generated under vacuum conditions, which is then exposed to a pulsed beam of D_2O seeded in He and/or Ar at a stagnant pressure of 250 Torr. Changes in m/z = 20 were monitored by a quadrupole mass spectrometer (QMS). The interaction of [C_4mim][NTf_2] with D_2O was assessed as a function of temperature and collision energy to determine energetics. The initial dissolution probability (S) was assessed as a function of temperature and collision energy. S decreases as a function of increasing temperature and increasing collision energy. From these data the initial dissolution enthalpy (ΔH) and entropy (ΔS) were determined to be -53 kJ mol^{-1} and -210 J mol^{-1} K^{-1}, respectively.

APXPS is similar to UHV XPS described previously, with the additional capability of probing surfaces under Torr level pressures (*43*). This is accomplished by bringing a small aperture (typically ~0.3 mm) close to the sample surface. Behind the aperture is a differentially pumped electrostatic lens system for the collection of photoelectrons. This setup requires the electrons to only travel through small submillimeter distances of elevated pressure, thereby reducing electron scattering between the sample surface and the electron energy analyzer. APXPS was recently utilized to examine the interaction of water vapor with [C_4mim][Ace] (*40*). A droplet of IL was deposited on gold and placed in a vacuum chamber, followed by exposing the sample to increasing water vapor pressures at room temperature. The probing depth was estimated to be 8.5 nm. Figure 7a shows APXPS O 1s spectra at 10^{-6} and 1.6 Torr water vapor, where growth in interfacial water (O_w) and gas phase water (O_g) increases at the higher pressure. A quantitative assessment of interfacial water is shown in Figure 7b, plotting the mole ratio of water to IL pair as a function of water vapor pressure. From high vacuum to 1 Torr there is a sharp rise to 2 waters per IL pair. Above 1 Torr the number of waters per IL pair increases roughly linearly with pressure. At 5.0 Torr, there is approximately 5 waters per IL pair. These results are the first quantitative assessment of water at the IL-gas interface as a function of surrounding water vapor pressure.

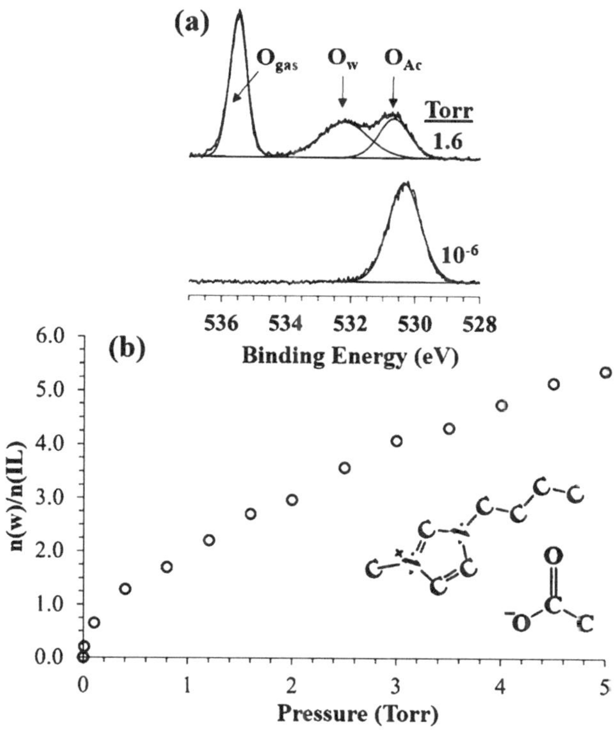

Figure 7. (a) APXPS O 1s spectra of [C_4mim][Ace] in the presence of 10^{-6} and 1.6 Torr water vapor. (b) The water to IL pair mole ration as a function of water vapor pressure. Data are from Reference (40)).

Ionic Liquid-Solid Interface

IL-solid interfaces in the presence of water have been examined by atomic force microscopy (AFM) (*44–51*) and surface enhanced infrared reflection absorption spectroscopy (SEIRAS) (*52*). AFM is a type of scanning force microscopy (SFM) that measures the force between a sharp tip and the sample surface to obtain topographic information about the sample (*28*). It operates in two modes: constant force (sample is adjusted vertical during measurements) or constant height (sample position is constant and cantilever tip deflection is recorded). AFM is highly sensitive with the ability to probe single atoms. This technique is typically performed in lab air, allowing experimental conditions of IL studies to include ambient water vapor or liquid phase water.

The effects of water on the ability of an IL to wet a mica surface have been examined by AFM (*53*). Such studies show that AFM is a valuable tool for the assessment of IL film morphology. Herein we focus on studies that examine the IL-solid interface which drive the tip into the IL and approach the solid surface to examine directly the IL-solid interface. The effects of water on the IL-solid interface via AFM have been examined on mica (*44, 45, 47, 48, 50*) and silica (*48*). Figure 8 shows results of AFM examining the [C_6mim][O_3SOC_2]-mica interface under dry and wet (~45 % RH) conditions (*45*). The dry sample (Figure 8a) shows two transition regimes, one with a thickness of ~1.1 nm adjacent to the interface and another with a thickness of ~0.7 nm away from the interface. The layer adjacent to the interface is attributed to positively charged cations interacting with the negatively charged mica surface (~1.1 nm), while the layer away from the interface is due to a mixed layer composed of cations and anions (~0.7 nm). When the IL is in the presence of ~45 % RH (Figure 8b) an additional transition regime appears with an average thickness of ~0.3 nm closest to the mica interface attributed to a monolayer of water. A slight expansion of ~0.05 nm is noted in the third film-thickness transition regime furthest from the interface attributed to water interacting with [O_3SOC_2]$^-$ anions.

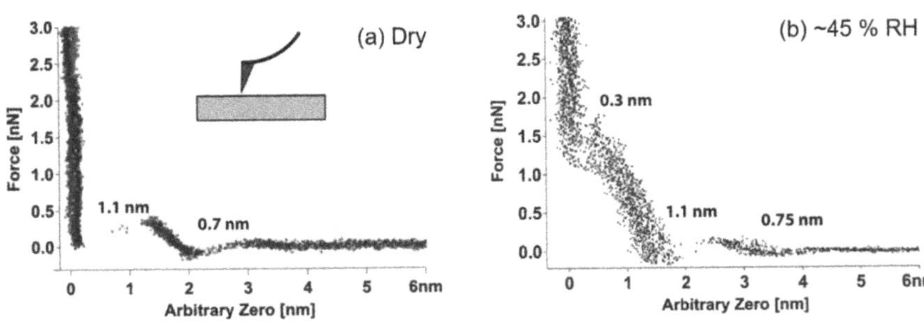

Figure 8. AFM of the [C_6mim][O_3SOC_2]-mica interface under (a) dry conditions and (b) in the presence of 45% RH. (Used with permission from Reference (45)).

The efficacy for water to disrupt IL structuring at the IL-solid interface depends on the size of the alkyl chain on the imidazolium ring, hydrogen bonding with the IL anion, and hydrogen bonding sites on the solid interface (*44, 45, 47, 48, 50*). The effect of surface electrical potential in the presence of water has also been examined at the IL-solid interface for HOPG (*49*) and Au (*50–52*) showing that the electrification of the interface can significantly disrupt IL structuring depending on the IL examined and the extent of applied electric potential under both positive and negative bias.

SEIRAS is a vibrational spectroscopy technique with the ability to probe liquid-solid interfaces with a probing depth of ~5 nm (*54*). SEIRAS has been used to investigate the [C_4mim][NTf_2]-Au electrode interface in vacuum and with samples prepared by exposure to water saturated Ar gas (*52*). An Au film was deposited onto a hemispherical Si prism and spectra were recorded in the Kretschmann attenuated total reflection configuration (*54*). The CF and OH vibrational modes were probed as a function of bias showing that presence of water at the IL-Au interface is potential dependent and interacts strongly with the anion. The amount of water at the IL-Au interface was greater at more positive potentials due to anions being more abundant at the Au electrode interface.

Molecular Dynamic Simulation Studies of IL Interfaces

Molecular Dynamic (MD) simulation studies have been utilized to study the IL-vacuum (*55–63*) and IL-solid interface in the presence of water (*64, 65*) and are summarized in Table 2. The IL-vacuum interface is the main interface studied with MD to date in the presence of water. For these studies the IL is surrounded by a vacuum in the MD simulation cell in the z-direction and typically the water is introduced in the vapor phase and ends up predominantly within the condensed phase IL throughout the simulations. The IL-vacuum interface has been examined using hydrophilic and hydrophobic ILs, and also in the presence of other gases including CO_2, N_2, O_2.

MD was used to examine the [C_4mim][NTf_2]-vacuum interface for an MD cell containing 368 IL pairs and 96 water molecules ($x_w = 0.21$) with a simulation run of 12 ns at 350 K (*61*). The density profiles are shown in Figure 9 for anion, cation, and water molecules (Figure 9a), water and selected carbon atoms of the cation (Figure 9b), and water and selected atoms of the anion (Figure 9c). Water is predominantly present in the condensed phase IL and absent from the vacuum. The density profile in Figure 9b shows the alkyl chain is enhanced at the IL-vacuum interface with the water molecules present in an interior layer below the cation. As seen from Figure 9c the anion is also present below the cation alkyl layer and is interacting with the water layer. This indicates a stronger interaction between [NTf_2]$^-$ and water compared to the cation alkyl chain. When introducing CO_2 into the system (data not shown) it is predominantly enhanced, unlike water, in the outermost region of the IL-vacuum interface above the alkyl chain. This strong propensity for CO_2 to be at the interface exists both in the absence and presence of water. This outer layer of CO_2 and inner layer of water is consistent for a number of ILs (*61–63*). In the case of [C_2mim][Gly], the presence of a water inner layer can lead to a decrease in the diffusion of CO_2 across the interface thus affecting bulk absorption kinetics (*62*).

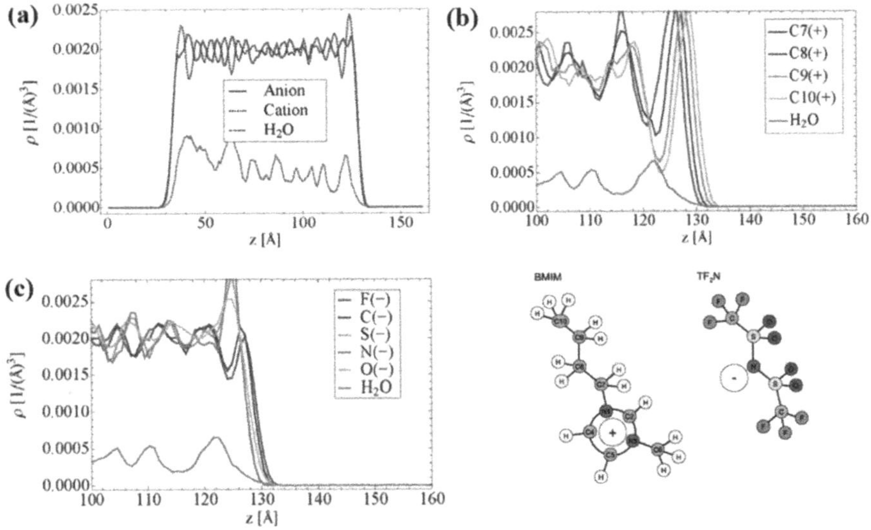

Figure 9. MD density profiles of [C4mim][NTf2]/water at x_w = 0.21 at the IL-vacuum interface at 350 K showing (a) anions, cations and water, (b) selected atoms in the cation, and (c) selected atoms in the anion. (Used with permission from Reference (61)).

Potential mean free force (PMF) calculations of the IL-vacuum interface examine the free energy for an individual water molecule as it crosses the interface going from vacuum towards the IL bulk. Deng et al. (*59*, *60*) show for the [C_4mim][BF_4], [C_4mim][NTf_2] and [C_4mim][PF_6] there is a monotonic decrease in the free energy. These results suggest that there is no energy minimum for water at the IL-vacuum interface for the ILs studied. Interestingly, while these PMF papers do not show a minimum energy at the interface, the experimental LOSMS studies from Deyko and Jones (*25*) do show a minimum in the enthalpy (Figure 3) suggesting more studies are needed to deconvolute enthalpy and entropy contributions.

MD has also been utilized to explore the effects of water in [C_4mim][PF_6] and [C_4mim][BF_4] at the IL-solid electrode interface (*64*). The system contained 656 IL pairs and 12 water molecules between two planar electrodes. Figure 10 shows the density profiles of the anions, cations and water for [C_4mim][PF_6] at negative, zero and positive charge densities (σ). Water resides near the electrode interface regardless of the potential applied, but has a larger accumulation around the positive electrode. This is due to its strong interaction with the anion, which is accumulated more at the positive electrode than the cation. A similar water enhancement at the interface is seen for positively charged graphene in the presence of [C_4mim][BF_4] (*65*).

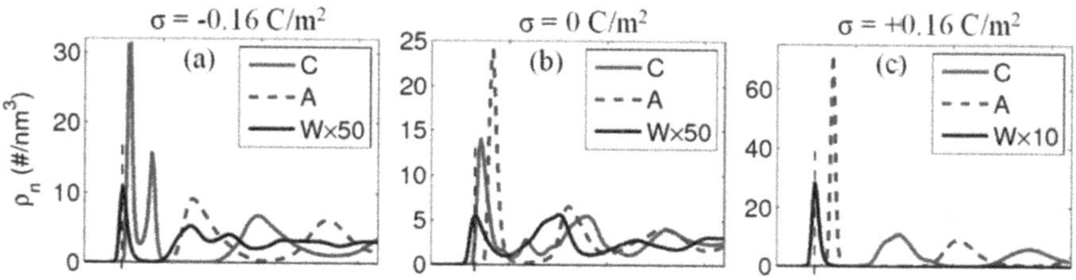

Figure 10. MD density profiles of anions, cations and water at three different current densities. (Used with permission from Reference (64)).

Summary and Future Outlook

Herein we have summarized experiments and MD simulations involving water at IL-vacuum, IL-gas and IL-solid interfaces. Studies to date have focused predominantly on imidazolium based ILs. The focus of experimental instrumentation covered in this review include surface microscopy, spectroscopy, mass spectrometry and scattering techniques with surface sensitivities ranging from submonolayer to several nanometers in order to give an overview of the molecular level understanding of IL interfaces as it pertains to the influence of water.

SFG has been the most widely used technique to examine the influence of water on the IL-gas interface. These molecular level investigations have used polarization dependent C-H vibrational modes to show that water can significantly impact IL cation structure at the outermost layer, depending on the IL used and amount of water present. A depth dependent picture of the IL-vacuum and IL-gas interface has been provided by NICISS, NR and XR studies. For example, NICISS results in the IL rich regime suggest that as x_w increases for [C_5mim][Cl] water remains as a layer just below the interfacially enhanced cation layer, whereas the [Cl]− anion gets pushed deeper into the bulk (*22*). These IL-vacuum and IL-gas results suggest that both the outer layer (via SFG) and sublayers (via NICISS, NR and XR) can be significantly altered by the presence of water.

The energetics of water interacting with ILs at the IL-vacuum and IL-gas interface has been explored using UHV XPS, LOSMS, FJSB-KW and MD simulations. Experimental studies have provided enthalpies for water interactions in the interfacial regime. For example, a UHV XPS study of water interacting with [C_8mim][BF_4] gave an adsorption enthalpy of -76 kJ mol^{-1}, which is greater than the enthalpy of bulk absorption (-34 kJ mol^{-1}) (*24*). These results suggest that water is stabilized in the interfacial regime, although possible entropic contributions also need to be determined. An energetic stabilization of water in the interfacial regime would be consistent with an enhanced concentration that is typically observed relative to the bulk. However, it should be noted that PMF MD simulation studies have shown that for water the free energy decreases monotonically as it crosses the interface, indicating that there is no free energy minimum for water at the interface relative to the bulk (*59*, *60*). Indeed, complementary studies are strongly needed between experimental investigations and MD simulations to assess this further.

In summary for the IL-gas interface, results to date show that water in most cases is found to significantly impact the IL structure within the top few layers. Theoretically, this would suggest that one can use water in known concentrations to control the cation interfacial structure. Moreover, in some cases water is enhanced at the outermost layer or within an underlayer relative to the bulk concentration. It is interesting to then ask what impact might these effects (cation orientation and development of enhanced water layer) have on the adsorption and uptake (ergo, uptake probability) of other gas phase sorbates. There is currently a lack of surface sensitive experiments assessing the effects of water on the uptake of other sorbates and is a potential avenue of exploration. MD simulations have suggested, for example, that a layer of water developing at the interface can impede CO_2 uptake (*62*).

The influence of water on the IL-solid interfaces of mica, silica, HOPG and Au have been assessed by AFM, both in the absence and presence of an electrical potential. When comparing the neat IL to an IL in the presence of water, clear disruptions in the interfacial layering of the IL have been observed. The extent to which this occurs is a function IL hydrogen bonding and alkyl chain length, the hydrogen bonding nature of the solid surface, and the presence of sufficient water. Like the IL-gas interface, the influence of water extends several layers into the IL away from solid interface. Given the strong interactions between IL ions and an electric potential, applying a bias in positive and negative regimes has a significant impact on layering which changes as a function of x_w. SEIRAS is chemical specific, allowing for the observation of water interacting more strongly with the anion compared to the cation for [C_4mim][NTf_2] at the IL-Au interface under an electric potential (*52*). While SFG has been used predominantly to study the IL-gas interface as it pertains to the influence of water, SFG is also a powerful tool to examine the IL-solid electrified interface (*66, 67*). Additional studies using SFG while varying x_w would provide valuable information on the influence of water on cation orientation at the IL-solid interface.

For the IL-vacuum interface, the few MD studies presented in Table 2 are vastly outnumbered by additional studies in the literature that assess the influence of water on bulk properties. Most MD simulations put periodic boundaries in three dimensions as compared to those that incorporate a vacuum above the IL in one of the three dimensions. Indeed, the experimental studies presented herein assessing water at IL interfaces significantly outnumber the MD simulations, and the surface science studies presented herein would greatly benefit from complementary MD simulations. This would include simulations of both IL-vacuum and IL-solid interfaces.

References

1. Zhang, S.; Sun, N.; He, X.; Lu, X.; Zhang, X. Physical Properties of Ionic Liquids: Database and Evaluation. *J. Phys. Chem. Ref. Data* **2006**, *35*, 1475–1517.
2. Javaid, A. Membranes for Solubility-Based Gas Separation Applications. *Chem. Eng. J.* **2005**, *112*, 219–226.

3. Zhang, Q.; Shreeve, J. M. Energetic Ionic Liquids as Explosives and Propellant Fuels: A New Journey of Ionic Liquid Chemistry. *Chem. Rev.* **2014**, *114*, 10527–10574.
4. Xin, B.; Hao, J. Imidazolium-Based Ionic Liquids Grafted on Solid Surfaces. *Chem. Soc. Rev.* **2014**, *43*, 7171–7187.
5. Theo, W. L.; Lim, J. S.; Hashim, H.; Mustaffa, A. A.; Ho, W. S. Review of Pre-Combustion Capture and Ionic Liquid in Carbon Capture and Storage. *Appl. Energy* **2016**, *183*, 1633–1663.
6. Xiong, L.; Compton, R. G. Amperometric Gas Detection: A Review. *Int. J. Electrochem. Sci.* **2014**, *9*, 7152–7181.
7. Gebresilassie Eshetu, G.; Armand, M.; Scrosati, B.; Passerini, S. Energy Storage Materials Synthesized from Ionic Liquids. *Angew. Chem., Int. Ed.* **2014**, *53*, 13342–13359.
8. Steinrück, H.; Wasserscheid, P. Ionic Liquids in Catalysis. *Catal. Lett.* **2015**, *145*, 380–397.
9. Tran, C.; Lacerda, S.; Oliveira, D. Absorption of Water by Room-Temperature Ionic Liquids: Effect of Anions on Concentration and State of Water. *Appl. Spectrosc.* **2003**, *57*, 152–157.
10. Seddon, K.; Stark, A.; Torres, M. Influence of Chloride, Water, and Organic Solvents on the Physical Properties of Ionic Liquids. *Pure Appl. Chem.* **2000**, *72*, 2275–2287.
11. Widegren, J.; Laesecke, A.; Magee, J. The Effect of Dissolved Water on the Viscosities of Hydrophobic Room-Temperature Ionic Liquids. *Chem. Commun.* **2005**, 1610–1612.
12. Jacquemin, J.; Husson, P.; Padua, A.; Majer, V. Density and Viscosity of several Pure and Water-Saturated Ionic Liquids. *Green Chem.* **2006**, *8*, 172–180.
13. Vila, J.; Gines, P.; Rilo, E.; Cabeza, O.; Varela, L. M. Great Increase of the Electrical Conductivity of Ionic Liquids in Aqueous Solutions. *Fluid Phase Equilib.* **2006**, *247*, 32–39.
14. Ziobrowski, Z.; Rotkegel, A. The Influence of Water Content in Imidazolium Based ILs on Carbon Dioxide Removal Efficiency. *Sep. Purif. Technol.* **2017**, *179*, 412–419.
15. Sturlaugson, A. L.; Fruchey, K. S.; Fayer, M. D. Orientational Dynamics of Room Temperature Ionic Liquid/Water Mixtures: Water-Induced Structure. *J. Phys. Chem. B* **2012**, *116*, 1777–1787.
16. Plechkova, N.; Seddon, K. *Ionic Liquids Uncoiled-Critical Expert Overviews*; Wiley: Hoboken, NJ, 2013; 413 pgs.
17. Gonfa, G.; Bustam, M.; Man, Z.; Mutalib, M. A. Unique Structure and Solute–solvent Interaction in Imidazolium Based Ionic Liquids: A Review. *Asian Trans. Eng.* **2011**, *1*, 24–34.
18. Lei, Z.; Dai, C.; Chen, B. Gas Solubility in Ionic Liquids. *Chem. Rev.* **2014**, *114*, 1289–1326.
19. Andersson, G.; Ridings, C. Ion Scattering Studies of Molecular Structure at Liquid Surfaces with Applications in Industrial and Biological Systems. *Chem. Rev.* **2014**, *114*, 8361–8387.

20. Lovelock, K. R. J. Influence of the Ionic Liquid/Gas Surface on Ionic Liquid Chemistry. *Phys. Chem. Chem. Phys.* **2012**, *14*, 5071–5089.
21. Bhargava, B.; Yasaka, Y.; Klein, M. L. Computational Studies of Room Temperature Ionic Liquid–water Mixtures. *Chem. Commun.* **2011**, *47*, 6228–6241.
22. Reichelt, M.; Hammer, T.; Morgner, H. Influence of Water on the Surface Structure of 1-Hexyl-3-Methylimidazolium Chloride. *Surf. Sci.* **2011**, *605*, 1402–1411.
23. Ridings, C.; Lockett, V.; Andersson, G. Significant Changes of the Charge Distribution at the Surface of an Ionic Liquid due to the Presence of Small Amounts of Water. *Phys. Chem. Chem. Phys.* **2011**, *13*, 21301–21307.
24. Lovelock, K. R. J.; Smith, E. F.; Deyko, A.; Villar-Garcia, I. J.; Licence, P.; Jones, R. G. Water Adsorption on a Liquid Surface. *Chem. Commun.* **2007**, *2007*, 4866–4868.
25. Deyko, A.; Jones, R. G. Adsorption, Absorption and Desorption of Gases at Liquid Surfaces: Water on [C_8C_1Im][BF_4] and [C_2C_1Im][Tf_2N]. *Faraday Discuss.* **2012**, *154*, 265–288.
26. Lovelock, K. R. J.; Villar-Garcia, I. J.; Maier, F.; Steinrück, H. P.; Licence, P. Photoelectron Spectroscopy of Ionic Liquid-Based Interfaces. *Chem. Rev.* **2010**, *110*, 5158–5190.
27. Steinrück, H. P. Recent Developments in the Study of Ionic Liquid Interfaces using X-Ray Photoelectron Spectroscopy and Potential Future Directions. *Phys. Chem. Chem. Phys.* **2012**, *14*, 5010–5029.
28. Bubert, H.; Jenett, H. *Surface and Thin Film Analysis: Principles, Instrumentation, Applications*; Wiley-VCH: Weinheim, 2002; xvii, 336 pgs.
29. Baldelli, S. Influence of Water on the Orientation of Cations at the Surface of a Room-Temperature Ionic Liquid: A Sum Frequency Generation Vibrational Spectroscopic Study. *J. Phys. Chem. B* **2003**, *107*, 6148–6152.
30. Rivera-Rubero, S.; Baldelli, S. Influence of Water on the Surface of Hydrophilic and Hydrophobic Room-Temperature Ionic Liquids. *J. Am. Chem. Soc.* **2004**, *126*, 11788–11789.
31. Sung, J.; Jeon, Y.; Kim, D.; Iwahashi, T.; Iimori, T.; Seki, K.; Ouchi, Y. Air-Liquid Interface of Ionic Liquid Plus H_2O Binary System Studied by Surface Tension Measurement and Sum-Frequency Generation Spectroscopy. *Chem. Phys. Lett.* **2005**, *406*, 495–500.
32. Rivera-Rubero, S.; Baldelli, S. Influence of Water on the Surface of the Water-Miscible Ionic Liquid 1-Butyl-3-Methylimidazolium Tetrafluoroborate: A Sum Frequency Generation Analysis. *J. Phys. Chem. B* **2006**, *110*, 15499–15505.
33. Deng, G.; Li, X.; Guo, Y.; Liu, S.; Lu, Z.; Guo, Y. Orientation and Structure of Ionic Liquid Cation at Air/[Bmim][BF_4] Aqueous Solution Interface. *Chin. J. Chem. Phys.* **2013**, *26*, 569–575.
34. Deng, G.; Guo, Y.; Li, X.; Zhang, Z.; Liu, S.; Lu, Z.; Guo, Y. Surface of Room Temperature Ionic Liquid [Bmim][PF_6] Studied by Polarization- and Experimental Configuration-Dependent Sum Frequency Generation Vibrational Spectroscopy. *Sci China Chem.* **2015**, *58*, 439–447.

35. Saha, A.; SenGupta, S.; Kumar, A.; Choudhury, S.; Naik, P. D. Vibrational Sum–frequency Generation Spectroscopy of Ionic Liquid 1-butyl-3-methylimidazolium Tris (Pentafluoroethyl) Trifluorophosphate at the Air–water Interface. *Chem. Phys.* **2016**, *475*, 14–22.
36. Deng, G.; Li, X.; Liu, S.; Zhang, Z.; Lu, Z.; Guo, Y. Successive Adsorption of Cations and Anions of Water–1-Butyl-3-Methylimidazolium Methylsulfate Binary Mixtures at the Air–Liquid Interface Studied by Sum Frequency Generation Vibrational Spectroscopy and Surface Tension Measurements. *J. Phys. Chem. C* **2016**, *120*, 12032–12041.
37. Lauw, Y.; Horne, M. D.; Rodopoulos, T.; Webster, N. A. S.; Minofar, B.; Nelson, A. X-Ray Reflectometry Studies on the Effect of Water on the Surface Structure of [C_4Mpyr][NTf_2] Ionic Liquid. *Phys. Chem. Chem. Phys.* **2009**, *11*, 11507–11514.
38. Goodchild, I.; Collier, L.; Millar, S. L.; Prokes, I.; Lord, J. C. D.; Butts, C. P.; Bowers, J.; Webster, J. R. P.; Heenan, R. K. Structural Studies of the Phase, Aggregation and Surface Behaviour of 1-Alkyl-3-Methylimidazolium Halide Plus Water Mixtures. *J. Colloid Interface Sci.* **2007**, *307*, 455–468.
39. Ohoyama, H.; Teramoto, T. Initial Dissolution of D2O at the Gas-Liquid Interface of the Ionic Liquid C4min]NTf2] Associated with Hydrogen-Bond Network Formation. *Phys. Chem. Chem. Phys.* **2016**, *18*, 28061–28068.
40. Broderick, A.; Khalifa, Y.; Shiflett, M. B.; Newberg, J. T. Water at the Ionic Liquid-Gas Interface Examined by Ambient Pressure X-Ray Photoelectron Spectroscopy. *J. Phys. Chem. C* **2017**, *121*, 7337–7343.
41. Sloutskin, E.; Lynden-Bell, R. M.; Balasubramanian, S.; Deutsch, M. The Surface Structure of Ionic Liquids: Comparing Simulations with X-Ray Measurements. *J. Chem. Phys.* **2006**, *125*, 174715.
42. Penfold, J.; Thomas, R. K. The Application of the Specular Reflection of Neutrons to the Study of Surfaces and Interfaces. *J. Phys.: Condens. Matter* **1990**, *2*, 1369–1412.
43. Bluhm, H. Photoelectron Spectroscopy of Surfaces Under Humid Conditions. *J. Electron. Spectrosc. Relat. Phenom.* **2010**, *177*, 71–84.
44. Cheng, H.-W.; Dienemann, J.-N.; Stock, P.; Merola, C.; Chen, Y.-J.; Valtiner, M. The Effect of Water and Confinement on Self-Assembly of Imidazolium Based Ionic Liquids at Mica Interfaces. *Sci. Rep.* **2016**, *6*, 30058.
45. Jurado, L. A.; Kim, H.; Rossi, A.; Arcifa, A.; Schuh, J. K.; Spencer, N. D.; Leal, C.; Ewoldt, R. H.; Espinosa-Marzal, R. M. Effect of the Environmental Humidity on the Bulk, Interfacial and Nanoconfined Properties of an Ionic Liquid. *Phys. Chem. Chem Phys.* **2016**, *18*, 22719–22730.
46. Espinosa-Marzal, R.; Arcifa, A.; Rossi, A.; Spencer, N. Ionic Liquids Confined in Hydrophilic Nanocontacts: Structure and Lubricity in the Presence of Water. *J. Phys. Chem. C* **2014**, *118*, 6491–6503.
47. Wang, Z.; Li, H.; Atkin, R.; Priest, C. Influence of Water on the Interfacial Nanostructure and Wetting of [Rmim][NTf_2] Ionic Liquids at Mica Surfaces. *Langmuir* **2016**, *32*, 8818–8825.
48. Sakai, K.; Okada, K.; Uka, A.; Misono, T.; Endo, T.; Sasaki, S.; Abe, M.; Sakai, H. Effects of Water on Solvation Layers of Imidazolium-Type

Room Temperature Ionic Liquids on Silica and Mica. *Langmuir* **2015**, *31*, 6085–6091.

49. Begic, S.; Li, H.; Atkin, R.; Hollenkamp, A. F.; Howlett, P. C. A Comparative AFM Study of the Interfacial Nanostructure in Imidazolium Or Pyrrolidinium Ionic Liquid Electrolytes for Zinc Electrochemical Systems. *Phys. Chem. Chem. Phys.* **2016**, *18*, 29337–29347.
50. Cheng, H.; Stock, P.; Moeremans, B.; Baimpos, T.; Banquy, X.; Renner, F. U.; Valtiner, M. Characterizing the Influence of Water on Charging and Layering at Electrified Ionic-Liquid/Solid Interfaces. *Adv. Mater. Interfaces* **2015**, *2*, 1500159.
51. Cui, T.; Lahiri, A.; Carstens, T.; Borisenko, N.; Pulletikurthi, G.; Kuhl, C.; Endres, F. Influence of Water on the Electrified Ionic Liquid/Solid Interface: A Direct Observation of the Transition from a Multilayered Structure to a Double-Layer Structure. *J. Phys. Chem. C* **2016**, *120*, 9341–9349.
52. Motobayashi, K.; Osawa, M. Potential-Dependent Condensation of Water at the Interface between Ionic Liquid [BMIM][TFSA] and an Au Electrode. *Electrochem. Commun.* **2016**, *65*, 14–17.
53. Gong, X.; Kozbial, A.; Li, L. What Causes Extended Layering of Ionic Liquids on the Mica Surface? *Chem. Sci.* **2015**, *6*, 3478–3482.
54. Motobayashi, K.; Minami, K.; Nishi, N.; Sakka, T.; Osawa, M. Hysteresis of Potential-Dependent Changes in Ion Density and Structure of an Ionic Liquid on a Gold Electrode: In Situ Observation by Surface-Enhanced Infrared Absorption Spectroscopy. *J. Phys. Chem. Lett.* **2013**, *4*, 3110–3114.
55. Lynden-Bell, R. Gas-Liquid Interfaces of Room Temperature Ionic Liquids. *Mol. Phys.* **2003**, *101*, 2625–2633.
56. Lynden-Bell, R.; Kohanoff, J.; Del Popolo, M. Simulation of Interfaces between Room Temperature Ionic Liquids and Other Liquids. *Faraday Discuss.* **2005**, *129*, 57–67.
57. Picalek, J.; Minofar, B.; Kolafa, J.; Jungwirth, P. Aqueous Solutions of Ionic Liquids: Study of the Solution/Vapor Interface using Molecular Dynamics Simulations. *Phys. Chem. Chem. Phys.* **2008**, *10*, 5765–5775.
58. Chaban, V. V.; Prezhdo, O. V. Water Phase Diagram is significantly Altered by Imidazolium Ionic Liquid. *J. Phys. Chem. Lett.* **2014**, *5*, 1623–1627.
59. Dang, L. X.; Wick, C. D. Anion Effects on Interfacial Absorption of Gases in Ionic Liquids. A Molecular Dynamics Study. *J. Phys. Chem. B* **2011**, *115*, 6964–6970.
60. Dang, L. X.; Chang, T. Molecular Mechanism of Gas Adsorption into Ionic Liquids: A Molecular Dynamics Study. *J. Phys. Chem. Lett.* **2012**, *3*, 175–181.
61. Perez-Blanco, M. E.; Maginn, E. J. Molecular Dynamics Simulations of Carbon Dioxide and Water at an Ionic Liquid Interface. *J. Phys. Chem. B* **2011**, *115*, 10488–10499.
62. Herrera, C.; García, G.; Alcalde, R.; Atilhan, M.; Aparicio, S. Interfacial Properties of 1-Ethyl-3-Methylimidazolium Glycinate Ionic Liquid regarding CO_2, SO_2 and Water from Molecular Dynamics. *J. Mol. Liq.* **2016**, *220*, 910–917.

63. Aparicio, S.; Atilhan, M. On the Properties of CO_2 and Flue Gas at the Piperazinium-Based Ionic Liquids Interface: A Molecular Dynamics Study. *J. Phys. Chem. C* **2013**, *117*, 15061–15074.
64. Feng, G.; Jiang, X.; Qiao, R.; Kornyshev, A. A. Water in Ionic Liquids at Electrified Interfaces: The Anatomy of Electrosorption. *ACS Nano* **2014**, *8*, 11685–11694.
65. Docampo-Álvarez, B.; Gómez-González, V.; Montes-Campcs, H.; Otero-Mato, J. M.; Méndez-Morales, T.; Cabeza, O.; Gallego, L.; Lynden-Bell, R. M.; Ivaništšev, V. B.; Fedorov, M. V. Molecular Dynamics Simulation of the Behaviour of Water in Nano-Confined Ionic Liquid–water Mixtures. *J. Phys.: Condens. Matter* **2016**, *28*, 464001.
66. García-Rey, N.; Dlott, D. D. Structural Transition in an Ionic Liquid Controls CO_2 Electrochemical Reduction. *J. Phys. Chem. C* **2015**, *119*, 20892–20899.
67. Baldelli, S. Surface Structure at the Ionic Liquid-Electrified Metal Interface. *Acc. Chem. Res.* **2008**, *41*, 421–431.

Chapter 11

Radiation and Radical Chemistry of Ionic Liquids for Energy Applications

James F. Wishart*

Chemistry Division, Brookhaven National Laboratory, Upton, New York 11973-5000, United States
***E-mail: wishart@bnl.gov**

Ionic liquids (ILs) are becoming important components of many advanced devices and technologies. Several such applications, such as photoelectrochemical solar cells, high-performance batteries and recycling of spent nuclear fuel, expose ILs to extreme conditions where they are subject to ionization or the injection of excess charges. It is thus important to understand what happens to ILs under those conditions. The unique combinations of IL properties, and their inherent binary (cation-anion) nature, lead to significant behavioral differences compared to conventional liquids. This chapter explores ionization processes in ionic liquids, the formation of reactive intermediates, and the influence of IL properties on the ensuing chemical reactivity, as revealed by radiation chemistry techniques. Mechanisms of radiolytic damage accumulation in ILs and strategies to avoid or control them are discussed.

Introduction

Due to their unique and tunable combinations of properties, ionic liquids (ILs) have found many uses in energy-related applications (*1–5*). Many of these applications expose ILs to stresses that challenge their long-term stability and performance. For example, in batteries and supercapacitors, ILs can be subjected to extremes of voltage and current that cause them to be chemically modified by oxidation or reduction reactions, and subsequent reactivity of the initial radical intermediates. ILs used in solar applications (photoelectrochemical or "Gratzel" cells, heat transfer fluids, etc.) can undergo photolysis and thermolysis. In open systems, stress-induced IL modification can be exacerbated by the presence of oxygen (*6*).

Ionizing radiation poses one of the greatest challenges for IL stability in certain key applications where their properties are especially attractive. Their generally low volatility and high conductivity are important factors for space-rated batteries (*7*) and for their use as ionic propellants in thruster modules for very small satellites ("cubesats") (*8–10*). Space applications necessarily entail extended exposure to space radiation in the form of energetic electrons, protons and heavier nuclei, and the cumulative impact on spacecraft systems using ILs as electrolytes, propellants or lubricants can be significant (*11*), not to mention the effects on the astronauts.

Closer to Earth, ionic liquids can play significant roles towards making nuclear energy safer, more efficient and less burdensome to the environment (*12*). They have desirable properties that enhance the safety and efficacy of liquid/liquid extraction systems for nuclear separations, such as tunability by design, low volatility, high combustion resistance, the flexibility of exploiting both neutral extraction and ion exchange extraction mechanisms (*13*), and if desired, ILs can incorporate boron to provide a wide margin of safety against criticality (*14*).

When ionic liquids are used for nuclear separations they are exposed to potentially high levels of ionizing alpha, beta and gamma radiation. The incident radiation can induce oxidation, reduction, bond breaking, radical attachment, excited-state chemistry and several other reactions, depending on the chemical composition of the IL (*15–30*). Thus, as radiation exposure increases, products of these reactions accumulate up to and beyond the point where they alter the performance of the IL for the intended purpose (e.g., liquid/liquid extraction, lubrication, electrolyte, propellant, etc.)

These problems are not unique to ionic liquids; the challenges posed by radiolysis of conventional water/organic separations systems for nuclear fuel processing, including radiation damage to extractant molecules (*31–34*), has been a major focus of research and engineering for many decades. However, ILs provide new opportunities to re-engineer traditional separations processes and new ways to deal with the radiation damage problems. The behaviors of ILs under radiation vary enormously over their exceedingly wide range of chemical compositions. IL composition is thus a tool that can be adjusted to control and minimize the radiation susceptibility by identifying particular cation and anion structural motifs that are resistant to radiation and combining cations and anions to synergize that resistance. Examples of such radiation-stable families of cations and anions will be described in this chapter, but to comprehend why they are more stable than certain other ions it is necessary to understand how ionizing radiation interacts with ILs.

Early Stages in the Radiolysis of Ionic Liquids

The Interaction of Radiation with Ionic Liquids

Radiolysis of a material (in this case a liquid) is initiated by the transit of a high-energy photon (x- or γ-ray) or particle (electron or heavy ion e.g., $^{1}H^{+}$,

$^{4}He^{2+}$, $^{12}C^{6+}$, etc.) (Figure 1) (*35–41*). (Neutron radiolysis is a special case that converts into charged particle radiolysis.) The incident radiation transfers varying increments of energy to the material by interacting electrostatically with its electrons, leading to the formation of excited states, and if sufficient energy is transferred, "ionizing" the liquid by ejecting electrons out of their bound states with excess kinetic energy, producing "secondary" electrons that can induce further ionizations in turn. The liquid molecules that have lost electrons are referred to as "holes", and they are also radicals because they have unpaired electrons. If the liquid molecules are neutral as in ordinary solvents, the holes are monocations (Eqn. 1, where the wavy arrow indicates the radiolysis event). However, if the liquid is an IL (with monovalent ions for simplicity's sake), ionization of a cation yields a dication radical (Eqn. 2) and ionization of an anion yields a neutral (or zwiterion) radical (Eqn. 3). Each electron in a molecule is a potential acceptor of ionizing energy; therefore in ILs Eqns. 2 and 3 both occur, in relative proportion to the respective electron counts of the cation and anion.

$$L \rightsquigarrow e^{-}_{excess} + L^{+\bullet} \qquad 1$$

$$C^{+} \rightsquigarrow e^{-}_{excess} + C^{2+\bullet} \qquad 2$$

$$A^{-} \rightsquigarrow e^{-}_{excess} + A^{\bullet} \qquad 3$$

In equations 1-3, $L^{+\bullet}$, $C^{2+\bullet}$ and $A^{\bullet}$ are depicted for simplicity as molecules lacking one electron from their normal complement. However, it is more realistic to consider the electron-deficient wavefunction extending over multiple molecules (and so could the excess electron) (*42–46*) until a relaxation or reaction event causes it to localize (*47*). Also, depending on the electronic structures of the ions making up the IL, holes can be transferred between different ions, i.e., $C^{2+\bullet}$ can oxidize A^{-} or $A^{\bullet}$ can oxidize C^{+}.

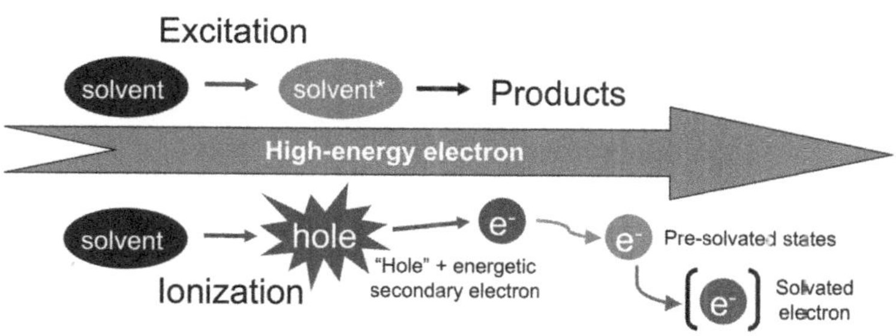

Figure 1. Excitation and ionization processes induced in a liquid by transit of ionizing radiation.

As mentioned above, secondary electrons are ejected from the molecules of the medium with varying amounts of kinetic energy. They lose their kinetic energy to the medium via subsequent excitation and ionization events until they come to kinetic (translational) rest. However, at that point the surrounding medium has not yet responded to the presence of an excess negative charge where there was none before. As the solvent reorganizes around the electron to minimize the total potential energy, the electron passes through one or more "pre-solvated" states before it becomes fully solvated (*40, 41*). Pre-solvated electrons have higher potential energies than solvated ones; consequently they can show different reactivity patterns toward electron acceptors, and higher mobility (diffusivity) because they are less deeply trapped by the solvent around them. In highly concentrated solutions, such as electrolytes or systems for extracting and separating metal ions, the different reactivity of pre-solvated electrons can have significant consequences, as discussed below.

Reactivity of Initially-Formed Species

Figure 2 depicts the processes and reactions that the initially formed electrons and holes undergo on picosecond and longer timescales. Holes and electrons may recombine to produce excited or ground states of their original parent molecules. Capture of the excess electrons by scavengers (S) occurs in competition with the solvation process. Some scavengers react more readily with energetic pre-solvated electrons than with solvated ones (indicated by crossed-out reaction arrows). Holes may oxidize solutes or they may fragment, producing various radical species. In the field of radiation chemistry, evidence is beginning to accumulate that holes also undergo relaxation or solvation processes just like electrons, with similar consequences for reactivity, but this aspect has not been as widely explored.

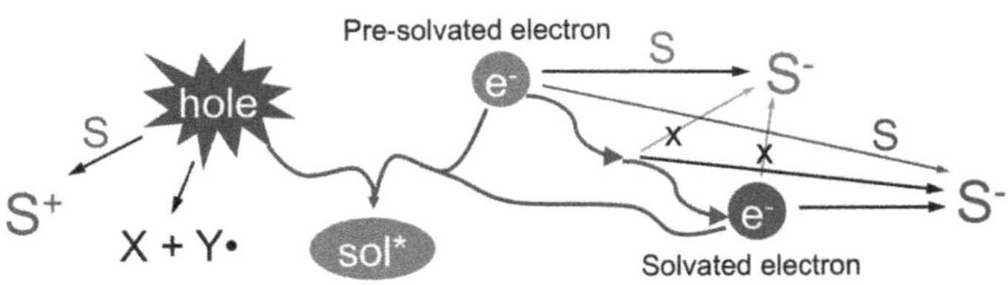

Figure 2. Solvation dynamics and reactivity of the initially formed excess electron and hole species after ionization. Adapted with permission from reference (48). Copyright 2010 American Chemical Society.

Electron solvation processes are very fast in conventional, lower-viscosity solvents such as water (average solvation time $<\tau> < 1$ ps) (*40, 41*) and short-chain alcohols, diols and even glycerol ($<\tau>$ on the order of tens of picoseconds) (*49–51*). In contrast, electron solvation in ILs of even modest viscosity can be significantly slower and dynamically heterogeneous, for example 1-butyl-1-methylpyrrolidinium bis(trifluoromethylsulfonyl)amide, $[C_4mpyr][NTf_2]$, where $<\tau>$ is 270 ps ($\tau_1 = 70$ ps, $\tau_2 = 574$ ps) for a viscosity of 90 cP at 21 °C (*52*). As solvation times get longer, lower scavenger concentrations become competitive with the solvation process and the reactivities of the pre-solvated electron states become more important factors in determining radiolytic product yields and subsequent degradation pathways for practical separations systems and other applications.

The significance of this effect became evident during early attempts to use competition kinetics according to the mechanism in Eqns. 4 and 5 to determine the reaction rate constants of H-atom attachment to aromatic solutes such as pyrene in the ionic liquid methyltributylammonium bis(trifluoromethylsulfonyl)amide, $[N_{1444}][NTf_2]$ (*53*). When rate constants for H-atom attachment to the arenes (Eqn. 5) that were measured directly by pulse radiolysis disagreed with the ratios predicted by competition kinetics, the reason was discovered to be that pre-solvated electron capture (*54*) was forming arene radical anions that were subsequently protonated to form the same product through a supplementary pathway (Eqns. 6, 7) (*53*).

$$e^-_{solv} + H^+ \rightarrow H^\bullet \qquad (4)$$

$$H^\bullet + \text{pyrene} \rightarrow \text{pyrene-}H^\bullet \qquad (5)$$

$$e^-_{pre} + \text{pyrene} \rightarrow \text{pyrene}^{-\bullet} \qquad (6)$$

$$\text{pyrene}^{-\bullet} + H^+ \rightarrow \text{pyrene-}H^\bullet \qquad (7)$$

Pre-solvated electron scavenging in ionic liquids has been empirically quantified on several occasions (*54–58*) by measuring the amount of solvated electrons that are "missing" as a function of scavenger concentration when the decay kinetics of the solvated electron are extrapolated to time zero. However, such studies cannot provide insight into the competition between solvation and scavenging. One set of experiments on the picosecond timescale (*52*) independently compared the electron solvation dynamics in $[C_4mpyr][NTf_2]$ with electron scavenging by duroquinone in the same IL and showed that the solvation process controlled the window for pre-solvated electron scavenging by duroquinone. Interestingly, that study also observed a rising fraction of "missing" pre-solvated electrons with increasing duroquinone concentration, indicating that a precursor to the pre-solvated electron was also being scavenged. That process is too fast to observe with existing pulse radiolysis instrumentation, but it was elegantly revealed by femtosecond photodetachment experiments using perchloric acid as the electron scavenger (*59*). It is important to note that the slower dynamics of ionic liquids, coupled with ultrafast techniques, critically enabled the observation that there were not just one, but two scavenging processes responsible for the "missing" solvated electrons. This never would have been inferred from traditional experiments, and thus ionic liquids have already made important contributions to fundamental radiation chemistry.

Common Reaction Pathways for Reactive Intermediates

Figure 2 offers a generic picture of the reactivity induced in ionic liquids by ionizing radiation, but the probabilities of the various pathways and the yields of radiolysis products inescapably depend on the composition of the IL (*17*). Numerous studies have examined IL radiation chemistry and radiation stability by breaking the problems down to families of related cations or anions. This is a necessary and important first step, but it must always be kept in mind that the radiation chemistry of a cation or anion in a given IL can be strongly influenced or even dominated by the identity of the counterion constituting the IL, because of its contribution to the IL's density of electronic states, most importantly the HOMO and LUMO (*42–46*, *60–62*). With that caveat stipulated, we can review the radiation chemistry of various ion families within ILs.

Aromatic Cations (imidazolium, pyridinium, and others)

ILs with 1,3-dialkylimidazolium ($[C_nC_mim]^+$) and 1-alkylpyridinium ($[C_npy]^+$) cations were at the forefront of the explosion of interest in ILs that started in the late 1990s, and they continue to be the most widely used and heavily studied families of ILs even today. Not surprisingly, the earliest radiation chemistry studies of ILs were performed on these salts (*63–66*). First and foremost, excess electrons in aromatic cation ILs are rapidly scavenged by the aromatic cations ($[C^+]$) to produce radicals ($[C^\bullet]$) or dimer (or multimer) radical cations ($[C_2^{+\bullet}]$). Preliminary measurements using the ultrafast Optical Fiber Single Shot (OFSS) detection system (*67*) at the BNL Laser-Electron Accelerator Facility (*68*) indicate that electron capture in $[C_2mim][NTf_2]$ is complete within 15 ps.

Upon electron capture, the chemistries of imidazolium and pyridinium cations diverge. When reduced, pyridinium cations form relatively stable neutral aromatic radicals (*66*) that may interact with other pyridinium cations to form charge-resonance-stabilized dimer or multimer radical cations (*27*, *46*). In these cases the six-member aromatic rings delocalize the excess charge and bond distortion in comparison to the pyridinium cation is minimal. These pyridinyl radicals can in turn reduce solutes by intermolecular electron transfer reactions (*66*, *69*).

On the other hand, electron capture by imidazolium and related five-member aromatic cations activates them towards radical chemistry by causing a pyramidal distortion of the imidazole ring at the C-2 position between the nitrogens, where a concentration of unpaired electron density is formed (*19*, *21*, *23*, *26*). The pyramidal imidazoyl radical then attacks another imidazolium cation (Figure 3), predominantly at the C-2 position, to form a covalent dimer radical cation that has been deduced from low-temperature EPR measurements and from product analysis by electrospray mass spectroscopy (*19*, *21*, *23*, *26*). Similar species have been produced by reaction of $[C_2mim][AlCl_4]$ with lithium metal (*70*). Evidence exists that $[C^\bullet]$ and $[C_2^{+\bullet}]$ are reductants (*71*), however if water is present in the ionic liquid $[C_2^{+\bullet}]$ can also be irreversibly hydrolyzed. As shown in Figure 3, loss of the aliphatic arm R′ through C-N bond fragmentation is another important pathway for cation damage. If R′ is a benzyl group, C-N fragmentation is the major pathway for reduced imidazolium and triazolium cations due to formation of a relatively stable benzyl radical (*26*). For thiazolium cations, formation of $[C_2^{+\bullet}]$ is the dominant pathway (*26*).

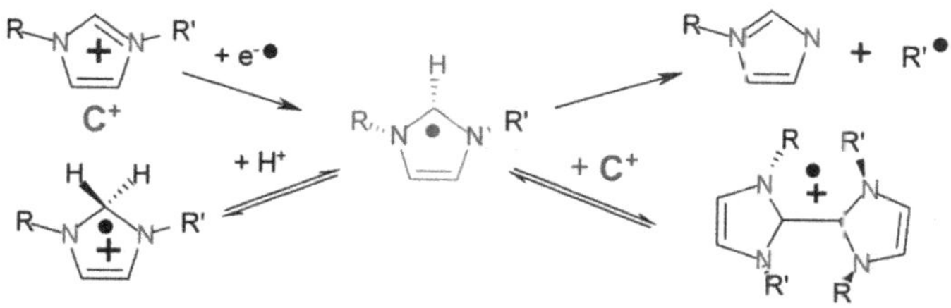

Figure 3. Reactions following reduction of imidazolium cations in ILs. Adapted with permission from reference (23). Copyright 2011 American Chemical Society.

Aliphatic Cations (ammonium, pyrrolidinium, phosphonium)

Quaternary ammonium and phosphonium cations are relatively poor acceptors of solvated electrons in ILs. Despite the fact that in water, hydrated electrons react with quaternary ammonium ions with rate constants on the order of 10^6 – 10^7 $M^{-1}s^{-1}$ (*72, 73*), solvated electrons in many quaternary ammonium ILs survive for hundreds of nanoseconds or longer (*30, 54, 56–58, 71, 74, 75*) despite the 2-5 M cation concentrations. Solvated electrons are very reactive species, but their diffusion-limited rate constants are relatively slow in comparison to common molecular solvents such as water or acetonitrile ($k > 10^{10}$ $M^{-1}s^{-1}$). In a relatively viscous IL such as $[N_{1444}][NTf_2]$ (viscosity 787 cP at 20 °C (*76*)) the diffusion-limited rate constants are ~2 x 10^8 $M^{-1}s^{-1}$ (*54*), while in lower viscosity ILs such as $[C_4mpyr][NTf_2]$ and $[N_{1113}][NTf_2]$ (viscosities 95 (*76*) and 103 (*77*) cP, at 20 °C respectively) they are 3-7 x 10^8 $M^{-1}s^{-1}$ (*56–58, 78*). Given that the electron is a charged species just like the ions of the IL, its diffusional movement is linked to the dynamics of the ionic motions and its diffusional properties resemble a molecular anion rather than a quantum particle that is capable of tunneling to diffuse. For example, diffusion-limited electron transfer rate constants for molecular radicals and anions are about the same as those for the solvated electron (*69*).

As shown in Eqn. 4, solvated electrons can react with various proton donors to produce hydrogen atoms. Even in "pure" ILs, proton donors can exist in the forms of water and trace impurities, and excess protons are generated from the radiolysis event itself when cationic holes on cation-bound aliphatic groups transfer protons (usually to the anions) to stabilize themselves as neutral radicals (Eqn. 8). In aliphatic ILs the H-atoms thus formed can abstract H-atoms from the alkyl chains to produce more aliphatic (or alkyl) radicals (Eqn. 9). Alkyl groups on anions, such as alkyl sulfates or alkylsulfonates, also undergo analogous reactions to Eqns. 8 and 9.

$$C^{2+\bullet} + A^{-} \rightarrow C(\text{-H})^{+\bullet} + HA \qquad 8$$

$$C^{+} + H^{\bullet} \rightarrow C(\text{-H})^{+\bullet} + H_2 \qquad 9$$

Alkyl radicals can also be formed through C-N bond fragmentation via dissociative electron attachment to cations, as shown in the upper right corner of Figure 3 in the case of imidazole. While solvated electrons react slowly with aliphatic cations, pre-solvated electrons may be more efficient (Eqn. 10).

$$e^-_{pre} + NR_4^+ \rightarrow NR_3 + R^\bullet \qquad 10$$

Evidence for the formation of both kinds of alkyl radicals ($C(-H)^{+\bullet}$ and $R^\bullet$) abounds in electrospray MS studies of irradiated imidazolium (*15*, *17*) and tetraalkylammonium (*16*) ILs. The $C(-H)^{+\bullet}$ radicals can react to cross-link cations, which increases the viscosity of the IL as the radiolysis products accumulate. The $R^\bullet$ radicals can attach to ions, modifying their properties, and they can react with each other or abstract H atoms to form volatile hydrocarbon products (*17*).

The production of molecular hydrogen from the radiolysis of ILs increases as the aliphatic content of the IL increases, through mechanisms akin to Eqns. 8 and 9. Table 1 shows the radiolytic yields (moles per Joule of absorbed energy) of molecular hydrogen for gamma radiolysis of several ILs and analogous neutral molecules. The H_2 yield for the aromatic molecule imidazole is very low, but the addition of one methyl group triples the yield. Aliphatic amines have much higher yields; in particular the yield for the secondary amine *N*-Methylbutylamine is higher than for the tertiary amine *N,N*-Dimethylbutylamine because of the amine proton that can scavenge electrons according to Eqn. 4, creating additional H-atoms that produce H_2 via Eqn. 9.

The hydrogen yields for the ionic liquids are lower than for the related neutral molecules, but it must be remembered for the purposes of comparison that the $[NTf_2]^-$ anion absorbs part of the incident radiation without producing H_2. Among the imidazolium salts the H_2 yield increases as the alkyl chain lengthens from ethyl to hexyl. Similarly, the H_2 yields in the aliphatic cation ILs increase with alkyl content, and there is no significant difference between ammonium and phosphonium cations. In applications where ILs will be exposed to significant irradiation, consideration should be placed on selecting IL components that will minimize H_2 production, for safety purposes.

Functionalized Cations for Radiation Stability

The concept of "Task-Specific Ionic Liquids" (*82*), or TSILs, refers to the addition of functional groups or modifications to the standard IL framework in order to impart a functional capability, such as metal extraction (*82*, *83*) or CO_2 capture (*84*, *85*). The addition of such functionalities diversifies the gamut of radiation-induced reactions that can occur in TSILs, which can lead to loss of task-specific performance if radiation exposure accumulates. The scope of such reactions is beyond the present review, however it has been shown that the TSIL concept can be used to improve the performance of IL-based systems under ionizing radiation, according to the goal desired.

There are two basic ways that functional modifications can be used to improve the radiation stability of an IL-based system. The first way is to add components that capture and stabilize the initially formed electrons and holes, in order to prevent them from causing irreversible chemical damage that leads to performance degradation. The second way is to deliberately introduce components that act as sacrificial agents to protect more critical parts of the system, i.e., extractants in liquid/liquid separations systems, from oxidative or reductive damage.

Table 1. Molecular hydrogen yields for gamma radiolysis

Compound	*H_2 Yield x 10^7 (mol/J)*	*Reference*
Imidazole	0.050	(*79*)
1-Methylimidazole	0.15	(*79*)
N-Methylbutylamine	6.02	(*79*)
N,N-Dimethylbutylamine	3.98	(*79*)
$[C_2mim][NTf_2]$	0.102	(*79*)
$[C_4mim][NTf_2]$	0.23	(*80*)
$[C_6mim][NTf_2]$	0.33	(*80*)
$[C_4mpyr][NTf_2]$	0.65	(*81*)
$[HN_{222}][NTf_2]$	0.72	(*81*)
$[N_{1114}][NTf_2]$	0.76	(*79*)
$[N_{2226}][NTf_2]$	1.72	(*80*)
$[P_{2226}][NTf_2]$	1.72	(*80*)
$[N_{2228}][NTf_2]$	2.14	(*80*)
$[P_{2228}][NTf_2]$	2.03	(*80*)
$[N_{222(12)}][NTf_2]$	2.23	(*80*)
$[P_{222(12)}][NTf_2]$	2.38	(*80*)
$[P_{888(14)}][NTf_2]$	2.5	(*81*)

A good way to prevent an excess charge from activating a chemical bond towards fragmentation is to stabilize it by delocalization over an extended conjugated system. As mentioned above, pyridinyl radicals formed from electron capture by pyridinium cations are examples of this effect. However, hole capture is just as important in preventing radiation damage. For example, radiolytic degradation of tetraalkyldiglycolamide extractants in imidazolium and pyridinium ILs proceeds through an oxidative pathway that is not effectively blocked by the IL (as well as via excited states), and not via reduction to any significant extent (*86*). Neutral aromatic systems can serve as hole traps; for example, long-lived pyrene cations produced by hole capture were observed by pulse radiolysis in $[N_{1444}][NTf_2]$ (*53*).

Pyrene-based ILs are not practical, but holes can be usefully trapped by the incorporation of benzyl groups into the IL. Benzyl groups can stabilize holes through the formation of cationic charge resonance (CR) states that delocalize the hole over two or more aromatic rings. Such cationic CR states have been known in the radiation chemistry of arenes for a long time (*87*) and the pyrene dimer radical cation was even observed in the IL radiolysis experiment mentioned just above (*53*).

Shkrob and coworkers investigated the excess charge-stabilizing properties of benzyl pendant groups in ILs with 5- and 6-member aromatic cations (*26*, *27*). The behaviors of the benzyl-derivatized imidazolium and pyridinium cations were different for the reducing (excess electron) side of the radiolysis process but the same for the oxidizing (hole) side. Upon reduction of benzylimidazolium cations, N-C bond cleavage produces a relatively stable benzyl radical and a substituted imidazole molecule (*26*). Reduction of benzylpyridinium cations instead produces a charge resonance state involving the excess electron and a pair of intact benzylpyridinium cations (*27*).

On the oxidizing side, benzyl-derivatized imidazolium, pyridinium and pyrrolidinium cations all produce cationic charge resonance states shared between benzyl rings of two or more cations. Localization of a positive charge between two cations would be energetically unfavorable in many conventional solvents; however in ILs this interaction is stabilized by the surrounding anions. Evidence for the existence of cationic and anionic CR states is shown by the pulse radiolysis transient absorption spectra shown in Figure 4 (*27*). The CR states of all the benzyl derivatives have broad and similar-shaped absorptions with peaks around 960-1000 nm that extend well into the NIR. This is reflective of the fact that the CR absorption features are the sums of populations with distributions of structural conformations. The CR absorption feature profiles are quite distinct from those of the solvated electron, as depicted in Figure 4 for [C_4mpyr][NTf_2].

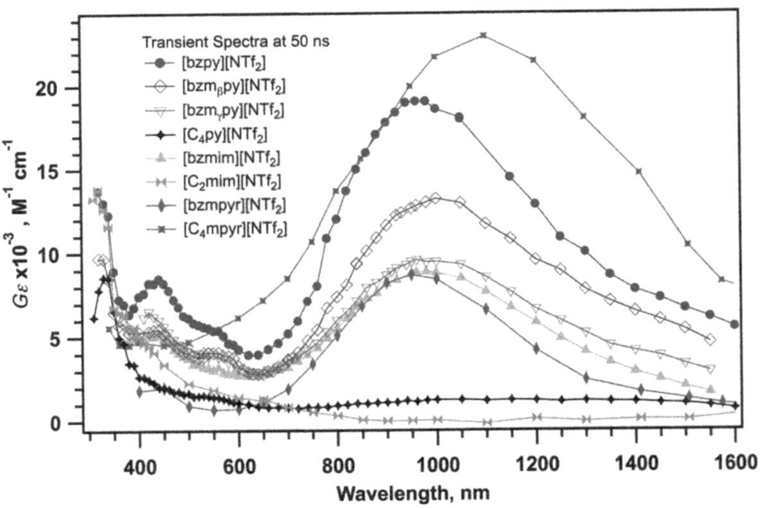

Figure 4. Pulse radiolysis transient absorption spectra for eight ILs, measured at 50 ns after the electron pulse. The cation partners of the [NTf_2]$^-$ anion are 1-benzylpyridinium, 1-benzyl-3-methylpyridinium, 1-benzyl-4-methylpyridinium, 1-butylpyridinium, 1-benzyl-3-methylimidazolium, 1-ethyl-3-methylimidazolium, 1-benzyl-1-methylpyrrolidinium, and 1-butyl-1-methylpyrrolidinium. Adapted with permission from reference (27). Copyright 2013 American Chemical Society.

At early times (5 ns), the addition of electron and hole scavengers had no influence on the yield of CR states (as detected by absorbance) because they could not compete with the high concentration of pre-formed CR traps for excess electrons and holes (*27*). Benzyl and related aromatic groups are therefore extremely useful for capturing and stabilizing excess charges before they can induce radiolytic damage.

Another way to protect separations systems from radiolytic damage is to introduce sacrificial components to absorb the oxidative or reductive damage to keep it from harming critical components such as extractants. In some cases the IL itself is sufficient to protect an extractant; as in the case of trialkylphosphates in imidazolium and pyrrolidinium ILs (*18*). In nuclear separation systems using hydrocarbon solvents to extract uranium and plutonium as complexes with tributylphosphate (TBP), dissociative electron attachment (DEA) to TBP causes loss of a butyl group, converting it into a counter-extractant that limits the useful life of the extraction system. In contrast, certain ILs protect trialkylphosphates from radiolytic degradation (*18*). A recent study showed that $[C_8mim][NTf_2]$ preserved the Th^{4+} extraction efficiency of four extractants under gamma radiolysis much better than did xylene (*88*).

It is well understood that even the simplest ILs are structurally different than conventional solvents, even before the issue of polar/non-polar structural heterogeneity is taken into account (*89*). This difference in structural organization also affects the spatial distribution of extractants in the system in way that could make them more susceptible to radiolytic damage in low-polarity organic solvents and less susceptible in ILs (*90*). For further protection, task-specific antioxidant cations can be introduced (*29*).

Radiation-Induced Reactivity of IL Anions

Although it could be argued that anions generally receive less attention among the IL community than cations, they are equally important contributors to the prompt and long-timescale radiation chemistry of ILs. For one thing, since many anions contain elements with higher atomic numbers (e.g., O, F, P, S) than cations typically do (H, C, N, sometimes P), a significant portion of the direct effect of high-energy radiation is absorbed by the anions (*17*), producing ionizations and excitations. Secondly, anions contribute to the electronic density of states in a given IL, as noted above, and therefore mediate the electron and hole chemistry. Attachment of excess electrons to anions such as nitrate (*24*, *43*) is not unusual. Indeed, since anions are surrounded by cations, an initially formed and unsolvated excess electron might be most stabilized on an anionic site until such time as the solvation process permits it to occupy its own cation-surrounded cavity (*43*, *44*, *47*).

Radiolysis of most of the anions that are commonly used in ILs results in fragmentation of the anion, forming a wide variety of radicals that are cataloged in Table 2. Anion fragmentation can be caused by DEA, oxidation or excited state chemistry (induced either by direct excitation or electron-hole recombination). Upon oxidation, halides and pseudohalides (e.g. $[N(CN)_2]^-$, $[SCN]^-$) form dimer radical anions (*22, 91*) that are well known from previous aqueous radiation chemistry work.

Table 2. Radical products derived from the constituent anions and identified in frozen ILs using EPR spectroscopy. Adapted with permission from reference (*22*). Copyright 2011 American Chemical Society.

Anion	*Radicals observed*
NTf_2^-	$\bullet CF_3$, $\bullet CF_2SO_2NTf^-$, $\bullet NTf_2$
NO_3^-	$NO_3^{2-\bullet}$, $NO_2^\bullet$, $NO_2^{2-\bullet}$
$CR_3CO_2^-$ [a]	$\bullet CR_3$, $\bullet CR_2CO_2^-$
$CR_3SO_3^-$ [a]	$\bullet CR_3$, $\bullet CR_2SO_3^-$, $CR_3SO_3^\bullet$
$CF_3CF_2CO_2^-$	$\bullet CF_2CF_2CO_2^-$, $CF_3{\bullet}CFCO_2^-$
$CH_3OSO_3^-$	$\bullet CH_2OH$, $\bullet CH_2OSO_3$-, $CH_3OSO_3^\bullet$
$(CH_3O)_2PO_2^-$	$\bullet CH_2OH$
$B(CN)_4^-$	$\bullet B(CN)_2$, $B(CN)_3^{-\bullet}$
$B(ox)_2^-$ [b]	$(ox)BO{\bullet}C{=}O$
X^- [c]	$X_2^{-\bullet}$
HSO_3^-	$H^\bullet$

[a] R= H or F. [b] ox=oxalato. [c] X = Br or $N(CN)_2$

Evidence for these radical intermediates produced from anion fragmentation is also abundantly available from mass spectroscopic and gas chromatographic studies of radiolysis products (*15–17*). Mass spectroscopy revealed multiple products of anion fragment attachment to cations and gaseous products produced by radical-radical reactions of anion and cation fragments. One study looked at radiolysis product distribution and rates (yields) of anion and cation loss (degradation) when the cation ($[C_4mim]^+$) was held constant and the anions ($[NTf_2]^-$, $[OTf]^-$, $[PF_6]^-$, $[BF_4]^-$) were varied. It was found that the yield of cation degradation varied depending on the anion used, but in all cases there was more damage to the cations and less to the anions than would have been predicted based on the cation/anion partition of the direct radiolysis effect. In other words, the electronic structure of the IL and subsequent chemistry partially directed damage away from the anions and towards the cations (*17*).

The radiolytic fragmentation of the $[NTf_2]^-$ anion produces other products besides the radicals listed in Table 2, chief among them SO_2 (*92*). In contact with air and moisture, which are common in the context of separations, SO_2 is first converted to $[SO_3]^{2-}$ and then eventually oxidized to $[SO_4]^{2-}$. The accumulation of sulfite and sulfate anions can interfere with extraction processes, particularly in the case of strontium where $SrSO_3$ and $SrSO_4$ form precipitates (*92, 93*). Fortunately, extraction efficiency can be recovered by washing the irradiated IL with water to remove the water-soluble anions (*92*).

It should be pointed out that the rates of radiolytic damage accumulation in ionic liquids incorporating the anions and cations mentioned above are within the range expected for conventional solvents, with the exception that less molecular hydrogen is produced in ILs (*17*). On the other hand, some of the IL anion radiolysis products might be more corrosive or acid-generating than those of normal solvents (*92*). However, just as in the case of radiation-resistant cations, certain classes of anions permit the design of ILs with lower overall susceptibility to radiation damage.

Radiation-Resistant Anions

The principle of using conjugated systems to delocalize excess charges and prevent them from stimulating reactivity and radiation damage works equally well for anions as it does for cations. A good example is the series of cyclic aromatic diamide anions shown in Figure 5.

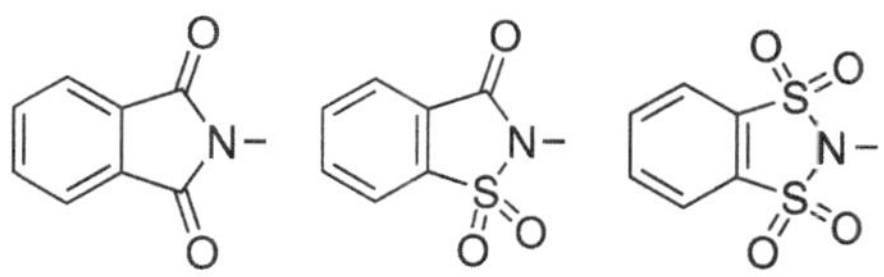

Figure 5. (Right to left): phthalimide, saccharinate (94) and 1,2 benzenedisulfonimide anions.

Each of these anions forms a stable neutral radical when oxidized (*95*). Electron addition to these anions is followed by protonation at the nitrogen to form a stable radical anion (*95*). Preliminary studies have shown that ILs incorporating saccharinate, which is an inexpensive, mass commodity anion, can stabilize the performance of extraction systems up to doses of 2 MGy.

Conjugated anionic systems to stabilize excess charges do not necessarily need to be cyclic. A series of ionic liquids with uncommon polynitrile anions containing 4-6 nitrile groups was shown to behave very similarly to the saccharinate family mentioned above. Despite not having the advantage of a cyclic structure, the polynitrile anions did not undergo fragmentation when oxidized to neutral radicals. When the anions were reduced, the resulting dianions were stabilized by protonation to form radical anions. This is an interesting class of ILs that has not been characterized in detail but might be useful in numerous applications.

A third significant family of radiation-resistant anions involves aromatic heterocyclic anions (AHAs), which have recently been developed for the absorption of CO_2 (*85*). Since the properties of these ILs change upon incorporation of CO_2, they can be considered "switchable" ILs and under the right conditions this switchability can be used to drive an extraction process in one direction or the other. Thus, there are compelling reasons to work out their radiation stability. Like the two anion classes mentioned above, AHAs resist fragmentation and form neutral imidyl radicals when oxidized and anionic H-atom adducts when reduced (reduction followed by protonation) (*96*).

New types of anions are constantly being added to the IL tool kit, and there will no doubt be several more families of conjugated, radiation-resistant anions in the pipeline.

Conclusions

Significant progress has been made in understanding the radiation chemistry of ionic liquids. The identities of reactive intermediates have been elucidated for many classes of cations and anions. In some cases product yields have been quantified. We are increasing our understanding of how to control and redirect radiolytic damage into more benign pathways. However, there is still much more work to be done along several lines of inquiry:

1) Ionic liquids provide a unique window into fundamental radiation chemistry on short timescales, due to their slower dynamics. In turn, these fundamental processes determine the identity and yields of reactive intermediates on longer timescales. The dynamics of excess charge solvation are particularly important for controlling reactivity in systems where solutes are concentrated, such as nuclear separations, batteries and systems for solar photoconversion and catalysis.

2) Radiation chemistry and pulse radiolysis would be excellent tools for studying general questions of chemical reactivity in ILs, especially redox-induced processes. They complement photochemical approaches by inducing reactions without he need of a chromophore and the constraints of excited-state lifetimes. They complement mechanistic electrochemistry because the redox equivalents can be delivered in a controlled manner (stepwise rather than in contact with an electrode) and the available time resolution (picosecond) far exceeds the best electrochemical system. However, before pulse radiolysis can be used for general reactivity studies in a particular solvent, the radiation chemistry of that solvent must be well characterized, including the yields and reactivities of the primary radiolytic species. These have been partially worked out in a limited number of ILs but the goal is to have fine control over reaction conditions, such as what has been worked out in water for many years. Advances will continue to be made along that front.

3) As alluded to in several places in this chapter, ionic liquids have significant contributions to make in the field of separations chemistry, in metallurgy, recycling and in the nuclear fuel cycle. In comparison to conventional organic solvents where neutral complex extraction is the only mechanism, ionic liquids have other pathways including ion exchange (*97*). While ion exchange was originally considered undesirable, controlled application of the ion exchange mechanism is now appreciated to be a useful lever for manipulating exchange processes. In addition, ILs provide new opportunities to re-invent extraction systems to take advantage of their unique properties. An example is using the IL to replace the aqueous phase rather than the organic one, as in the case of

an IL-based inverse TALSPEAK system that effectively separates lanthanides from actinides without the narrow pH constraints of the original aqueous-based TALSPEAK system (*98*). The system, which is based on a mixture of choline and betainium [NTf_2] salts (*30*), incorporates one of the extractants into the cationic matrix. This very innovative use of ILs as unconventional extraction solvents merits further development of other systems along the same concept and underscores the need for further IL radiation chemistry studies.

Acknowledgments

The author wishes to thank Ilya Shkrob, Sheng Dai, Mark Dietz, Andrew Cook and David Grills for fruitful collaborations. This work was supported by the US-DOE Office of Science, Division of Chemical Sciences, Geosciences and Biosciences under contract DE-SC0012704.

References

1. Wishart, J. F. Energy Applications of Ionic Liquids. *Energy Environ. Sci.* **2009**, *2*, 956–961.
2. Zhang, S. J.; Sun, J.; Zhang, X. C.; Xin, J. Y.; Miao, Q. Q.; Wang, J. J. Ionic Liquid-Based Green Processes for Energy Production. *Chem. Soc. Rev.* **2014**, *43*, 7838–7869.
3. Smiglak, M.; Pringle, J. M.; Lu, X.; Han, L.; Zhang, S.; Gao, H.; MacFarlane, D. R.; Rogers, R. D. Ionic Liquids for Energy, Materials, and Medicine. *Chem. Commun.* **2014**, *50*, 9228–9250.
4. MacFarlane, D. R.; Tachikawa, N.; Forsyth, M.; Pringle, J. M.; Howlett, P. C.; Elliott, G. D.; Davis, J. H., Jr.; Watanabe, M.; Simon, P.; Angell, C. A. Energy Applications of Ionic Liquids. *Energy Environ. Sci.* **2014**, *7*, 232–250.
5. Eshetu, G. G.; Armand, M.; Scrosati, B.; Passerini, S. Energy Storage Materials Synthesized from Ionic Liquids. *Angew. Chem., Int. Ed.* **2014**, *53*, 13342–13359.
6. Fox, E. B.; Smith, L. T.; Williamson, T. K.; Kendrick, S. E. Aging Effects on the Properties of Imidazolium-, Quaternary Ammonium-, Pyridinium-, and Pyrrolidinium-Based Ionic Liquids Used in Fuel and Energy Production. *Energy Fuels* **2013**, *27*, 6355–6361.
7. Yamagata, M.; Tanaka, K.; Tsuruda, Y.; Sone, Y.; Fukuda, S.; Nakasuka, S.; Kono, M.; Ishikawa, M. The First Lithium-Ion Battery with Ionic Liquid Electrolyte Demonstrated in Extreme Environment of Space. *Electrochemistry* **2015**, *83*, 918–924.
8. Krejci, D.; Lozano, P. Scalable Ionic Liquid Electrospray Thrusters for Nanosatellites. In *Guidance, Navigation, and Control 2016*; Chart, D. A., Ed., Univelt Inc.: San Diego, 2016; Vol. 157, pp 801−810.
9. Tolstogouzova, A. B.; Belykh, S. F.; Gurov, V. S.; Lozovan, A. A.; Taganov, A. I.; Teodoro, O.; Trubitsyn, A. A.; Chenakin, S. P. Ion-Beam Sources Based on Room-Temperature Ionic Liquids for Aerospace

Applications, Nanotechnology, and Microprobe Analysis (Review). *Instrum. Exp. Tech.* **2015**, *58*, 1–14.
10. Courtney, D. G.; Li, H. Q.; Lozano, P. Emission Measurements from Planar Arrays of Porous Ionic Liquid Ion Sources. *J. Phys. D, Appl. Phys.* **2012**, *45*, 485203.
11. Zeitlin, C.; Hassler, D. M.; Cucinotta, F. A.; Ehresmann, B.; Wimmer-Schweingruber, R. F.; Brinza, D. E.; Kang, S.; Weigle, G.; Böttcher, S.; Böhm, E.; Burmeister, S.; Guo, J.; Köhler, J.; Martin, C.; Posner, A.; Rafkin, S.; Reitz, G. Measurements of Energetic Particle Radiation in Transit to Mars on the Mars Science Laboratory. *Science* **2013**, *340*, 1080.
12. Sun, X. Q.; Luo, H. M.; Dai, S. Ionic Liquids-Based Extraction: A Promising Strategy for the Advanced Nuclear Fuel Cycle. *Chem. Rev.* **2012**, *112*, 2100–2128.
13. Wankowski, J. L.; Dietz, M. L. Ionic Liquid (IL) Cation and Anion Structural Effects on Metal Ion Extraction into Quaternary Ammonium-based ILs. *Solvent Extr. Ion Exch.* **2016**, *34*, 48–59.
14. Harmon, C. D.; Smith, W. H.; Costa, D. A. Criticality Calculations for Plutonium Metal at Room Temperature in Ionic Liquid Solutions. *Radiat. Phys. Chem.* **2001**, *60*, 157–159.
15. Berthon, L.; Nikitenko, S. I.; Bisel, I.; Berthon, C.; Faucon, M.; Saucerotte, B.; Zorz, N.; Moisy, P. Influence of Gamma Irradiation on Hydrophobic Room-Temperature Ionic Liquids [BuMeIm]PF_6 and [BuMeIm]$(CF_3SO_2)_2N$. *Dalton Trans.* **2006**, 2526–2534.
16. Bosse, E.; Berthon, L.; Zorz, N.; Monget, J.; Berthon, C.; Bisel, I.; Legand, S.; Moisy, P. Stability of [$MeBu_3N$][Tf_2N] under Gamma Irradiation. *Dalton Trans.* **2008**, 924–931.
17. Le Rouzo, G.; Lamouroux, C.; Dauvois, V.; Dannoux, A.; Legand, S.; Durand, D.; Moisy, P.; Moutiers, G. Anion Effect on Radiochemical Stability of Room-Temperature Ionic Liquids under Gamma Irradiation. *Dalton Trans.* **2009**, 6175–6184.
18. Shkrob, I. A.; Chemerisov, S. D.; Wishart, J. F. The Initial Stages of Radiation Damage in Ionic Liquids and Ionic Liquid-Based Extraction Systems. *J. Phys. Chem. B* **2007**, *111*, 11786–11793.
19. Shkrob, I. A.; Wishart, J. F. Charge Trapping in Imidazolium Ionic Liquids. *J. Phys. Chem. B* **2009**, *113*, 5582–5592.
20. Wishart, J. F.; Shkrob, I. A. The Radiation Chemistry of Ionic Liquids and its Implications for their Use in Nuclear Fuel Processing. In *Ionic Liquids: From Knowledge to Application*; Plechkova, N. V., Rogers, R. D., Seddon, K. R., Eds.; American Chemical Society: Washington, DC, 2009; Vol. 1030, pp 119–134.
21. Shkrob, I. A. Deprotonation and Oligomerization in Photo-, Radiolytically, and Electrochemically Induced Redox Reactions in Hydrophobic Alkylalkylimidazolium Ionic Liquids. *J. Phys. Chem. B* **2010**, *114*, 368–375.
22. Shkrob, I. A.; Marin, T. W.; Chemerisov, S. D.; Wishart, J. F. Radiation Induced Redox Reactions and Fragmentation of Constituent Ions in Ionic Liquids. 1. Anions. *J. Phys. Chem. B* **2011**, *115*, 3872–3888.

23. Shkrob, I. A.; Marin, T. W.; Chemerisov, S. D.; Hatcher, J. L.; Wishart, J. F. Radiation Induced Redox Reactions and Fragmentation of Constituent Ions in Ionic Liquids. 2. Imidazolium Cations. *J. Phys. Chem. B* **2011**, *115*, 3889–3902.
24. Shkrob, I. A.; Marin, T. W.; Chemerisov, S. D.; Wishart, J. F. Radiation and Radical Chemistry of NO_3^-, HNO_3, and Dialkylphosphoric Acids in Room-Temperature Ionic Liquids. *J. Phys. Chem. B* **2011**, *115*, 10927–10942.
25. Shkrob, I. A.; Wishart, J. F. Free Radical Chemistry in Room-Temperature Ionic Liquids. In *Encyclopedia of Radicals in Chemistry, Biology and Materials*; Chatgilialoglu, C., Studer, A., Eds.; John Wiley & Sons, Ltd.: Chichester, U.K., 2012; pp 433−448.
26. Shkrob, I. A.; Marin, T. W.; Luo, H. M.; Dai, S. Radiation Stability of Cations in Ionic Liquids. 1. Alkyl and Benzyl Derivatives of 5-Membered Ring Heterocycles. *J. Phys. Chem. B* **2013**, *117*, 14372–14384.
27. Shkrob, I. A.; Marin, T. W.; Hatcher, J. L.; Cook, A. R.; Szreder, T.; Wishart, J. F. Radiation Stability of Cations in Ionic Liquids. 2. Improved Radiation Resistance through Charge Delocalization in 1-Benzylpyridinium. *J. Phys. Chem. B* **2013**, *117*, 14385–14399.
28. Shkrob, I. A.; Marin, T. W.; Bell, J. R.; Luo, H. M.; Dai, S. Radiation Stability of Cations in Ionic Liquids. 3. Guanidinium Cations. *J. Phys. Chem. B* **2013**, *117*, 14400–14407.
29. Shkrob, I. A.; Marin, T. W. Radiation Stability of Cations in Ionic Liquids. 4. Task-Specific Antioxidant Cations for Nuclear Separations and Photolithography. *J. Phys. Chem. B* **2013**, *117*, 14797–14807.
30. Shkrob, I. A.; Marin, T. W.; Wishart, J. F.; Grills, D. C. Radiation Stability of Cations in Ionic Liquids. 5. Task-Specific Ionic Liquids Consisting of Biocompatible Cations and the Puzzle of Radiation Hypersensitivity. *J. Phys. Chem. B* **2014**, *118*, 10477–10492.
31. Mincher, B. J.; Modolo, G.; Mezyk, S. P. Review Article: The Effects of Radiation Chemistry on Solvent Extraction: 1. Conditions in Acidic Solution and a Review of TBP Radiolysis. *Solvent Extr. Ion Exch.* **2009**, *27*, 1–25.
32. Mincher, B. J.; Modolo, G.; Mezyk, S. P. Review Article: The Effects of Radiation Chemistry on Solvent Extraction: 2. A Review of Fission-Product Extraction. *Solvent Extr. Ion Exch.* **2009**, *27*, 331–353.
33. Mincher, B. J.; Modolo, G.; Mezyk, S. P. Review Article: The Effects of Radiation Chemistry on Solvent Extraction 3: A Review of Actinide and Lanthanide Extraction. *Solvent Extr. Ion Exch.* **2009**, *27*, 579–606.
34. Mincher, B. J.; Modolo, G.; Mezyk, S. P. Review: The Effects of Radiation Chemistry on Solvent Extraction 4: Separation of the Trivalent Actinides and Considerations for Radiation-Resistant Solvent Systems. *Solvent Extr. Ion Exch.* **2010**, *28*, 415–436.
35. Wishart, J. F. Photochemistry and Radiation Chemistry: A Perspective. In *Photochemistry and Radiation Chemistry*; Wishart, J. F., Nocera, D. G., Eds.; American Chemical Society: Washington, DC, 1998; Vol. 254, pp 1−4.
36. Richter, H. W. Radiation Chemistry: Principles and Applications. In *Photochemistry and Radiation Chemistry*; American Chemical Society: 1998; Vol. 254, pp 5−33.

37. Spotheim-Maurizot, M.; Mostafavi, M.; Douki, T.; Belloni, J. *Radiation Chemistry: From Basics to Applications in Material and Life Sciences*; EDP Sciences: Paris, 2008.
38. Hatano, Y.; Katsumura, Y.; Mozumder, A. *Charged Particle and Photon Interactions with Matter: Recent Advances, Applications, and Interfaces*; CRC Press: Boca Raton, 2010.
39. Wishart, J. F.; Rao, B. S. M. *Recent Trends in Radiation Chemistry*; World Scientific Publishing Co.: Singapore, 2010.
40. De Waele, V.; Lampre, I.; Mostafavi, M. Time-Resolved Study on Nonhomogeneous Chemistry Induced by Ionizing Radiation with Low Linear Energy Transfer in Water and Polar Solvents at Room Temperature. In *Charged Particle and Photon Interactions with Matter*; CRC Press: 2010; pp 289–324.
41. Alizadeh, E.; Sanche, L. Precursors of Solvated Electrons in Radiobiological Physics and Chemistry. *Chem. Rev.* **2012**, *112*, 5578–5602.
42. Liu, J. X.; Cukier, R. I.; Bu, Y. X. Bending Vibration-Governed Solvation Dynamics of an Excess Electron in Liquid Acetonitrile Revealed by Ab Initio Molecular Dynamics Simulation. *J. Chem. Theory Comput.* **2013**, *9*, 4727–4734.
43. Margulis, C. J.; Annapureddy, H. V. R.; De Biase, P. M.; Coker, D.; Kohanoff, J.; Del Popolo, M. G. Dry Excess Electrons in Room-Temperature Ionic Liquids. *J. Am. Chem. Soc.* **2011**, *133*, 20186–20193.
44. Xu, C. H.; Margulis, C. J. Solvation of an Excess Electron in Pyrrolidinium Dicyanamide Based Ionic Liquids. *J. Phys. Chem. B* **2015**, *119*, 532–542.
45. Wang, Z. P.; Zhang, L.; Chen, X. H.; Cukier, R. I.; Bu, Y. X. Excess Electron Solvation in an Imidazolium-Based Room-Temperature Ionic Liquid Revealed by Ab Initio Molecular Dynamics Simulations. *J. Phys. Chem. B* **2009**, *113*, 8222–8226.
46. Wang, Z. P.; Zhang, L.; Cukier, R. I.; Bu, Y. X. States and Migration of an Excess Electron in a Pyridinium-Based, Room-Temperature Ionic Liquid: an *ab initio* Molecular Dynamics Simulation Exploration. *Phys. Chem. Chem. Phys.* **2010**, *12*, 1854–1861.
47. Xu, C. H.; Durumeric, A.; Kashyap, H. K.; Kohanoff, J.; Margulis, C. J. Dynamics of Excess Electronic Charge in Aliphatic Ionic Liquids Containing the Bis(trifluoromethylsulfonyl)amide Anion. *J. Am. Chem. Soc.* **2013**, *135*, 17528–17536.
48. Wishart, J. F. Ionic Liquids and Ionizing Radiation: Reactivity of Highly Energetic Species. *J. Phys. Chem. Lett.* **2010**, *1*, 3225–3231.
49. Toigawa, T.; Gohdo, M.; Norizawa, K.; Kondoh, T.; Kan, K.; Yang, J.; Yoshida, Y. Examination of the Formation Process of Pre-solvated and Solvated Electron in *n*-Alcohol Using Femtosecond Pulse Radiolysis. *Radiat. Phys. Chem.* **2016**, *123*, 73–78.
50. Lampre, I.; Pernot, P.; Bonin, J.; Mostafavi, M. Comparison of Solvation Dynamics of Electrons in Four Polyols. *Radiat. Phys. Chem.* **2008**, *77*, 1183–1189.

51. Lampre, I.; Bonin, J.; Soroushian, B.; Pernot, P.; Mostafavi, M. Formation and Solvation Dynamics of Electrons in Polyols. *J. Mol. Liq.* **2008**, *141*, 124–129.
52. Wishart, J. F.; Funston, A. M.; Szreder, T.; Cook, A. R.; Gohdo, M. Electron Solvation Dynamics and Reactivity in Ionic Liquids Observed by Picosecond Radiolysis Techniques. *Faraday Discuss.* **2012**, *154*, 353–363.
53. Grodkowski, J.; Neta, P.; Wishart, J. F. Pulse Radiolysis Study of the Reactions of Hydrogen Atoms in the Ionic Liquid Methyltributylammonium Bis[(Trifluoromethyl)sulfonyl]imide. *J. Phys. Chem. A* **2003**, *107*, 9794–9799.
54. Wishart, J. F.; Neta, P. Spectrum and Reactivity of the Solvated Electron in the Ionic Liquid Methyltributylammonium Bis(trifluoromethylsulfonyl)imide. *J. Phys. Chem. B* **2003**, *107*, 7261–7267.
55. Katoh, R.; Yoshida, Y.; Katsumura, Y.; Takahashi, K. Electron Photodetachment from Iodide in Ionic Liquids through Charge-Transfer-to-Solvent Band Excitation. *J. Phys. Chem. B* **2007**, *111*, 4770–4774.
56. Asano, A.; Yang, J.; Kondoh, T.; Norizawa, K.; Nagaishi, R.; Takahashi, K.; Yoshida, Y. Molar Absorption Coefficient and Radiolytic Yield of Solvated Electrons in Diethylmethyl(2-methoxy)ammonium Bis(trifluoromethanesulfonyl)imide Ionic Liquid. *Radiat. Phys. Chem.* **2008**, *77*, 1244–1247.
57. Takahashi, K.; Sato, T.; Katsumura, Y.; Yang, J.; Kondoh, T.; Yoshida, Y.; Katoh, R. Reactions of Solvated Electrons with Imidazolium Cations in Ionic Liquids. *Radiat. Phys. Chem.* **2008**, *77*, 1239–1243.
58. Kondoh, T.; Asano, A.; Yang, J. F.; Norizawa, K.; Takahashi, K.; Taguchi, M.; Nagaishi, R.; Katoh, R.; Yoshida, Y. Pulse Radiolysis Study of Ion-Species Effects on the Solvated Electron in Alkylammonium Ionic Liquids. *Radiat. Phys. Chem.* **2009**, *78*, 1157–1160.
59. Molins i Domenech, F.; FitzPatrick, B.; Healy, A. T.; Blank, D. A. Photodetachment and Electron Reactivity in 1-Methyl-1-butyl-pyrrolidinium Bis(trifluoromethylsulfonyl)amide. *J. Chem. Phys.* **2012**, *137*, 034512.
60. Dhungana, K. B.; Faria, L. F. O.; Wu, B.; Liang, M.; Ribeiro, M. C. C.; Margulis, C. J.; Castner, E. W. Structure of Cyano-Anion Ionic Liquids: X-ray Scattering and Simulations. *J. Chem. Phys.* **2016**, *145*, 024503.
61. Ilawe, N. V.; Fu, J.; Ramanathan, S.; Wong, B. M.; Wu, J. Z. Chemical and Radiation Stability of Ionic Liquids: A Computational Screeding Study. *J. Phys. Chem. C* **2016**, *120*, 27757–27767.
62. Ong, S. P.; Andreussi, O.; Wu, Y. B.; Marzari, N.; Ceder, G. Electrochemical Windows of Room-Temperature Ionic Liquids from Molecular Dynamics and Density Functional Theory Calculations. *Chem. Mat.* **2011**, *23*, 2979–2986.
63. Allen, D.; Baston, G.; Bradley, A. E.; Gorman, T.; Haile, A.; Hamblett, I.; Hatter, J. E.; Healey, M. J. F.; Hodgson, B.; Lewin, R.; Lovell, K. V.; Newton, B.; Pitner, W. R.; Rooney, D. W.; Sanders, D.; Seddon, K. R.; Sims, H. E.; Thied, R. C. An Investigation of the Radiochemical Stability of Ionic Liquids. *Green Chem.* **2002**, *4*, 152–158.

64. Marcinek, A.; Zielonka, J.; Gebicki, J.; Gordon, C. M.; Dunkin, I. R. Ionic liquids: Novel Media for Characterization of Radical Ions. *J. Phys. Chem. A* **2001**, *105*, 9305–9309.
65. Behar, D.; Gonzalez, C.; Neta, P. Reaction Kinetics in Ionic Liquids: Pulse Radiolysis Studies of 1-Butyl-3-methylimidazolium Salts. *J. Phys. Chem. A* **2001**, *105*, 7607–7614.
66. Behar, D.; Neta, P.; Schultheisz, C. Reaction Kinetics in Ionic Liquids as Studied by Pulse Radiolysis: Redox Reactions in the Solvents Methyltributylammonium Bis(trifluoromethylsulfonyl)imide and *N*-Butylpyridinium Tetrafluoroborate. *J. Phys. Chem. A* **2002**, *106*, 3139–3147.
67. Cook, A. R.; Shen, Y. Z. Optical Fiber-Based Single-Shot Picosecond Transient Absorption Spectroscopy. *Rev. Sci. Instrum.* **2009**, *80*, 073106.
68. Wishart, J. F.; Cook, A. R.; Miller, J. R. The LEAF Picosecond Pulse Radiolysis Facility at Brookhaven National Laboratory. *Rev. Sci. Instrum.* **2004**, *75*, 4359–4366.
69. Skrzypczak, A.; Neta, P. Diffusion-Controlled Electron-Transfer Reactions in Ionic Liquids. *J. Phys. Chem. A* **2003**, *107*, 7800–7803.
70. Sherren, C. N.; Mu, C. H.; Webb, M. I.; McKenzie, I.; McCollum, B. M.; Brodovitch, J. C.; Percival, P. W.; Storr, T.; Seddon, K. R.; Clyburne, J. A. C.; Walsby, C. J. Merging the Chemistry of Electron-Rich Olefins with Imidazolium Ionic Liquids: Radicals and Hydrogen-Atom Adducts. *Chem. Sci.* **2011**, *2*, 2173–2177.
71. Wishart, J. F. Ionic Liquid Radiation Chemistry. In *Ionic Liquids Further UnCOILed: Critical Expert Overviews*; Plechkova, N., Seddon, K. R., Eds.; Wiley, Ltd.: Chichester, U.K., 2014; pp 259–274.
72. Bobrowski, K.; Grodkowski, J.; Zagorski, Z. P. Rate Constants of the Reactions of Tetraalkylammonium Cations with $e_{(aq)}^-$ Determined by Pulse Radiolysis Method. *Radiochem. Radioanal. Lett.* **1979**, *40*, 329–337.
73. Buxton, G. V.; Greenstock, C. L.; Helman, W. P.; Ross, A. B. Critical Review of Rate Constants for Reactions of Hydrated Electrons, Hydrogen Atoms and Hydroxyl Radicals (·OH/·O⁻) in Aqueous Solution. *J. Phys. Chem. Ref. Data* **1988**, *17*, 513–886.
74. Wishart, J. F.; Lall-Ramnarine, S. I.; Raju, R.; Scumpia, A.; Bellevue, S.; Ragbir, R.; Engel, R. Effects of Functional Group Substitution on Electron Spectra and Solvation Dynamics in a Family of Ionic Liquids. *Radiat. Phys. Chem.* **2005**, *72*, 99–104.
75. Kimura, A.; Taguchi, M.; Kondoh, T.; Yang, J. F.; Nagaishi, R.; Yoshida, Y.; Hirota, K. Decomposition of Halophenols in Room-Temperature Ionic Liquids by Ionizing Radiation. *Radiat. Phys. Chem.* **2010**, *79*, 1159–1164.
76. Funston, A. M.; Fadeeva, T. A.; Wishart, J. F.; Castner, E. W. Fluorescence Probing of Temperature-Dependent Dynamics and Friction in Ionic Liquid Local Environments. *J. Phys. Chem. B* **2007**, *111*, 4963–4977.
77. Katsuta, S.; Shiozawa, Y.; Imai, K.; Kudo, Y.; Takeda, Y. Stability of Ion Pairs of Bis(trifluoromethanesulfonyl)amide-Based Ionic Liquids in Dichloromethane. *J. Chem. Eng. Data* **2010**, *55*, 1588–1593.

78. Lall-Ramnarine, S. I.; Castano, A.; Subramaniam, G.; Thomas, M. F.; Wishart, J. F. Synthesis, Characterization and Radiolytic Properties of Bis(oxalato)borate Containing Ionic Liquids. *Radiat. Phys. Chem.* **2009**, *78*, 1120–1125.
79. Dhiman, S. B.; Goff, G. S.; Runde, W.; LaVerne, J. A. Hydrogen Production in Aromatic and Aliphatic Ionic Liquids. *J. Phys. Chem. B* **2013**, *117*, 6782–6788.
80. Dhiman, S. B.; Goff, G. S.; Runde, W.; LaVerne, J. A. Gamma and Heavy Ion Radiolysis of Ionic Liquids: A Comparative Study. *J. Nucl. Mater.* **2014**, *453*, 182–187.
81. Tarabek, P.; Liu, S. Y.; Haygarth, K.; Bartels, D. M. Hydrogen Gas Yields in Irradiated Room-Temperature Ionic Liquids. *Radiat. Phys. Chem.* **2009**, *78*, 168–172.
82. Visser, A. E.; Swatloski, R. P.; Reichert, W. M.; Mayton, R.; Sheff, S.; Wierzbicki, A.; Davis, J. H.; Rogers, R. D. Task-Specific Ionic Liquids for the Extraction of Metal Ions from Aqueous Solutions. *Chem. Commun.* **2001**, 135–136.
83. Sun, X.; Do-Thanh, C.-L.; Luo, H.; Dai, S. The Optimization of an Ionic Liquid-Based TALSPEAK-like Process for Rare Earth Ions Separation. *Chem. Eng. J.* **2014**, *239*, 392–398.
84. Bates, E. D.; Mayton, R. D.; Ntai, I.; Davis, J. H. CO_2 Capture by a Task-Specific Ionic Liquid. *J. Am. Chem. Soc.* **2002**, *124*, 926–927.
85. Seo, S.; Quiroz-Guzman, M.; DeSilva, M. A.; Lee, T. B.; Huang, Y.; Goodrich, B. F.; Schneider, W. F.; Brennecke, J. F. Chemically Tunable Ionic Liquids with Aprotic Heterocyclic Anion (AHA) for CO_2 Capture. *J. Phys. Chem. B* **2014**, *118*, 5740–5751.
86. Shkrob, I. A.; Marin, T. W.; Bell, J. R.; Luo, H. M.; Dai, S.; Hatcher, J. L.; Rimmer, R. D.; Wishart, J. F. Radiation-Induced Fragmentation of Diamide Extraction Agents in Ionic Liquid Diluents. *J. Phys. Chem. B* **2012**, *116*, 2234–2243.
87. Kira, A.; Imamura, M. Absorption Spectra of Dimer Cations and Other Cationic Species Produced by Warming of Gamma-Irradiated Glassy Solutions of Aromatic Hydrocarbons. *J. Phys. Chem.* **1979**, *83*, 2267–2273.
88. Singh, M.; Sengupta, A.; Murali, M. S.; Kadam, R. M. Comparative Study on the Radiolytic Stability of TBP, DHOA, Cyanex 923 and Cyanex 272 in Ionic Liquid and Molecular Diluent for the Extraction of Thorium. *J. Radioanal. Nucl. Chem.* **2016**, *309*, 615–625.
89. Canongia Lopes, J. N.; Padua, A. A. H. Nanostructural Organization in Ionic Liquids. *J. Phys. Chem. B* **2006**, *110*, 3330–3335.
90. Shkrob, I. A.; Marin, T. W.; Wishart, J. F. Radiation Induced Reactions and Fragmentation in Room Temperature Ionic Liquids. In *Applications of EPR in radiation research*; Lund, A., Shiotani, M., Eds.; Springer, Ltd.: Singapore, 2014; pp 453−488.
91. Grodkowski, J.; Nyga, M.; Mirkowski, J. Formation of $Br_2^{\bullet -}$, $BrSCN^{\bullet -}$ and $(SCN)_2^{\bullet -}$ Intermediates in the Ionic Liquid Methyltributylammonium Bis[(trifluoromethyl)sulfonyl]imide. Pulse Radiolysis Study. *Nukleonika* **2005**, *50*, S35–S38.

92. Ao, Y. Y.; Peng, J.; Yuan, L. Y.; Cui, Z. P.; Li, C.; Li, J. Q.; Zhai, M. L. Identification of Radiolytic Products of C_4mim NTf_2 and their Effects on the Sr^{2+} Extraction. *Dalton Trans.* **2013**, *42*, 4299–4305.
93. Yuan, L. Y.; Peng, J.; Zhai, M. L. Radiation Effects on Imidazolium Ionic Liquids and Their Extraction Systems. *Prog. Chem.* **2011**, *23*, 1469–1477.
94. Carter, E. B.; Culver, S. L.; Fox, P. A.; Goode, R. D.; Ntai, I.; Tickell, M. D.; Traylor, R. K.; Hoffman, N. W.; Davis, J. H. Sweet Success: Ionic Liquids Derived from Non-Nutritive Sweeteners. *Chem. Commun.* **2004**, 630–631.
95. Shkrob, I. A.; Marin, T. W.; Chemerisov, S. D.; Hatcher, J. L.; Wishart, J. F. Toward Radiation-Resistant Ionic Liquids. Radiation Stability of Sulfonyl Imide Anions. *J. Phys. Chem. B* **2012**, *116*, 9043–9055.
96. Shkrob, I. A.; Marin, T. W. The AHA Moment: Assessment of the Redox Stability of Ionic Liquids Based on Aromatic Heterocyclic Anions (AHAs) for Nuclear Separations and Electric Energy Storage. *J. Phys. Chem. B* **2015**, *119*, 14766–14779.
97. Garvey, S. L.; Hawkins, C. A.; Dietz, M. L. Effect of Aqueous Phase Anion on the Mode of Facilitated Ion Transfer into Room-Temperature Ionic Liquids. *Talanta* **2012**, *95*, 25–30.
98. Shkrob, I. A.; Marin, T. W.; Jensen, M. P. Ionic Liquid Based Separations of Trivalent Lanthanide and Actinide Ions. *Ind. Eng. Chem. Res.* **2014**, *53*, 3641–3653.

Chapter 12

Experimental Study of the Interactions of Fullerene with Ionic Liquids

M. F. Costa Gomes,* L. Pison, and A. A. H. Padua

Institut de Chimie de Clermont-Ferrand, CNRS and Université Clermont-Auvergne, F-63000 Clermont-Ferrand, France
***E-mail: margarida.costa-gomes@uca.fr**

The interactions between fullerenes and room temperature ionic liquids can be investigated experimentally using titration calorimetry. In the present work, we have added the ionic liquid 1-decyl-3-methylimidazolium bis-(trifluoromethylsulfonyl)imide, [$C_{10}C_1$Im][NTf_2], into a solution of C_{60} in 1,2-dichlorobenzene and the heat effect was recorded. The background heat of mixing of the two liquids was measured separately. It was observed that the enthalpy of mixing of the ionic liquid with the organic solvent in presence of C_{60} is more negative by approximately 4 J·mol^{-1} than in the absence of fullerene.

Introduction

Fullerenes are carbon materials that are not chemically inert and thus can be transformed, most often when present in liquid media, to find applications in materials chemistry, electrochemistry or catalysis (*1*). Therefore, the quantitative characterization and the understanding of the nature of fullerene solutions are important, the principal thermodynamic property to be studied being solubility. Fullerenes are scarcely soluble in any solvent, the highest reported solubilities at 298 K being 0.07 and 0.069 mol·L^{-1} in 1-chloronaphthalene or 1-phenyl-naphthalene, respectively (*2*). Other good solvents for C_{60} include 1,2-dichlorobenzene and o-xylene, with reported solubilities of 0.032−0.037 (*2, 3*) and 0.011−0.013 mol·L^{-1} (*4, 5*), respectively. Toluene or benzene only dissolve around 10^{-3} mol·L^{-1} of C_{60} at 298 K (*1*). A general observation is that solubility is enhanced in highly polarizable solvents with relatively low polarity and low cohesive energy density (*6*).

Ionic liquids have been considered as possible solvents for carbon materials (*7*) namely carbon nanotubes (*8*, *9*) or C_{60} (*6*, *10*) and also as effective media for the exfoliation of graphite (*11*). These purely ionic, salt-like materials that are composed solely of cations and anions form structured phases because of charge ordering and also because ionic liquids with sufficiently long nonpolar side chains show a heterogeneous structure composed of ionic and nonpolar domains (*12*). It has been shown that different molecular solutes can interact with these domains and thus be solvated in different molecular environments (*13*).

There have been few studies reported in the literature concerning the solvation of fullerenes in ionic liquids as the ionic nature of these solvents, like the polarity of organic solvents, is believed to cause a low solubility of C_{60} (*14*). Ionic liquids have, nevertheless, shown promising properties as solvents for the synthesis of functionalized fullerenes (*15*) and show remarkable properties for different applications in materials science and electrochemistry (*16*, *17*).

Following a previous computational study on the solvation of fullerene and fluorinated fullerene in molecular and ionic liquids (*10*), in this work we present for the first time an experimental study on the energy required to replace a good solvent of C_{60} – 1,2-dichlorobenzene, DCB – by an ionic liquid – 1-decyl-3-methylimidazolium bis-(trifluoromethylsulfonyl)imide [$C_{10}C_1$Im]-[NTf_2]. We have used an experimental protocol previously tested for the study of metallic nanoparticules in ionic liquids (*18*) that allows to obtain an estimation of the difference in interaction energy between C_{60} in 1,2-dichlorobenzene and C_{60} in [$C_{10}C_1$Im][NTf_2]. Our aim is to contribute with experimental data to improve the understanding about the mechanisms of solvation and stabilization of this kind of carbon nanomaterial in ionic liquids.

Experimental

Materials

The ionic liquid 1- decyl-3-methylimidazolium bis(trifluoromethylsulfonyl)imide [$C_{10}C_1$Im][NTf_2] was purchased from Iolitec with a purity better than 98%. The liquid was used without further purification but was dried under primary vacuum at room temperature during 24 h. 1,2-dichlorobenzene (DCB) from Fluka, grade purum ≥98% from GC, was used in the measurements. Fullerene C_{60} was purchased at Sigma-Aldrich with a 99.9% purity and was used as received from the manufacturer.

Apparatus and Operation

Calorimetric measurements were performed at 313 K and atmospheric pressure using an isothermal titration nanocalorimeter (TA Instruments) equipped with 1 mL stainless steel cells housed in a Thermal Activity Monitor TAM III thermostat (TA Instruments). The temperature of the thermostat is controlled precisely within 10^{-5} K. An electrical calibration was done before each experiment, and the instrument was chemically calibrated several times by titration of a 0.01 M aqueous solution of 18-crown-6 ethers with an 0.2 M aqueous solution of $BaCl_2$ as described in references (*19*, *22*). The enthalpies of binding of Ba^{2+} ions to 18-crown-6 were found to be 2% higher than those reported in the literature. No correction attributable to these differences was introduced in the raw data and, based on these values, the authors estimate the overall uncertainty of the present calorimetric determinations as ±2%.

For the determination of the enthalpy of mixing of $[C_{10}C_1Im][NTf_2]$ (IL) and DCB, $\Delta_{mix}H_{IL+DCB}$, approximately 0.8 mL of degassed ionic liquid or degassed DCB were introduced into the 1 mL glass measuring and reference cells. The liquid in the measuring cell was stirred by a gold-platted propeller at 80 rpm and volumes of 9 μL of the second liquid (different from the one in the cells) were injected during 40 s using a motor driven pump (Thermometric 3810 Syringe Pump) equipped with a 100 μL gastight Hamilton Syringe. The intervals between consecutive injections were 40 min, which allowed for a good thermal stabilization of the solution and the return to a stable baseline. Each experiment consisted in 10 injections. Undesirable effects of diffusion of the liquid in the cell into the injection canula were avoided by immersing the canula only about 3 mm prior to the first injection. A peak with an area proportional to the resulting heat effect Q_i translated to the thermal effect due to each injection of liquid. The integration of the peaks from the recorded calorimetric plots was performed using the TAM III Assistant software.

For the measurement of the interaction energy between C_{60} and $[C_{10}C_1Im][NTf_2]$, the sample cell of the calorimeter was filled with approximately 0.8 mL of a solution of 0.0311 mol·L^{-1} of C_{60} in DCB and the reference cell was filled with the same amount of solution. The solution in the sample cell was stirred at 80 rpm and volumes of 9 μL of the ionic liquid were injected following the procedure described above. The heat effect measured corresponds to the enthalpy of mixing of the two liquids in the presence of C_{60}, $\Delta_{mix}H_{IL+DCB(C_{60})}$. This corresponds to the contributions due to the new interactions between C_{60} and the ionic liquid, the loss in the interactions between the C_{60} and DCB, and the gain of interactions between the ionic liquid and DCB. This last contribution is determined by "blank" experiment, without C_{60}, where the enthalpy mixing of the two liquids, $\Delta_{mix}H_{IL+DCB}$, is determined.

Data Reduction

The heat effects involved in injections of small quantities of $[C_{10}C_1Im][NTf_2]$ in DCB (or of DCB in $[C_{10}C_1Im][NTf_2]$) , Q_{IL} (or Q_{DCB}), are directly related to the partial molar excess enthalpy of the IL in DCB, $\bar{H}^E_{IL}$ (or of DCB in the IL, $\bar{H}^E_{DCB}$), according to eq 1.

$$\bar{H}^E_{DCB} = \left(\frac{\partial (n_{IL} + n_{DCB}) \Delta_{mix} H_{IL+DCB}}{\partial n_{DCB}} \right)_{p,T,n_{IL}} \approx \frac{Q_{DCB}}{\Delta n_{DCB}} \text{ and}$$

$$\bar{H}^E_{IL} = \left(\frac{\partial (n_{IL} + n_{DCB}) \Delta_{mix} H_{IL+DCB}}{\partial n_{IL}} \right)_{p,T,n_{DCB}} \approx \frac{Q_{IL}}{\Delta n_{IL}} \quad (1)$$

where n_{IL} and n_{DCB} denote the amounts of $[C_{10}C_1Im][NTf_2]$ and 1,2-dichlorobenzene, respectively. Δn_{IL} and Δn_{DCB} are the quantities of $[C_{10}C_1Im][NTf_2]$ and DCB per injection calculated from the injected volumes and from the densitites reported in the literature (*20*, *21*). If the enthalpy of mixing, $\Delta_{mix}H_{IL+DCB}$, is represented by a Redlich-Kister equation, eq 2, where x_{IL} and x_{DCB} are mole fractions of $[C_{10}C_1Im][NTf_2]$ and DCB in the mixture, respectively, then $\bar{H}^E_{IL}$ and $\bar{H}^E_{DCB}$ are obtained by appropriate derivatives with respect to composition, yielding eq 3 and 4 as described in references (*18*) and (*22*, *25*).

$$\begin{aligned} \Delta_{mix} H_{IL+DCB} &= (1 - x_{DCB}) x_{DCB} \sum_{i=0}^{n} A_i (1 - 2x_{DCB})^i \\ &= (1 - x_{IL}) x_{IL} \sum_{i=0}^{n} A_i (2x_{IL} - 1)^i \end{aligned} \quad (2)$$

$$\bar{H}^E_{DCB} = \Delta_{mix} H_{IL+DCB} + (1 - x_{DCB}) \left(\frac{\partial \Delta_{mix} H_{IL+DCB}}{\partial x_{DCB}} \right)_{p,T,x_{IL}} \quad (3)$$

$$\bar{H}^E_{IL} = \Delta_{mix} H_{IL+DCB} + (1 - x_{IL}) \left(\frac{\partial \Delta_{mix} H_{IL+DCB}}{\partial x_{IL}} \right)_{p,T,x_{DCB}} \quad (4)$$

The enthalpy of mixing of IL with DCB in the presence of C_{60}, $\Delta_{mix}H_{IL+DCB(C_{60})}$, is calculated the system in presence of C_{60} being simply treated as a binary system as the concentration was small, $x_{C60} < 10^{-2}$.

$$\bar{H}^E_{DCB(C60)} = (x_{DCB} - 1)^2 \left(x_{DCB} \sum_{i=0}^{n} -2iA_{i,C60} (1 - 2x_{DCB})^{-1+i} + \sum_{i=0}^{n} A_i (1 - 2x_{DCB})^i \right) \quad (5)$$

$$\Delta_{mix} H_{IL+DCB(C60)} = (1 - x_{DCB}) x_{DCB} \sum_{i=0}^{n} A_{i,C60} (1 - 2x_{DCB})^i \quad (6)$$

The different in the interaction energy between C_{60} in DCB and C_{60} in $[C_{10}C_1Im][NTf_2]$ can be considered as the difference in the enthalpy of mixing of the two liquids in presence and in absence of fullerene using eq 7

$$\Delta\Delta H_{DCB(C60)-IL(C60)} = \Delta_{mix} H_{IL+DCB(C_{60})} - \Delta_{mix} H_{IL+DCB} \quad (7)$$

Results and Discussion

The enthalpy of mixing of $[C_{10}C_1Im][NTf_2]$ with 1,2-dichlorobenzene was calculated from the experimental calorimetric data through the partial molar excess enthalpies of the two liquids in the mixture using eq 2. The experimental points were fitted to a Redlich-Kister function with three coefficients. The enthalpies of mixing as a function of composition, at 313 K, are represented in Figure 1. Because the experimental points were determined only at extreme mole fraction compositions, the Redlich-Kister fit at intermediary compositions, although still valid (*25*), is less significant. The mixing enthalpy is negative with a minimum of -164 J·mol^{-1} deviated to ionic liquid mole fractions lower than equimolar (at x_{IL} = 0.5 $\Delta_{mix}H_{IL+DCB} = -84.2$ J·mol^{-1}), reflecting the energetically favourable interactions in the mixing process. These values are significantly less negative than the ones found for mixtures of 1-alkyl-3-methylimidazolium bis-(trifluoromethylsulfonyl)imide ionic liquids with polar compounds and are closer to the reported values of the enthalpy of mixing with apolar liquids like toluene (*e.g.* $\Delta_{mix}H \approx -0.2$ kJ·mol^{-1} for toluene + $[C_2C_1Im][NTf_2]$ at 298 K (*23*) or $\Delta_{mix}H \approx -0.6$ kJ·mol^{-1} for toluene + $[C_4C_1Im][NTf_2]$ at 363 K (*24*)).

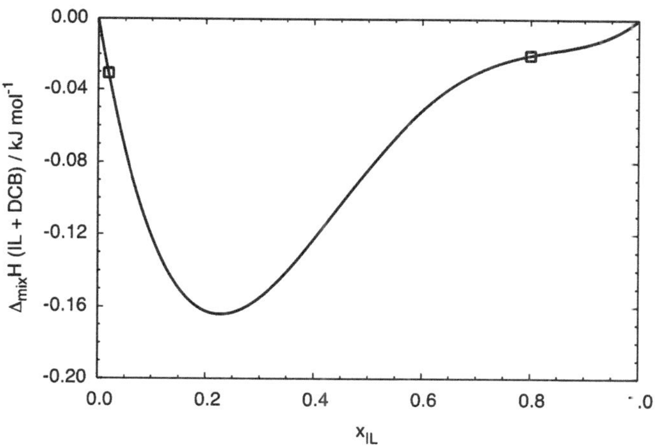

Figure 1. Enthalpy of mixing of $[C_{10}C_1Im][NTf_2]$ + 1,2-dichlorobenzene as a function of the mole fraction of $[C_{10}C_1Im][NTf_2]$, x_{IL}, at 313 K. The Redlich-Kister fit at intermediary compositions is less significant as the experimental determinations are only at extreme compositions.

In the presence of C_{60}, the measured values of the partial molar excess enthalpies were fitted to a Redlich-Kister polynomial as was done for the mixtures of $[C_{10}C_1Im][NTf_2]$ with 1,2-dichlorobenzene. The values are depicted in Figure 2 and it is observed that, when the ionic liquid $[C_{10}C_1Im][NTf_2]$ is added to the solution of C_{60} in 1,2-dichlorobenzene, the heat of mixing is more negative than when $[C_{10}C_1Im][NTf_2]$ is added to the solution pure 1,2-dichlorobenzene.

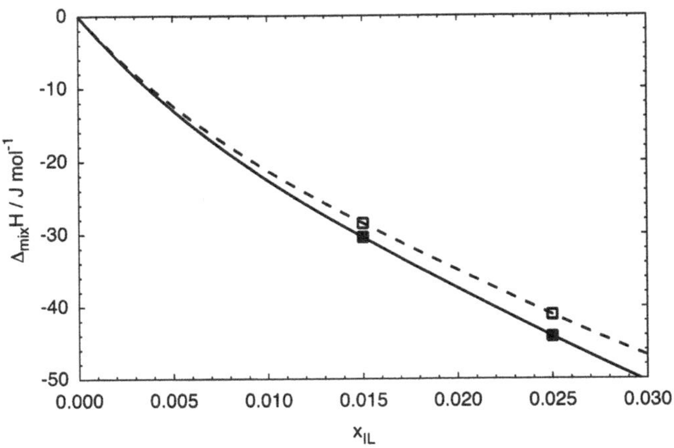

Figure 2. Enthalpies of mixing of [$C_{10}C_1Im$][NTf_2] + 1,2-dichlorobenzene (empty symbols and dashed lines) and of [$C_{10}C_1Im$][NTf_2] + C_{60} in 1,2-dichlorobenzene (full symbols and full lines).

The differences encountered are already apparent in the heats directly measured by the calorimeter as seen in Figure 3.

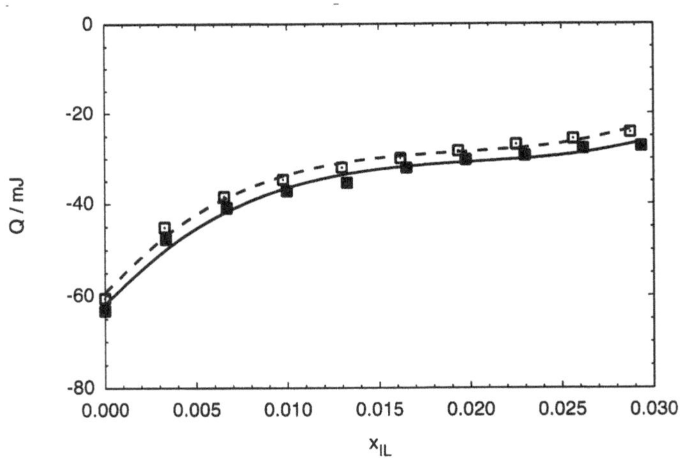

Figure 3. Experimental heat effects of the additions of [$C_{10}C_1Im$][NTf_2] to 1,2-dichlorobenzene (open symbols) and of [$C_{10}C_1Im$][NTf_2] to C_{60} in 1,2-dichlorobenzene (full symbols).

The difference between the two curves represented in Figure 2 can be understood as the difference in the interaction energy of C_{60} with 1,2-dichlorobenzene and of C_{60} with [$C_{10}C_1$Im][NTf_2], $\Delta\Delta H_{DCB(C60)-IL(C60)}$. As can be seen in Figure 4, the values of $\Delta\Delta H_{DCB(C60)-IL(C60)}$ are negative, indicating an exothermal heat effect due to interactions of the ionic liquid with C_{60}, even if the solubility of the fullerene is much lower in [$C_{10}C_1$Im][NTf_2] (*10*) than in 1,2-dichlorobenzene (*2, 3*).

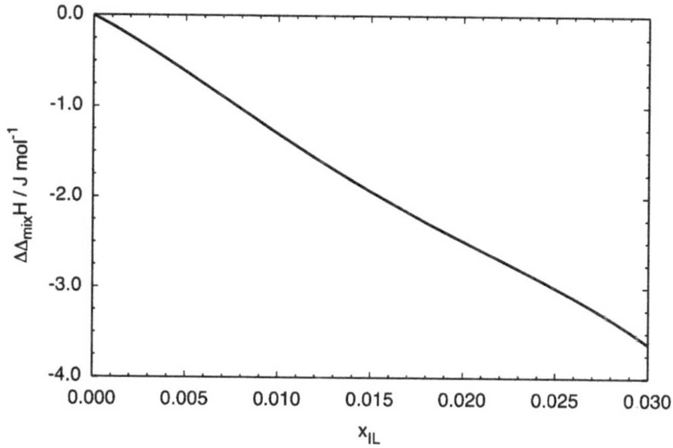

Figure 4. Difference in the enthalpy of mixing of [$C_{10}C_1$Im][NTf_2] + 1,2-dichlorobenzene and [$C_{10}C_1$Im][NTf_2] + C_{60} in 1,2- dichlorobenzene obtained from eq 7.

Conclusion

We present an experimental study on the thermodynamics of mixing of an imidazolium based ionic liquid with a solution of C_{60} in 1,2-dichlorobenzene. The authors show that isothermal titration calorimetry techniques are sufficiently precise to enable experimental access to the difference in the interaction energy between carbon-based materials (herein C_{60}) and different solvents, in particular ionic liquids. It has been shown in this work that the addition of an imidazolium based ionic liquid to a solution of fullerene in an organic solvent (1,2-dichlorobenzene in the present case) is more exothermic than the addition of the ionic liquid to the pure organic solvent. The difference, low but significant (approximately −4 J·mol^{-1} of ionic liquid for x_{IL} = 0.03 in a solution of concentration 0.0311 mol L^{-1} of C_{60} in DCB), can be attributed to the interaction between the ionic liquid and the fullerene.

References

1. Mechedlov-Petrossyan, N. O. Fullerenes in Liquid Media: An Unsettling Intrusion Into the Solution Chemistry. *Chem. Rev.* **2013**, *113*, 5149–5193.
2. Ruoff, R. S.; Tse, D. S.; Malhotra, R.; Lorents, D. C. Solubility of Fullerene (C60) in a Variety of Solvents. *J. Phys. Chem.* **1993**, *97*, 3379–3383.
3. Doome, R. J.; Dermaut, S.; Fonseca, A.; Hammida, M.; Nagy, J. B. New Evidences for the Anomalous Temperature-Dependent Solubility of C_{60} and C_{70} Fullerenes in Various Solvents. *Fullere Sci. Technol.* **1997**, *5*, 1593–1606.
4. Scrivens, W. A.; Tour, J. M. Potent Solvents for C_{60} and Their Utility for the Rapid Acquisition of ^{13}C NMR Data for Fullerenes. *J. Chem. Soc., Chem. Commun.* **1993**, 1207–1209.

5. Beck, M. T.; Mańdi, G. Solubility of C_{60}. *Fullerene Sci. Technol.* **1997**, *5*, 291–310.
6. Liu, H.; Tao, G.; Evans, D. G.; Kou, Y. Solubility of C_{60} in Ionic Liquids. *Carbon* **2005**, *43*, 1782–1785.
7. Fukushima, T.; Kosaka, A.; Ishimura, Y.; Yamamoto, T.; Takigawa, T.; Ishii, N.; Aida, T. Molecular Ordering of Organic Molten Salts Triggered by Single-Walled Carbon Nanotubes. *Science* **2003**, *300*, 2072–2074.
8. Wang, J.; Chu, H.; Li, Y. Why Single-Walled Carbon Nanotubes Can Be Dispersed in Imidazolium-Based Ionic Liquids. *ACS Nano* **2008**, *2*, 2540–2546.
9. Lu, F.; Zhang, S.; Zheng, L. Dispersion of Multi-Walled Carbon Nanotubes (MWCNTs) by Ionic Liquid-Based Phosphonium Surfactants in Aqueous Solution. *J. Mol. Liq.* **2012**, *173*, 42–46.
10. Szala-Bilnik, J.; Costa Gomes, M. F.; Padua, A. A. H. Solvation of C_{60} Fullerene and $C_{60}F_{48}$ Fluorinated Fullerene in Molecular and Ionic Liquids. *J. Phys. Chem C* **2016**, *120*, 19396–19408.
11. Matsumoto, M.; Saito, Y.; Park, C.; Fukushima, T.; Aida, T. Ultrahigh-Throughput Exfoliation of Graphite into Pristine 'Single-Layer' Graphene Using Microwaves and Molecularly Engineered Ionic Liquids. *Nat. Chem.* **2015**, *7*, 730–736.
12. Padua, A. A. H.; Costa Gomes, M. F.; Canongia Lopes, J. N. A. Molecular Solutes in Ionic Liquids: A Structural Perspective. *Acc. Chem. Res.* **2007**, *40*, 1087–1096.
13. Canongia Lopes, J. N.; Costa Gomes, M. F.; Padua, A. A. H. Nonpolar, Polar, and Associating Solutes in Ionic Liquids. *J. Phys. Chem. B* **2006**, *110*, 16816–16818.
14. Yoshihara, T.; Murai, M.; Tamaki, Y.; Furube, A.; Katoh, R. Trace Analysis by Transient Absorption Spectroscopy: Estimation of the Solubility of C_{60} in Polar Solvents. *Chem. Phys. Lett.* **2004**, *394*, 161–164.
15. Zhu, Y. H. Application of Ultrasound Technique in the Synthesis of Methanofullerene Derivatives. *J. Phys. Chem. Solids* **2004**, *65*, 349–353.
16. Welton, T. Room-Temperature Ionic Liquids. Solvents for Synthesis and Catalysis. *Chem. Rev.* **1999**, *99*, 2071–2084.
17. Deutsch, D.; Tarabek, J.; Krause, M.; Janda, P.; Dunsch, L. Nanostructuring of C_{60} Fullerene Thin Films. *Carbon* **2004**, *42*, 1137–1141.
18. Podgorsek, A.; Pensado, A. S.; Santini, C. C.; Costa Gomes, M. F.; Padua, A. A. H. Interaction Energies of Ionic Liquids with Metallic Nanoparticles: Solvation and Stabilization Effects. *J. Phys. Chem. C* **2013**, *117*, 3537–3547.
19. Campbell, P. S.; Podgorsek, A.; Gutel, T.; Santini, C. C.; Padua, A. A. H.; Costa Gomes, M. F.; Bayard, F.; Fenet, B.; Chauvin, Y. How do Physical-Chemical Parameters Influence the Catalytic Hydrogenation of 1,3-Cyclohexadiene in Ionic Liquids? *J. Phys. Chem. B* **2010**, *114*, 8156–8165.
20. Tariq, M.; Serro, A. P.; Mata, J. L.; Saramago, B.; Esperança, J. M. S. S.; Canongia Lopes, J. N.; Rebelo, L. P. N. High-Temperature Surface Tension and Density Measurements of 1-Alkyl-3-Methylimidazolium Bistriflamide Ionic Liquids. *Fluid Phase Equilib.* **2010**, *294*, 131–138.

21. Goralski, P.; Piekarski, H. Heat Capacities and Densities of Some Liquid Chloro-, Bromo-, and Bromochloro-Substituted Benzenes. *J. Chem. Eng. Data* **2007**, *52*, 655–659.
22. Podgorsek, A.; Jacquemin, J.; Padua, A. A. H.; Costa Gomes, M. F. Mixing Enthalpy for Binary Mixtures Containing Ionic Liquids. *Chem. Rev.* **2016**, *116*, 6075–6106.
23. Marczak, W.; Verevkin, S. P.; Heintz, A. Enthalpies of Solution of Organic Solutes in the Ionic Liquid 1-Methyl-3-ethyl-imidazolium Bis-(trifluoromethyl-sulfonyl) Amide. *J. Solution Chem.* **2003**, *32*, 519–526.
24. Nebig, S.; Bölts, R.; Gmehling, J. Measurement of Vapor-Liquid Equilibria (VLE) and Excess Enthalpies (H^E) of Binary Systems with 1-Alkyl-3-Methylimidazolium Bis(trifluoromethylsulfonyl)imide and Prediction of These Properties and γ^∞ Using Modified UNIFAC (Dortmund). *Fluid Phase Equilib.* **2007**, *258*, 168–178.
25. Matteoli, E.; Lepori, L. Determination of the Excess Enthalpy of Binary Mixtures From the Measurements of the Heat of Solution of the Components: Application to the Perfluorohexane + Hexane Mixture. *Fluid Phase Equilib.* **2000**, *174*, 115–131.

Chapter 13

Biphasic Extraction, Recovery and Identification of Organic and Inorganic Compounds with Ionic Liquids

Rico E. Del Sesto,[1,*] Andrew T. Koppisch,[2,*] David T. Fox,[3] Mattie R. Jones,[1] Katherine S. Lovejoy,[3] Tyler E. Stevens,[4] and Todd C. Monson[4]

[1]Department of Chemistry, Dixie State University, 225 S. University Ave., St. George, Utah 84770, United States
[2]Department of Chemistry, Northern Arizona University, P.O. Box 5698, Flagstaff, Arizona 86011, United States
[3]Los Alamos National Laboratory, PO Box 1663, Los Alamos, New Mexico 87545, United States
[4]Nanoscale Sciences, Sandia National Laboratories, P.O. Box 5800, MS 1415, Albuquerque, New Mexico 87185, United States
[*]E-mail: delsesto@dixie.edu; andy.koppisch@nau.edu

New methods aimed at extraction and identification of contaminants and trace compounds are critical for a number of technological areas that greatly impact our society. Current methods to extract bulk or trace compounds from various sources often use volatile organic compounds or exhaustive treatment processes that also tend to destroy important chemical signatures needed for identification of contaminants. We have utilized ionic liquid extractants that are non-toxic and non-flammable, and that can quantitatively remove dissolved organic and inorganic contaminants from aqueous systems and solids. Additionally, complete recovery of trace analytes can be immediately followed by isolation and identification through standard analytical methods. The versatility of these new liquid extractants therefore provides a high-yielding recovery and single-pot approach to enhance identification of water contaminants and aid in forensic investigation.

Introduction

Room temperature ionic liquids (RTILs) have been explored to address numerous applications for over thirty years, ranging from electrochemical applications in batteries and energy storage to more obscure applications such as embalming and insect feeding deterrent (*1*, *2*). One major driver for their continued exploration is their primarily organic nature that allows for tunabaility of functional groups and therefore properties. In particular, many applications have specific needs in terms of properties such as thermal stability, viscosity, conductivity, and water miscibility. Due to the ionic and organic nature of RTILs, they can achieve performance standards that are not possibile with typical inorganic molten salts or conventional organic solvents; rather, RTILs tend to bring the advantages of each individual class to yield more powerful materials.

One of the significant advantages of ionic liquids used for extractions is the ability to introduce functional groups with the potential for strong intermolecular interactions (e.g. dipole-dipole, hydrogen bonding, etc.) while still maintaining water immiscibility. Using water-immiscible ionic liquids with the ideal properties, we are able to extract organic, inorganic, and highly water-soluble compounds from aqueous solutions and solid materials through a biphasic extraction. RTILs have been utilized for extraction of solids for a number of years, including dissolution of natural compounds for processing into energy-related materials, e.g. cellulose and nuclear material (*3–7*) and novel functional materials, e.g. hybrid textiles (*8–15*). Additionally, there are a number of teams exploring the use of RTILs for extraction of contaminants from water-sources for remediation purposes, e.g. water reclamation from sewage (*16*, *17*). From this prior work, along with our recent results (*18–20*), some information regarding structure property relationships of RTILs are becoming illuminated. For example, dissolution of strongly hydrogen-bonded, recalcitrant biomaterials such as cellulose and keratin into RTILs often depends on the nature of the anion, where strong hydrogen-bond acceptor anions such as halides and carboxylates coupled with weakly interacting cations often lead to greater solubilities (*18–22*).

One application of extractions with RTILs relates to quantitative and forensic needs. The extraction of target analytes into RTILs from aqueous systems facilitates their analysis through a number of instrumental methods. In many standard extraction-analysis protocols, further processing of samples post isolation is often necessary to prepare them for examination following extraction. One unique advantage to RTILs is that they are often non-interfering with analytical methods – most RTILs lack spectroscopic signatures in the UV-visible absorption and fluorescence regions, and their thermal/chemical stability and non-volatility can limit their interference with the analyte in many mass spectroscopy methods (*18–20*, *23*). In this report, we discuss the use of water-immiscible RTILs for the extraction of compounds from solids and aqueous solutions, and the ensuing analysis of the target analytes via optical and mass spectroscopic methods, among other analytical methods. While there are numerous classes of ionic liquids available for such methods, we have focused on tetralkylphosphonium cation-based RTILs, Figure 1, due to their observed broad-spectrum extraction ability and their propensity to readily form water-immiscible RTILs with nearly any anion (*24*, *25*). In particular, the tri(hexyl)tetradecyl phosphonium cation, $[P_{6\,6\,6\,14}]^+$, has proven extremely useful in the methods described herein as a result of RTIL formation with anions ranging from the halides to fluorinated organics. This versatility has allowed for some degree of tunability for extraction of specific analytes. More importantly, the extreme hydrophobic nature of the cation along with the highly charged, essentially "free" anion, introduce a wide range of properties into a single RTIL.

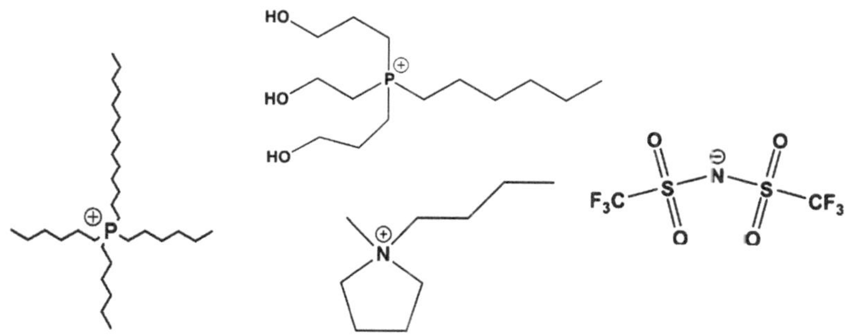

Figure 1. Typical phosphonium and pyrrolidinium cations - $[P_{6\,6\,6\,14}]^+$, $[P(C_3H_6OH)_3C_6]^+$, $[C_4Mpyr]^+$ - and $[Tf_2N]$ anion used for extraction and analysis.

Results and Discussion

Extractions using RTILs were performed generally by adding water immiscible ILs to aqueous solutions or solids containing the compounds of interest to demonstrate proof-of-principle. In some cases, particularly with extractions from solid samples, industrially relevant materials were used to test the extraction and identification of specific compounds (*18–20*). The general scheme for this process is shown in Figure 2. In a typical experiment, the solutions or solids were allowed to mix with the RTIL extractant for a fixed amount of time (depending on mixing ability of the two phases), then the phases separated. Characterization was then carried out on either the RTIL or aqueous layer (or both, as needed), by various spectroscopic techniques to demonstrate the efficiency of the process.

Below, we discuss our efforts to extract various compounds into RTIL, with different end-results based on the application. These processes successfully obtained high efficiency extractions of organic dyes, large hydrocarbons, and transition and lanthanide metals, with subsequent identification through direct analysis by optical spectroscopy, mass spectrometry, and magnetometry.

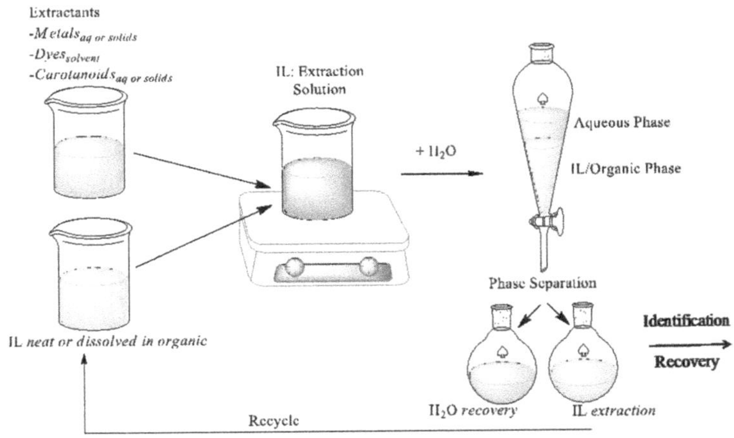

Figure 2. Method for extraction and identification using water-immiscible RTILs.

Organic Dyes

RTIL extraction methods were used to address a specific need within the area of trace extraction and identification of organic dyes from textile materials. In particular, dyes integrated into wool textiles are extremely difficult to remove without destroying the chemical signature of those dyes. Extreme or caustic conditions, such as concentrated acid or base, are typically required to disrupt the robust structure of the keratin protein in the wool in order to release the organic dyes (*26, 27*). This dye extraction and isolation process often encounters competing hydrolysis or other decomposition reactions facilitated by the extreme conditions. These competing, degradative pathways destroy the chemical makeup of the dyes, and thus reduce the effectiveness of the extraction needed for the analytical techniques. It has been well established that RTILs are able to disrupt and denature biopolymers such as cellulose, silk, chitin, and keratin (*3–15, 28–30*). We had also discovered that some ionic liquid cations, particularly phosphonium-based, are able to extract and solubilize organic dyes from aqueous solutions (*24, 25*). Building from this previously established work, we utilized RTILs to help denature the keratin-based structure of wool fibers, and in the process, extract dyes from the textiles.

The extraction efficiency of several common dyes of varying charge, including acid dyes that are often utilized in the coloring of wool, were determined using a biphasic system containing one of several phosphonium-based RTILs and an aqueous solution containing 2-5 mM dye at varying pH. Efficiency was determined using UV-visible absorption measurements of the aqueous solutions prior to and post extraction. Extraction of the dyes into the RTIL layer were fairly efficient, especially with the [$P_{6\,6\,6\,14}$]Cl ionic liquid, and generally ranged from 70-98% efficiency with positively and negatively charged dyes, as well as neutral dyes, Table 1. As isolation occurred with dyes of varying charge, it is postulated that extraction from the aqueous layer into the RTIL was by dissolution rather than any type of ion exchange process. Near 100% efficiency could be achieved by controlling the pH of the aqueous solution to ensure single speciation of the dyes being extracted.

Table 1. Extraction efficiency of aqueous organic dyes using $[P_{66614}]Cl$ RTIL. Adapted with permission from *Anal. Chem.* 2011, 83, 2921–2930. Copyright 2011 American Chemical Society.

Dye	Extraction Efficiency (no pH control)	Extraction Efficiency (pH 7)	Extraction Efficiency (pH 10)	Partition Coefficient (pH 7)	Charge at pH 2	Charge at pH 7.5	Charge at pH 10
malachite green	97.77%	n.d.	n.d.	44[a]	+2	+1	+1
Nile blue	74.14%	87.29%	n.d.	7	+1	+1	0
Nile red	76.27%	99.37%	n.d.	157	0	0	0
fluorescein	92.07%	97.94%	n.d.	48	+1	-2	-2
bromothymol blue	0.00%	0.00%	100.00%	n.d.	0	-1	-1
kiton red	98.31%	99.86%	n.d.	730	-1	-1	-1

[a]value given for the case without pH control. n.d. indicates not determined

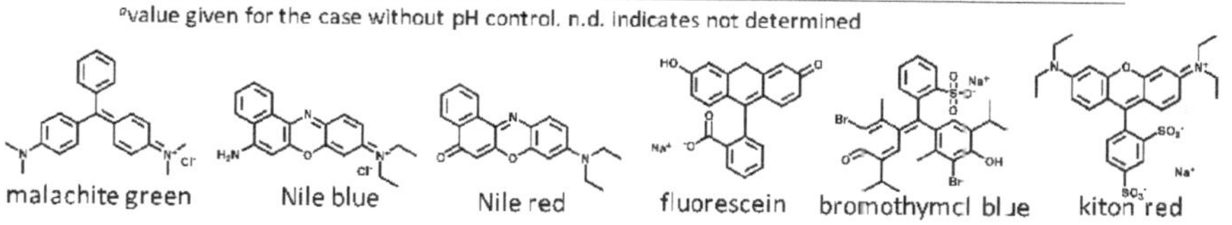

In addition to the ability of RTILs to extract and dissolve organic dyes, it has also been demonstrated that RTILs are able to denature biopolymers such as wool and silk. The general hypothesis is that the anions of the RTILs, in particular halide and carboxylate anions, tend to be particularly influential in disruption of biopolymer structures. This disruption ostensibly occurs through the strong hydrogen-bond interactions of the halide anion that serves to interfere with the network of hydrogen bonds that exist in the protein's native fold. Several RTILs were effective at completely dissolving keratin at up to 1% w/w. Merging the concepts of dye extraction with keratin denaturation or solubilization, we utilized RTILs to extract dyes from wool textiles, including commercially dyed wool yarns and commercially available wool clothing. In many cases, RTILs extracted enough of the dyes from keratin textiles to perform analysis of the dye components.

Conveniently, using RTILs for extraction of dyes from any source yielded an ideal matrix for the direct characterization of dissolved analytes. Specifically, the non-volatile solvent (the RTIL) containing dissolved species capable of absorbing UV energy (the textile dyes), provided ideal conditions for analysis of the dissolved dyes directly from the extraction solution via MALDI (matrix-assisted laser desorption ioniziaton) mass spectrometry. Extensive work has shown that RTILs as matrixes promote the ionization of analytes for MALDI-time of flight (TOF) mass spectrometry (*31–35*). The use of RTILs as MALDI matrixes intrinsically addresses a disadvantage of commonly used solid MALDI matrixes such as dihydroxybenzoic (DHB) or cyano-4-hydroxycinnamic acid (CHCA), namely that the solid matrixes frequently form heterogeneous deposits on the MALDI plate and result in poor spot-to-spot reproducibility. RTIL-based samples, which have negligible vapor pressure and are liquid at RT, do not require evaporation prior to use, thereby preventing non-homogeneous crystalline spots. This attribute of RTILs improves data reproducibility, reduces the need for many laser shots at a single sample, and improves the possibility of obtaining quantitative MALDI-MS data (*36–38*). Furthermore, given the negligible vapor pressure associated with RTIL, analysis from these solutions also afford minimal interference from the extraction solution to the analyte itself. As a consequence, MALDI analysis from the RTIL provided an added benefit of extending the spectroscopic window to molecular mass ranges commonly obscured by matrix ionization in standard MALDI analyses.

Using the RTIL $[P_{6\,6\,6\,14}]Cl$ to extract the dyes from commercially dyed wool fibers, followed by introducing the extraction solution directly into the MALDI-TOF, we were able to identify the dyes that comprise the complex mixture giving rise to the visible textile colors associated with the materials themselves. For example, this method identified that an "orange" wool fiber was actually a dye mixture containing acid yellow 49, acid orange 7, acid red 151 and acid red 4, Figure 3. Similarly, a "brown" fiber comprised a dye mixture contained acid orange 7, acid blue 62, and acid red 151. To show the effectiveness of this extraction-MALDI process as compared to other techniques, the same extraction solution was diluted in solvent and analyzed by ESI-TOF mass spectrometry. Using ESI, only the acid orange 7 signal was observable when analyzing the same orange fiber extraction solution. The rest of the component dyes in the mixture were not observed by ESI, possibly as a result of poor ionization or poor signal-noise ratios that often plague mass spectrometry observations in negative ionization mode.

A further advantage of this method is that the ionic liquid extraction solvents are not fragmented in the soft MALDI-MS technique, which further minimize interferences from the RTIL matrix in the spectrum of the analytes of interest. The limit of detection for wool dyes analyzed by this procedure is approximately 25 pmol (*19*, *20*). This limit can be achieved directly from neat sample mixtures without further processing or the addition of traditional ionization-promoting MALDI matrixes (e.g. CHCA), provided optimal signal-to-noise ratios. This technique will improve forensic analysis of dyed wool fibers by providing a "fingerprint" of the dyes used, achieved using fiber samples that are as short as 1 mm. The fingerprint is often characteristic of a particular dye batch, and in many cases is suitable for identifying two fibers that originated from the same garment or other device containing dyes or pigments.

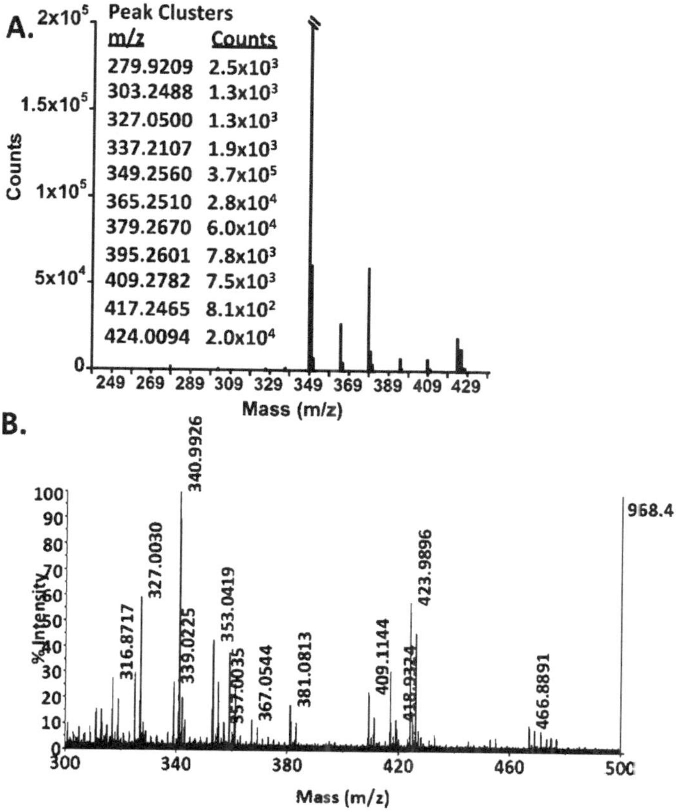

Figure 3. Negative mode ESI-TOF (top, A) and MALDI-MS (bottom, B) spectra of orange dye extracted from wool. Peaks in (A) are identified (m/z) as: 327.0500, Acid Orange 7 $[M]^-$; 424.0098, Acid Yellow 49 $[M]^-$. Peaks in (B) are identified (m/z) as acid orange 7 $[M- SO_3]^-$, 248.18; acid red 4 $[M - SO_3]^-$, 278.21; acid orange 7 $[M]^-$, 327.00; acid red 151 $[M - SO_3]^-$, 352.16; and acid yellow 49 $[M]^-$, 423.99. Reprinted with permission from Anal. Chem. 2012, 84, 9169-9175. Copyright 2012 American Chemical Society.

While this method provides an ideal single-pot method for extraction and identification of dyes, the complex mixtures that make up most dyed materials can sometimes make characterization extremely challenging. In order to optimize detection and fingerprinting dye mixtures, separation of the individual dyes prior to characterization would simplify analysis of MALDI spectra (though this does introduce an additional step for sample preparation prior to analysis). Unfortunately, a standard separation method for RTILs has not been well defined due to the charged and sometimes hydrophobic nature of the RTIL. Furthermore, separation methods often require tailoring to individual ionic liquids due to their wide ranging properties. Using a standard silica column, the RTIL $[P_{6\,6\,6\,14}]Cl$ containing a mixture of three acid dyes - kiton red, fluorescein, and fluorescent brightner - at 100mM concentrations (each) was developed on a silica column using a chloroform and methanol eluting solvent. Ideally, a solvent should be avoided for this added sample preparation step; however, due to the extremely high viscosity of many phosphonium-based ionic liquids (>1000 cP), a solvent was necessary to move the RTIL-dye mixture through the silica column. During elution, bands of RTIL-dye were clearly observable under UV-light, Figure 4, though complete separation of the bands was not achieved. Each band was collected into individual fractions, and absorption and fluorescence spectra were recorded. These spectra indicate that the fractions of the RTIL-dye mixtures were fairly well resolved, with indication of slight contamination by the other fractions. In the fluorescence spectra, the fluorescent brightner dye was not visible in the mixture due to its emission wavelength (495 nm) quenched by the other two dyes (absorptions at 498 and 560 nm). Recovery of each dye from the mixture was fairly low, generally in the 40-60% range from the original dye mixture. The dyes were observed to travel through the column as ion pairs with the phosphonium cations – infrared spectroscopy of the eluted fractions were similar to the spectrum of a mixture of $[P_{6\,6\,6\,14}]Cl$ and the dye sodium salt. The process to separate the RTIL-dye mixtures into individual dye components is currently being optimized.

Non-Lethal Hydrocarbon Biofuel Extractions

At this time, there is much effort within the renewable biofuel industry focused on production of either fatty acid based (biodiesel) or hydrocarbon based (terpene) fuel targets. The advantage of terpene based renewable fuels relative to fatty acid counterparts is that they are deoxygenated and are applicable to transportation engines that do not use diesel fuel. Most hydrocarbons that can be used as transportation fuels are sparingly soluble to insoluble in water, and biphasic extraction methods are therefore not explored often by the petroleum industry. However, isolation of hydrocarbon molecules from biofuel producing microorganisms, that are inherently aqueous systems themselves, is a critically important process for the economic viability of the biofuel.

An ideal extraction would be one that is both effective at isolating the biofuel away from the complex array of metabolites within, or on the surface of the producing organism, and also one that provides conditions that impart minimal deleterious effects to the culture health of the biofuel producing organism. Furthermore, microorganisms that have been engineered to metabolically produce heterologous biofuels may experience cell toxicity associated with intracellular accumulation of these water insoluble materials (*39*). Product toxicity associated with accumulation generally results in culture inviability at fairly low production yields. On the contrary, relief of this toxicity with biocompatible product extractions could conceivably represent a means to extend culture viability and hydrocarbon production. We explored the utilization of water-immiscible ionic liquids to extract biofuels in a non-lethal manner from a variety of microorganisms. The ultimate goal of these efforts were to develop a method to aid biofuel production via enabling a continuous flow process to maximize hydrocarbon production and isolation.

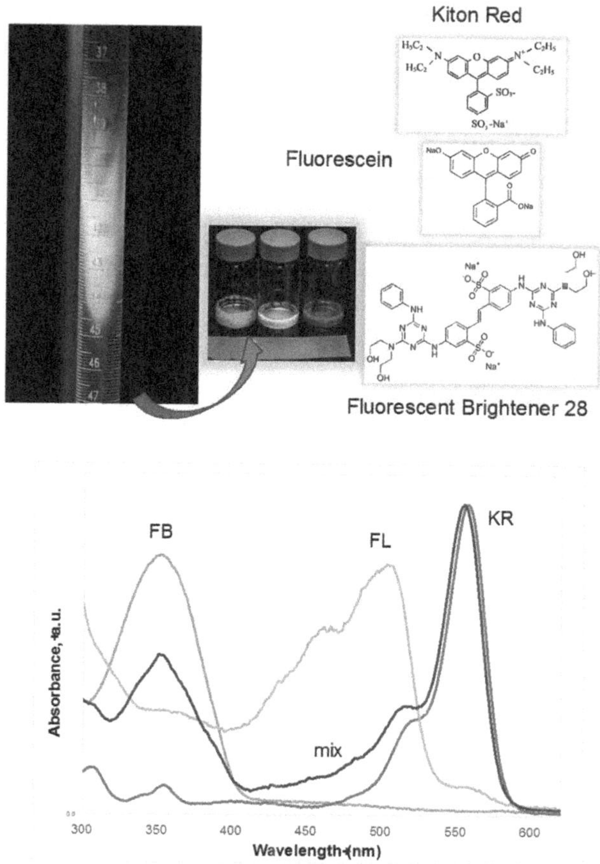

Figure 4. (top) Separation of a mixture of dyes in $[P_{6\,6\,6\,14}]Cl$ extraction solution on silica, and (bottom) UV-Visible spectra of recovered dyes following. FB = Fluorescent Brightener 28; FL = fluorescein, KR = Kiton Red 620; mix = extraction mixture.

To maximize extraction efficiency of hydrocarbons produced in real-time by microorganisms, the ideal RTIL must have optimized solubility of those hydrocarbons. With water-immiscible RTILs, the solubility of the hydrocarbons must be great enough such that the brief interaction between RTIL and microorganism at the biphasic interface is sufficient for phase transfer of the hydrocarbon to the RTIL layer. This can be significant challenge for production platforms that involve the biosynthesis of biofuel hydrocarbons in heterologous hosts (e.g. bacteria or yeast). While the produced hydrocarbons would otherwise phase separate in water, when produced in the context of a living organism, most will adhere to the outer membranes of the cell through van der Waals interactions. The colonial green algae *Botrycoccus braunii* var Showa (and other race B isolates) produces and accumulate triterpene biofuel hydrocarbons outside of the cell itself (*40*). The hydrocarbons produced by *B. braunii* are a mixture of triterpenes known as the "botyrococcenes", which include linear triterpenes such as squalene or C_{30}-botryococcene (Figure 5), as well as methylated (such as C_{32}-dimethyl or C_{34}-tetramethylbotryococcenes) and cyclized triterpenes. The methylated botryococcenes are attractive biofuel targets as they have been shown able to be processed with existing hydrocracking infrastructure to yield a variety of usable transportation fuels, such as gasoline, kerosene, aviation fuel and diesel, Figure 5 (*41*). *B. braunii* accumulates an unusually high amount of these hydrocarbons (frequently 30-40% dry cell weight), so much so that mature algal colonies examined with light microscopy may be induced to exude botryococcene based hydrocarbons simply by applying pressure to the glass slides they are mounted upon, Figure 6.

The drawback to the use of *B. braunii* as a production biofuel host is that it is relatively challenging to grow, however continuous extraction (e.g. "milking") of algae for biofuels is recognized as a useful mechanisms to increase productivity in other strains (*42*, *43*) - we investigated the potential of RTIL in this regard. Hydrocarbon solubility was determined in RTILs using the representative terpenoids of squalene (a structural analog of C_{30}-botryococcene) and beta-carotene. The solubility was determined using optical absorption in the UV and visible regions – the beta-carotene has a maximum absorption at 450 nm, and the squalene, which is not a conjugated system, can be determined indirectly using the established colorimetric procedure (*44*). For solubility measurements, the RTILs chosen were well known, severely hydrophobic RTILs, including the $[P_{6\,6\,6\,14}]Cl$, $[P_{6\,6\,6\,14}][Tf_2N]$, and $[C_4MPyr][Tf_2N]$, as all three are immiscible with water. All three exhibited low solubility of beta-carotene, dissolving approximately 1% w/w, Table 2. With squalene, which is structurally similar to botyrococcene, the phosphonium based RTILs exhibited solubility limits reaching nearly 10% w/w at room temperature, as compared to less than 1% w/w for the pyrrodinium RTIL with shorter alkyl chains (pyrrolidinium RTILs with longer alkyl subsitutions were not determined). These solubility limits in the phosphonium RTILs is comparable to, or greater than, the solubility of squalene in standard organic solvents such as hexane.

The toxicity of several RTILs towards the biofuel producing algae as well as several cyanobacterial strains were also tested, using RTILs ranging from the severely hydrophobic to completely water miscible. Toxicity of the RTIL on *B. braunii* was determined by the ability of the culture to produce oxygen via photosynthesis relative to a control strain, whereas cyanobacterial cell viability post-extraction was assessed with flow cytometry. The *B. braunii* was susceptible to nearly all RTILs with the exception of the $[C_4MPyr][Tf_2N]$. This poor survivability is not surprising, as *B. braunii* is well known for its sensitivity to slight environmental changes (which has also proven to be a hurdle for large scale production efforts of this strain relative to other candidate strains) (*45*). The cyanobacteria proved to be much more resistant to RTIL exposure, including both water miscible and water immiscible RTILs. The increased tolerance toward RTIL exposure demonstrated by the cyanobacteria relative to *B. braunii* suggested that using these organisms may ultimately be more promising as heterologous production hosts for botryococcene than the native producer. Similarly, efforts to reconstitute hydrocarbon production in RTIL-tolerant bacterial hosts may also be a desirable strategy. Surprisingly, the severely hydrophobic $[P_{6\,6\,6\,14}]^+$ RTILs displayed limited toxicity. There are some reports that suggest the toxicity of RTILs is correlated to the hydrophobicity of the RTIL cations (*46–50*). We believe that the fairly nontoxic properties of these RTILs is due to minimization of the mixing and interface with the cell surface of the producing microorganisms as a result of the hydrophobic nature and higher viscosity. While this may present hurdles for the practical implementation of continuous extraction of biofuels from planktonic algae and bacteria, we believe there is still sufficient mixing to accomplish this goal. Specifically, when the *B. braunii* are exposed to the RTIL $[P_{6\,6\,6\,14}][Cl]$, depigmentation occurs rapidly and the chlorophyll is readily apparent to be extracted from the aqueous layer, which becomes colorless, into the RTIL layer, which becomes a green solution. In terms of bacteria, the $[P_{6\,6\,6\,14}][Cl]$ and $[P_{6\,6\,6\,14}][Tf_2N]$ also seem to be the least toxic, with greater than 90% survivability after exposure to the RTIL in a biphasic system for 36 hours. Therefore, the $[P_{6\,6\,6\,14}][Cl]$ presents itself as a strong candidate for utilization of extracting biofuels from microorganisms with high extraction efficiency and low toxicity to the biocatalyst. Current work exploring this RTIL in a simulated system of continuous extraction of biologically produced compounds, using *E. coli* engineered to produce biofuels of interest is underway.

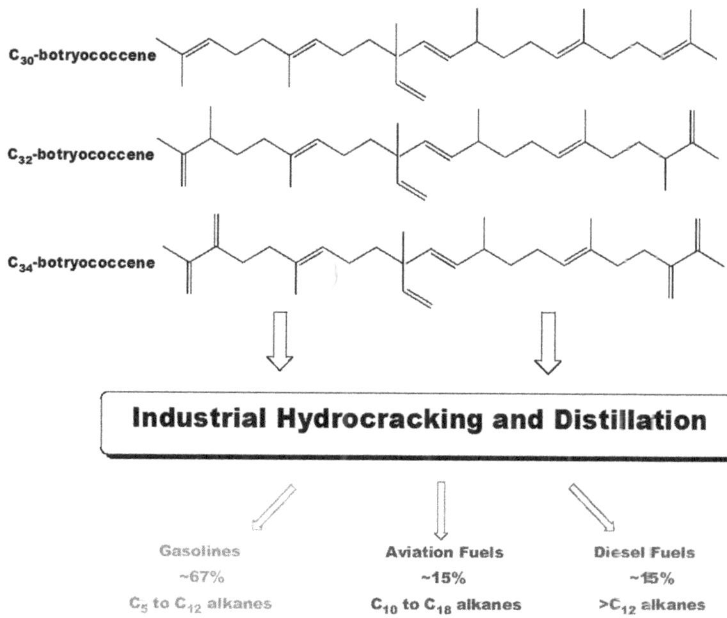

Figure 5. The branched hydrocarbon family of botryococcenes can readily feed into hydrocracking infrastructure to yield transportation fuels.

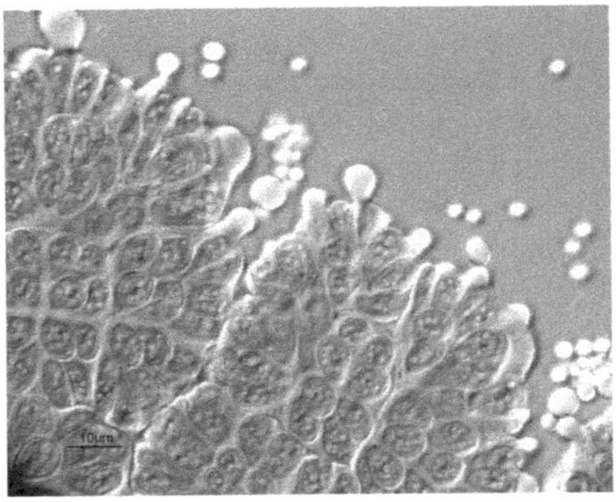

Figure 6. The branched hydrocarbon family of botryococcenes is produced in high yields by B. braunii var Showa. Photo Credit: Stan Vitha, TAMU Microscopy Center, Taylor L. Weiss and Timothy P. Devarenne, TAMU Department of Biochemistry & Biophysics

Table 2. Summary of solubility and toxicity results of ILs. With permission of Springer, *J. Applied Phycology*, 2013, 973-981.

Ionic Liquid	Solubility of β-carotene	Solubility of Squalene	Survival B. braunii	Percent Survival PCC6803 25 vol%[a]	Percent Survival PCC6803 50 vol%[a]
$[PC_6C_6C_6C_{14}][Cl]$	0.17 wt%	9.5 wt%	0%	90%	88%
$[PC_6C_6C_6C_{14}][Tf_2N]$	0.11 wt%	11.3 wt%	0%	90%	68%
$[C_4Mpyr][Tf_2N]$	0.2-0.3 wt%	1.6 wt%	56 ± 4%	92%	52%
$[PC_4C_4C_4C_4][Cl]$	1.0 wt%	2.5 wt%	0%	96%	77%
$[P(C_3OH)_3C_6][Br]$	< 0.1 wt%	3.5 wt%	0%	79%	n.d.
[Chol][Cl]	none (solid)	none (solid)	30 ± 6%	92%	81%
$[Chol][Tf_2N]$	< 0.1 wt%	1.1 wt%	35 ± 7%	67%	n.d.
hexane	0.14 wt%	miscible	88 ± 8%	n.d.	n.d.

n.d. not determined; [a]Normalized to control with BG-11 media.

Metal Salt Extractions

In addition to organic compounds, RTILs also have the potential to extract, recover, and identify inorganic compounds through a biphasic process using water-immiscible RTILs. Metals have played a critical role in the development of RTILs, with chloro-aluminates, -aurates, and -ferrates comprising some of the first room-temperature, air- and water-stable ionic liquids. Currently, RTILs have been used in numerous applications that have incorporated metals across nearly the entire periodic table, with applications in batteries, electrodeposition, and optical materials (*51*). In many RTILs, metal extractions are often driven by the interaction or coordination of the RTIL anion or cation components with the metal center. Our initial interest in metal-based RTILs was to take advantage of unique optical and magnetic characteristics of the metal centers when incorporated into viscous, room temperature, non-volatile liquids. In addition to extracting metals into RTILs, either by solvation (extraction) or coordination, we have attempted to recover the metals, a critical aspect to recycling and reuse of high-value metals, and identify metal species through mass spectrometry for forensic applications.

The extraction efficiency of metals into several RTILs was tested, with the $[P_{6\,6\,6\,14}][Cl]$ RTIL exhibiting higher extraction efficiencies and more importantly, from the materials perspective, maintained its liquid nature at room temperature following near stoichiometric extraction ratios. The extraction of the metals tested is primarily driven by chloride anion coordinaton to the metal centers, effectively yielding the new ionic liquids with $[MCl_x]^{z-}$ anions. The efficiency of extraction from aqueous solutions into the RTIL was near 100% for most metals tested as determined by microwave digestion of the RTILs in concentrated nitric acid, followed by characterization with ICP-AES. One exception to the near ideal extractions were the lanthanides, which were in the 60-80% range, likely due to the higher coordination numbers required for the larger f-element metals, Figure 7. Recovery of those metals from the ionic liquid was tested by exposing the metal-RTIL extraction materials to 6 M nitric acid. The nitric acid solutions were then diluted and characterized by ICP-AES. At least partial recovery of the metals from the RTIL was possible into the 6 M nitric acid, with the exception of Fe(III) and Bi(III) – nearly all of the metals remained in the RTIL even after exposure to these highly acidic conditions. The Fe(III)-based RTIL was extremely stable, and remained in the ionic liquid even after exposing the RTIL to an aqueous solution at pH 7 for over two years, with no evidence of diffusion of metal back into the aqueous layer.

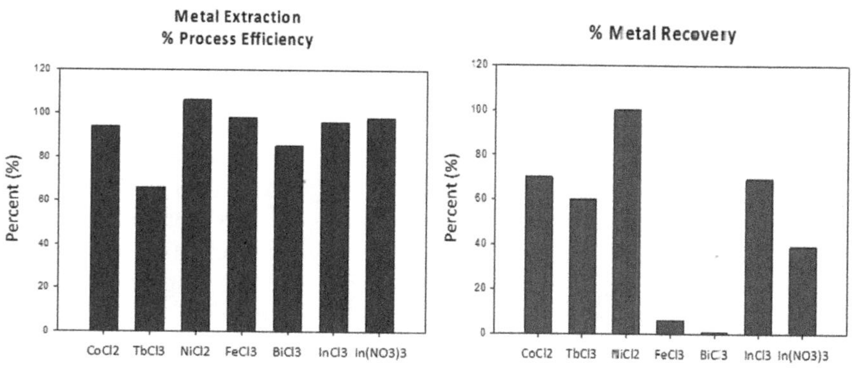

Figure 7. Extraction and recovery efficiencies of various metals using the $[P_{6\,6\,6\,14}][Cl]$ RTIL in a biphasic process.

As with the dye extraction solutions, the metal-based RTIL extractions were also highly absorbing in the 300-350 nm region, and therefore could be analyzed directly by MALDI-TOF without requiring additional processing. A RTIL containing equimolar amounts (100 mM) of three lanthanide metals – Ce(III), Er(III), and Dy(III), was prepared by extracting the metals from an aqueous solution. The RTIL layer containing all three metals was removed from the biphasic system and analyzed directly by MALDI-MS without requiring further processing.

In negative mode MALDI, the mass peaks and expected isotopic patterns for each metal were easy to distinguish from each other and from background noise, Figure 8. The Ce(III) was observed as $[CeCl_4]^-$ with peaks centered around 279.78 (expected 279.81). Likewise, $[DyCl_4]^-$ was observed with peaks at 303.80 and 305.81 amu (expected is 303.68 and 305.60, respectively); and $[ErCl_4]^-$ was observed at 308.8, 309.8 and 310.8 with the expected relative abundances. The higher coordination anions, e.g. $[LnCl_6]^{3-}$, were difficult to identify due to the higher charge and increased noise in that region of the MALDI. Due to the RTIL being silent in the negative mode of MALDI, it serves as an ideal matrix for this overall process of direct extraction and identification of analytes.

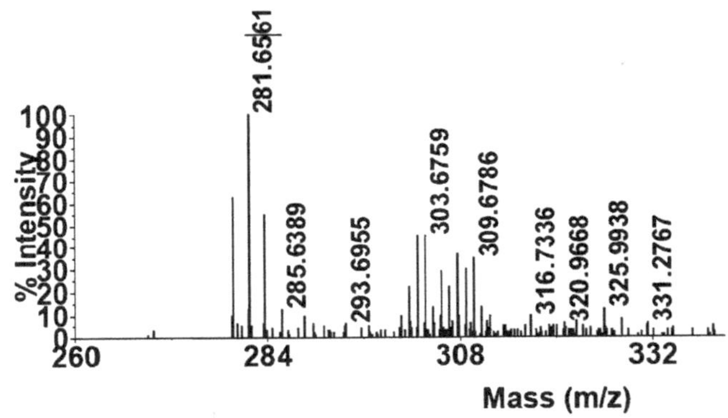

Figure 8. Characterization of lanthanide anion-based RTILs with the $[P_{6\,6\,6\,14}]^+$ cation through MALDI-MS of neat RTILs.

After extraction of the metals into the RTIL layer, it was observed that the RTIL appeared to flow, or be attracted, towards an external magnetic field applied using a rare-earth magnet. This may suggest that the new RTILs were "magnetic", meaning that some magnetic ordering was occurring via spin-spin coupling between metal centers. However, the magnetic properties, as measured with a Magnetic Property Measurement System (MPMS) magnetometer with SQUID detector, shows that the ionic liquids exhibit simple paramagnetic behavior above 50 K, showing only weak antiferromagnetic interactions at these temperatures (theta values were -5 to -10 K for the lanthanides tested, Ln = Tb(III), Dy(III), Ho(III), Er(III)). Therefore, these liquids are not magnetic, as per the standard definition of a magnetic material, but they are indeed solutions highly concentrated metal solutions, and therefore flow behavior towards magnetic fields is similar to that seen in highly concentrated aqueous systems in which the paramagnetic centers are simply attracted to an applied magnetic field. While this behavior is not necessarily unique to RTILs, recently one group has merged the magnetic properties and extraction capabilities of these RTILs to perform magnetic separations (*52–54*).

Interesting magnetic behavior is observed below 40 K with the $[P_{6\,6\,6\,14}]_3[LnCl_6]$ RTILs. There appears to be glassy magnetic behavior in zero-field cool/field cool (ZFC-FC) experiments, likely as a result of the high viscosity RTIL inabilty to achieve an equilibrated structure under the rapid cooling in this experiment, thus trapping out non-compensating clusters or domains of the metals in the RTIL structural glass, Figure 9. This anomaly occurs curiously close to where the magnetic phase transition for oxygen occurs, but the oxygen does not exhibit this deviating behavior that we observe in the ZFC-FC experiments. Another potential contributor to this strange behavior could be contamination from lanthanide oxides; however, all of lanthanide-based RTILs analyzed show this strange glassy behavior at near the same temperature with only slight deviations, and do not occur where lanthanide oxides would appear and do not vary as a function of an impurity of the oxides of different lanthanide metals.

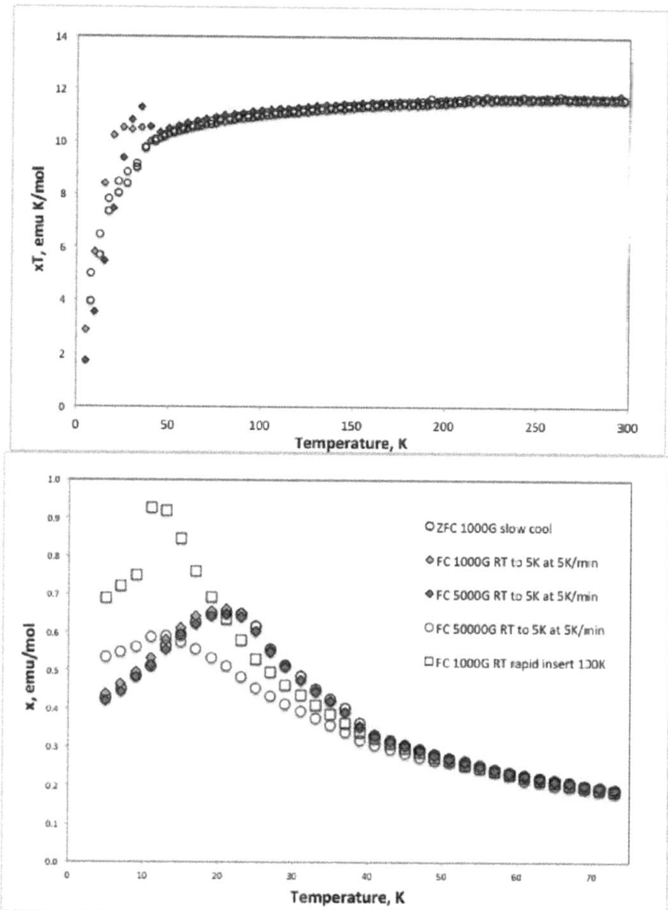

Figure 9. Magnetic properties of the $[P_{6\,6\,6\,14}]_3[TbCl_6]$ RTIL. (top) Magnetic moment, χT, and (bottom) magnetization, showing deviation from typical Curie-Weiss behavior below 50K, and the effects of cooling in zero field and in the presence of an external magnetic fields (ZFC-FC).

This anomaly could be due to magnetic glass behavior at low temperature, which might be driven by the rapid cooling of the highly viscous sample to well below the structural glass temperature (Tg for most of the $[P_{6\,6\,6\,14}]^+$ RTILs is generally in the 200 to 220 K temperature range). As the sample cools, the rapidly increasing viscosity of the ionic liquid prevents equilibration of the cation and anion structures, possibly trapping out domains or clusters of metal centers with uncompensated spins and/or spin-spin coupling within $[Ln_xCl_y]^{(y-3x)-}$ clusters that form during the cooling and vitrification process. We are currently exploring the potential for these clusters or glassy domains through low temperature structural characterization including SAXS and neutron scattering.

Conclusions

In summary, our work has demonstrated that RTILs have shown encouraging potential for application in biphasic extraction and recovery processes. In particular, the RTIL $[P_{6\,6\,6\,14}]Cl$ exhibits broad-spectrum extraction capabilities, and is effective at extracting small and large organic compounds, biomolecules, and metal salts, from aqueous solutions with fairly high efficiencies. The ability for this RTIL to function so well in extraction is likely due to the numerous intermolecular forces contained within the single RTIL. Additionally, the RTIL extractions of dyes, metals, and nearly any analyte that absorbs in the 300-350 nm range, can be utilized in a process of direct extraction-analysis with MALDI-MS spectroscopy. This non-destructive extraction method, coupled with the softer MALDI ionization process, is ideal for forensic analysis of trace compounds that may also be sensitive to extensive sample processing or harsher chemical environments. The extraction solutions described here may also lead to new materials application as well, as a result of their interesting optical and magnetic properties within a non-volatile, room-temperature liquid or glass. With the vast library of RTILs that currently exist, and those that are yet to be designed, numerous tasks are still being targeted by these liquid materials. As the field continues to grow and extend into more interdisciplinary fields, the potential impact that RTILs have on advancing technologies is vast and exciting.

Acknowledgments

This material is based upon work supported by the National Science Foundation under projects no. CHE-1410219 and CHE-1412648. We are grateful to Stan Vitha, Taylor L. Weiss, and Timothy P. Devarenne, for providing an image of botryococcene-producing *B. braunii*. Sandia National Laboratories is a multimission laboratory managed and operated by National Technology and Engineering Solutions of Sandia LLC, a wholly owned subsidiary of Honeywell International Inc. for the U.S. Department of Energy's National Nuclear Security Administration under contract DE-NA0003525. We gratefully acknowledge the Intelligence Community Postdoctoral Research Fellowship Program, the National Alliance for Advanced Biofuels and Bioproducts, and the Los Alamos National Flow Cytometry Resource. Los Alamos National Laboratory is operated by Los

Alamos National Security, LLC, for the National Nuclear Security Administration of the US Department of Energy under contract DE-AC52-06NA25396.

References

1. Majewski, P.; Pernak, A.; Grzymislawski, M.; Iwanik, K.; Pernak, J. Ionic Liquids in Embalming and Tissue Preservation. Can Traditional Formalin-fixation be Replaced Safely? *Acta Histochem.* **2003**, *105* (2), 135–142.
2. Pernak, J.; Syguda, A.; Mirska, I.; Pernak, A.; Nawrot, J.; Pradzynska, A.; Griffin, S. T.; Rogers, R. D. Choline-Derivative-Based Ionic Liquids. *Chem. Eur. J.* **2007**, *13*, 6817–6827.
3. Wang, H.; Gurau, G.; Rogers, R. D. Ionic Liquid Processing of Cellulose. *Chem. Soc. Rev.* **2012**, *41*, 1519–1537.
4. Berton, P.; Kelly, S.; Rogers, R. D. In *Metal Sustainability: Global Challenges, Consequences, and Prospects*; John Wiley & Sons: 2016; pp 295–316.
5. Shen, X.; Shamshina, J.; Berton, P.; Gurau, G.; Rogers, R. D. Hydrogels Based on Cellulose and Chitin: Fabrication, Properties, and Applications. *Green Chem.* **2016**, *18*, 53–75.
6. Kang, X.; Kugaa, S.; Wang, L.; Wu, M.; Huang, Y. Dissociation of Intra/inter-molecular Hydrogen Bonds of Cellulose Molecules in the Dissolution Process: A Mini Review. *J. Biores. Bioprod.* **2016**, *1* (2), 58–63.
7. Sivapragasam, M.; Moniruzzaman, M.; Goto, M. Recent Advances in Exploiting Ionic Liquids for Biomolecules: Solubility, Stability and Applications. *Biotechnol. J.* **2016**, *11*, 1000–1013.
8. Zhu, C.; et al. High Modulus Regenerated Cellulose Fibers Spun from a Low Molecular Weight Microcrystalline Cellulose Solution. *ACS Sustainable Chem. Eng.* **2016**, *4* (9), 4545–4553.
9. Li, R.; Wang, D. Preparation of Regenerated Wool Keratin Films From Wool Keratin–Ionic Liquid Solutions. *J. Appl. Polym. Sci.* **2013**, *127*, 2648–2653.
10. Zhu, C.; Chen, J.; Koziol, K. K.; Gilman, J.; Trulove, P.; Rahatekar, S. Effect of Fibre Spinning Conditions on the Electrical Properties of Cellulose and Carbon Nanotube Composite Fibres Spun Using Ionic Liquid as a Benign Solvent. *eXPRESS Polym. Lett.* **2014**, *8* (3), 154–163.
11. Wang, H.; Gurau, G.; Rogers, R. D. Ionic Liquid Processing of Cellulose. *Chem. Soc. Rev.* **2012**, *41*, 1519–37.
12. Phillips, D.; Drummy, L.; Conrady, D.; Fox, D.; Naik, R.; Stone, M.; Trulove, P.; De Long, H.; Mantz, R. Dissolution and Regeneration of *Bombyx mori* Silk Fibroin Using Ionic Liquids. *J. Am. Chem. Soc.* **2004**, *126*, 14350–14351.
13. Jost, K.; Durkin, D.; Haverhals, L.; Brown, K.; Langenstein, M.; De Long, H.; Trulove, P.; Gogotsi, Y.; Dion, G. Natural Fiber Welded Electrode Yarns for Knittable Textile Supercapacitors. *Adv. Energy Mater.* **2015**, *5*, 1401286.
14. Haverhals, L.; Reichert, W.; De Long, H.; Trulove, P. Natural Fiber Welding. *Macromol. Mater. Eng.* **2010**, *295*, 425–430.

15. Fahey, P.; Fox, E.; Durkin, D.; De Long, H.; Trulove, P. Combining Mechanical Tension with Natural-Fiber-Welding Treatments of Cellulose Yarns. *ECS Trans.* **2016**, *75*, 669–676.
16. Olkiewicz, M.; Plechkovab, N.; Earleb, M.; Fabregata, A.; Stübera, F.; Fortunyc, A.; Fonta, J.; Bengoa, C. Efficient Extraction of Lipids From Primary Sewage Sludge Using Ionic Liquids for Biodiesel Production. *Sep. Purif. Technol.* **2015**, *153*, 118–125.
17. Jeong, J.-H.; Takaoka, M.; Ichiura, H.; Oshita, K.; Fujimori, T. Evaluation of Metals in the Residue of Paper Sludge After Recovery of Pulp Components Using an Ionic Liquid. *J. Mater. Cycles Waste Manage.* **2016**, *18*, 215–221.
18. Lovejoy, K.; Davis, L.; McClellan, L.; Lillo, A.; Welsh, J.; Schmidt, E.; Sanders, C.; Lou, A.; Fox, D.; Koppisch, A.; Del Sesto, R. Evaluation of Ionic Liquids on Phototrophic Microbes and Their Use in Biofuel Extraction and Isolation. *J. Appl. Phycol.* **2013**, 973–981.
19. Lovejoy, K.; Lou, A.; Davis, L.; Sanchez, T.; Iyer, S.; Corley, C.; Wilkes, J.; Feller, R.; Fox, D.; Koppisch, A.; Del Sesto, R. Single-Pot Extraction-Analysis of Dyed Wool Fibers with Ionic Liquids. *Anal. Chem.* **2012**, *84*, 9169–9175.
20. Lovejoy, K.; Purdy, G.; Iyer, S.; Sanchez, T.; Robertson, A.; Koppisch, A.; Del Sesto, R. Tetraalkylphosphonium-Based Ionic Liquids for a Single-Step Dye Extraction/MALDI-MS Analysis Platform. *Anal. Chem.* **2011**, *83*, 2921–2930.
21. Xie, H.; Li, S.; Zhang, S. Ionic Liquids as Novel Solvents for the Dissolution and Blending of Wool Keratin Fibers. *Green Chem.* **2005**, *7*, 606–608.
22. Pinkert, A.; Marsh, K.; Pang, S.; Staiger, M. Ionic Liquids and Their Interaction with Cellulose. *Chem. Rev.* **2009**, *109*, 6712–6728.
23. Burrell, A.; Del Sesto, R.; Baker, S.; McCleskey, T.; Baker, G. The Large Scale Synthesis of Pure Imidazolium and Pyrrolidinium Ionic Liquids. *Green Chem.* **2007**, *9*, 449–454.
24. Del Sesto, R.; Corley, C.; Robertson, A.; Wilkes, J. Tetraalkylphosphonium-Based Ionic Liquids. *J. Organomet. Chem.* **2005**, *690*, 2536–2542.
25. Corley, C.; Del Sesto, R.; Wilkes, J. In *Molten Salts XIV: Proceedings of the International Symposium*; Electrochemical Society: 2006; Vol. 2004-24, pp 326–339.
26. Manhita, A.; Ferreira, T.; Candeias, A.; Dias, C. B. Extracting Natural Dyes From Wool - An Evaluation of Extraction Methods. *Anal. Bioanal. Chem.* **2011**, *400*, 1501–1514.
27. Pielesz, A.; Włochowicz, A.; Biniaś, W. Wool Fibre Keratin Treated with Azo Dyes by Fourier Transform Infrared Spectroscopy. *Spectrochim. Acta, Part A* **2000**, *56*, 1409–1420.
28. Shen, X.; Shamshina, J.; Berton, P.; Gurau, G.; Rogers, R. Hydrogels Based on Cellulose and Chitin: Fabrication, Properties, and Applications. *Green Chem.* **2016**, *18*, 53–75.
29. Brandt, A.; Gräsvik, J.; Hallett, J.; Welton, T. Deconstruction of Lignocellulosic Biomass with Ionic Liquids. *Green Chem.* **2013**, *15*, 550–583.

30. Prado, R.; Erdocia, X.; De Gregorio, G.; Labidi, J.; Welton, T. Willow Lignin Oxidation and Depolymerization Under Low Cost Ionic Liquid. *ACS Sustainable Chem. Eng.* **2016**, *4*, 5277–5288.
31. Armstrong, D.; Zhang, L.; He, L.; Gross, M. Ionic Liquids as Matrixes for Matrix-Assisted Laser Desorption/Ionization Mass Spectrometry. *Anal. Chem.* **2001**, *73*, 3679–3686.
32. Crank, J.; Armstrong, D. Towards a Second Generation of Ionic Liquid Matrices (ILMs) for MALDI-MS of Peptides, Proteins, and Carbohydrates. *J. Am. Soc. Mass Spectrom.* **2009**, *20*, 1790–1800.
33. Soukup-Hein, R.; Warnke, M.; Armstrong, D. Ionic Liquids in Analytical Chemistry. *Annu. Rev. Anal. Chem.* **2009**, *2*, 145–168.
34. Baker, G.; Baker, S.; Pandey, S.; Bright, F. An Analytical View of Ionic Liquids. *Analyst* **2005**, *130*, 800–808.
35. Berthod, A.; Ruiz-Angel, M.; Carda-Broch, S. Ionic Liquids in Separation Techniques. *J. Chromatogr. A* **2008**, *1184*, 6–18.
36. Mank, M.; Stahl, B.; Boehm, G. 2,5-Dihydroxybenzoic Acid Butylamine and Other Ionic Liquid Matrixes for Enhanced MALDI-MS Analysis of Biomolecules. *Anal. Chem.* **2004**, *76*, 2938–2950.
37. Li, T.; Gross, M. Ionic-Liquid Matrices for Quantitative Analysis by MALDI-TOF Mass Spectrometry. *J. Am. Soc. Mass Spectrom.* **2004**, *15*, 1833–1837.
38. Zabet-Moghaddam, M.; Heinzle, E.; Tholey, A. Qualitative and Quantitative Analysis of Low Molecular Weight Compounds by Ultraviolet Matrix-Assisted Laser Desorption/Ionization Mass Spectrometry Using Ionic Liquid Matrices. *Rapid Commun. Mass Spectrom.* **2004**, *18*, 141–148.
39. Dunlop, M.; Dossani, Z.; Szmidt, H.; Chu, H.; Lee, T.; Keasling, J.; Hadi, M.; Mukhopadhyay, A. Engineering Microbial Biofuel Tolerance and Export Using Efflux Pumps. *Mol. Syst. Biol.* **2011**, *7*, 487.
40. Wolf, F. R.; Nonomura, A. M.; Bassham, J. A. Growth and Branched Hydrocarbon Production in a Strain of *Botryococcus braunii* (Chlorophyta). *J. Phycol.* **1985**, *21*, 388–395.
41. Hillen, L.; Pollard, G.; Wake, L.; White, N. Hydrocracking of the Oils of *Botryococcus braunii* to Transport Fuels. *Biotechnol. Bioeng.* **1982**, *24*, 193–205.
42. Ventura, S.; Silva, F.; Quental, M.; Mondal, D.; Freire, M.; Coutinho, J. Ionic-Liquid-Mediated Extraction and Separation Processes for Bioactive Compounds: Past, Present, and Future Trends. *Chem. Rev.* **2017**, *117* (10), 6984–7052.
43. Singh, J.; Gu, S. Commercialization Potential of Microalgae for Biofuels Production. *Renew. Sust. Energy Rev.* **2010**, *14*, 2596–2610.
44. Rothblat, G.; Martak, D.; Kritchevsky, D. A Quantitative Colorimetric Assay for Squalene. *Anal. Biochem.* **1962**, *4*, 52–56.
45. Unkefer, C.; et al. Review of the Algal Biology Program Within the National Alliance for Advanced Biofuels and Bioproducts. *Algal Res.* **2017**, *22*, 187–215.
46. Wen, Q.; Chen, J.; Tang, Y.; Wang, J.; Yang, Z. Assessing the Toxicity and Biodegradability of Deep Eutectic Solvents. *Chemosphere* **2015**, *132*, 63–69.

47. Pernak, J.; Borucka, N.; Walkiewicz, F.; Markiewicz, B.; Fochtman, P.; Stolte, S.; Steudte, S.; Stepnowski, P. Synthesis, Toxicity, Biodegradability and Physicochemical Properties of 4-benzyl-4-methylmorpholinium-Based Ionic Liquids. *Green Chem.* **2011**, *13*, 2901–2910.
48. Stolte, S.; Matzke, M.; Arning, J.; Boschen, A.; Pitner, W.; Welz-Biermann, U.; Jastorff, B.; Ranke, J. Effects of Different Head Groups and Functionalised Side Chains on the Aquatic Toxicity of Ionic Liquids. *Green Chem.* **2007**, *9*, 1170–1179.
49. Petkovic, M.; Seddon, K.; Rebelo, L.; Pereira, C. Ionic liquids: A Pathway to Environmental Acceptability. *Chem. Soc. Rev.* **2011**, *40*, 1383–1403.
50. Cieniecka-Roslonkiewicz, A.; Pernak, J.; Kubis-Feder, J.; Ramani, A.; Robertson, A.; Seddon, K. Synthesis, Anti-Microbial Activities and Anti-Electrostatic Properties of Phosphonium-Based Ionic Liquids. *Green Chem.* **2005**, *7*, 855–862.
51. Berton, P.; Kelley, S.; Rogers, R. Stripping Uranium From Seawater-Loaded Sorbents with the Ionic Liquid Hydroxylammonium Acetate in Acetic Acid for Efficient Reuse. *Ind. Eng. Chem. Res.* **2016**, *55*, 4321–4327.
52. Yu, H.; Merib, J.; Anderson, J. Faster Dispersive Liquid-Liquid Microextraction Methods Using Magnetic Ionic Liquids as Solvents. *J. Chromatogr. A* **2016**, *1463*, 11–19.
53. Clark, K.; Nacham, O.; Purslow, J.; Pierson, S.; Anderson, J. Magnetic Ionic Liquids in Analytical Chemistry: A Review. *Anal. Chim. Acta* **2016**, *934*, 9–21.
54. Chatzimitakos, T.; Binellas, C.; Maidatsi, K.; Stalikas, C. Magnetic Ionic Liquid in Stirring-Assisted Drop-Breakup Microextraction: Proof-of-Concept Extraction of Phenolic Endocrine Disrupters and Acidic Pharmaceuticals. *Anal. Chim. Acta* **2016**, *910*, 53–59.

Editors' Biographies

Aaron M. Scurto

Professor Aaron M. Scurto obtained his B.S. in Chemical Engineering at The University of Delaware (1997) and his Ph.D. in Chemical Engineering at The University of Notre Dame (2002) under the direction of Professor Joan F. Brennecke. He is currently an Associate Professor at the University of Kansas in the Department of Chemical and Petroleum Engineering and is associated with the Center for Environmentally Beneficial Catalysis and the Department of Chemistry. He was an NSF post-doctoral Fellow at RWTH in Aachen, Germany under the guidance of Professor Dr. Walter Leitner's group (2003) and a post-doctoral associate at MIT working with Professor Alexander Klibanov (2004). Recently during a sabbatical (2015-2016), he was a visiting scientist at DuPont in Central Research and Development and a visiting professor at the University of Delaware in the Chemical and Biomolecular Engineering department. He has published over 50 papers and has 4 patents/patent applications. His research interests include sustainable chemistry and engineering, alternative solvents, phase equilibrium thermodynamics, biomass processing, compressed CO_2 applications, homogeneous catalysis, enzyme catalysis, and fermentation separations.

Mark B. Shiflett

Professor Mark B. Shiflett obtained his B.S. in Chemical Engineering at North Carolina State University (1989) and both his M.S. (1998) and Ph.D. (2001) in Chemical Engineering at The University of Delaware under the direction of Professor Henry C. Foley. He is currently a Foundation Distinguished Professor of Engineering at the University of Kansas in the Department of Chemical and Petroleum Engineering and is associated with the Center for Environmentally Beneficial Catalysis. He is an inventor on 44 U.S. Patents and has published over 70 papers. He worked for the DuPont Company in Wilmington, Delaware at the Experimental Station for 28 years on a variety of challenging projects including the development of non-ozone depleting refrigerants, a thermal plasma reactor, nanoporous carbon membranes, hydrogen storage materials, a next generation TiO_2 process, new applications for ionic liquids, and novel gas separation processes. His research interests include ionic liquids, green chemistry and engineering, energy efficient chemical processes, product design, desalination, thermodynamics, kinetics, and reaction engineering.

Indexes

Author Index

Subject Index

A

B

C

E

F

I

M

O

W

Printed in the USA/Agawam, MA
May 3, 2018

674041.001